MRI/DTI Atlas of the Human Brainstem in Transverse and Sagittal Planes

MRI/DTI Atlas of the Human Brainstem in Transverse and Sagittal Planes

George Paxinos
Neuroscience Research Australia and The University of New South Wales, Australia

Teri M Furlong
Neuroscience Research Australia and The University of New South Wales, Australia

Ken Ashwell
The University of New South Wales, Australia

Kristie Smith
Neuroscience Research Australia and The University of New South Wales, Australia

Evan Calabrese
Department of Radiology, Duke University Medical Centre, Durham, NC 27710, USA

G. Allan Johnson
Centre for In Vivo Microscopy, Department of Radiology
Duke University Medical Centre, Durham, NC 27710, USA

ELSEVIER

ACADEMIC PRESS
An imprint of Elsevier

Academic Press is an imprint of Elsevier
125 London Wall, London EC2Y 5AS, United Kingdom
525 B Street, Suite 1800, San Diego, CA 92101-4495, United States
50 Hampshire Street, 5th Floor, Cambridge, MA 02139, United States
The Boulevard, Langford Lane, Kidlington, Oxford OX5 1GB, United Kingdom

ISBN: 978-0-323-91583-0

For information on all Academic Press publications
visit our website at https://www.elsevier.com/books-and-journals

Publisher: Nikki Levy
Senior Acquisitions Editor: Megan McManus
Editorial Project Manager: Kathy Padilla
Production Project Manager: Julie Taylor
Designers: Yvette Paxinos and Matthew Limbert

Printed in the USA

Dedicated to

Cem
Neve
Cameron and Samantha
Jean
Jim Lagopoulos
Patricia Furlong (in memory of)

Contents

Preface

This atlas uses high resolution MR images juxtaposed with detailed diagrams. It is relevant to radiologists, radiotherapists, neurosurgeons, pathologists, anatomists, neurophysiologists, and graduate students in neuroscience. It provides neuroscientists and clinicians with MR images like those they see on their screens, but with the advantage of detailed delineations. Scientists working with brain connectivity will be able to follow tracts in colorized fractional anisotropy (FAC) images in both the transverse and the sagittal planes.

The key features of the present atlas are:

- TRANSVERSE (AXIAL) PLANE: Presents 60 axial levels at 1 mm intervals from the pyramidal decussation to the posterior commissure, with each open-book page featuring a colorized fractional anisotropy (FAC), diffusion-weighted (DWI), T2*-weighted gradient echo (GRE), and the accompanying fully labelled diagram.
- SAGITTAL PLANE: Presents 26 sagittal levels at 0.5 mm intervals, with each open-book page featuring the FAC, DWI, GRE, and the corresponding fully labelled diagram with the abbreviations and their explanations.
- Delineations primarily made based on direct observations of the MRI contrasts.
- The orientation of sections corresponds to that of Human Brainstem: Cytoarchitecture, Chemoarchitecture, Myeloarchitecture (Paxinos et al., 2019).
- Electronic diagrams available to purchasers of this book via a password-protected website.

How to cite this atlas: Paxinos G, Furlong TM, Ashwell KWS, Smith K, Calabrese E, and Johnson GA, MRI/DTI Atlas of the Human Brainstem, San Diego, Elsevier Academic Press, 2023.

Acknowledgements

We thank Christodoulos Skliros for help at various stages of this atlas. We also thank Mustafa (Steve) Kassem for brilliance in assisting page layout and image composition, Yvette Paxinos (http://www.paxifilm.com/) for the artistic cover design and the talented staff at Academic Press Elsevier, Natalie Farra, Anna Valutkevich, Kathy Padilla, and Julie Taylor, for delivering the book. This work was supported by NHMRC grants (APP1086083, APP1086643) and in part by ARC Center of Excellence for Integrative Brain Function (CE140100007) to G. Paxinos who is an NHMRC Senior Principal Research Fellow, Investigator Grant 2017282), as well as NIH P41 EB015897 and NIH 1S10OD010683 to G Allan Johnson.

Introduction

Atlases can be likened to scientific theories. Like theories they provide a means for finding our way in strange domains. Like theories, they are social constructs, the culmination of a community effort (original analogy by Toulmin, 1953). Our atlas relies on prior work by anatomists, physiologists, pharmacologists, psychologists, psychiatrists, neurologists and radiologists, engineers and physicists.

The development of brain atlases over the past 100 years can be mapped upon the successive introduction of new staining and imaging technologies and the development of more accurate stereotaxic coordinates. Olszewski and Baxter (1954) constructed the first comprehensive study of the cell groups in the human brainstem. Using Nissl and myelin stains, they produced a high-quality atlas which formed a starting point for the work of Paxinos and Huang (1995) who employed, in addition, histochemical staining (acetylcholinesterase, AChE). Subsequently, Paxinos et al., (2019) used three times as many levels of the brainstem as previous atlases to cover it in greater depth.

The next step in atlas development was the introduction of a wide range of histochemical and immunohistochemical stains to assist anatomical delineation (Paxinos and Watson, 1982; Paxinos, 1983, 1985). The most comprehensive presentation of this method is represented in Chemoarchitectonic Atlas of the Rat Brain of Paxinos et al., (1999, 2021). The recent editions of The Rat Brain in Stereotaxic Coordinates made use of additional immunohistochemical stains, even though these sections are not pictured in the atlas (Paxinos and Watson, 2005, 2007, 2014).

Since the discovery of gene targeting in mice (Capecchi, 1989), data on gene expression in the rodent brain have contributed to anatomical mapping, through the use of mouse lineages based on site-specific recombinases. This has led to an era of what Alexandra Joyner called 'genetic neuroanatomy' in which the developmental origins of neuron groups can be visualized in postnatal brains (Joyner and Zervas, 2006).

In the present work, we make another departure and use MR images to segment the human brainstem. The delineations in the present book were guided by the histological atlas Human Brainstem, Cytoarchitecture, Chemoarchitecture, Myeloarchitecture (Paxinos et al., 2019).

Over the past few decades, there have been predictions of the demise of classical histological neuroanatomical atlases and their imminent replacement by imaging atlases. However, there is no sign yet that these atlases are in fact replacing the classical paper atlases.

We believe the reason for this is that the resolution and contrast of the images offered have not been satisfactory, and the neuroanatomical delineations were even less. Hence, here we are justified in using the power of histological atlases to inform the delineation of an imaging one.

The present atlas offers two major enhancements when compared with earlier attempts to make MRI human brain atlases. Firstly, the MRI resolution/contrast obtained at the Duke Center for In Vivo Microscopy is considerably higher than that obtained in conventional clinical MRI. Secondly, the comprehensive set of contrasts provided using DTI has enabled us to identify more than 80% of structures identified in Human Brainstem, Cytoarchitecture, Chemoarchitecture, Myeloarchitecture (Paxinos et al., 2019). MRI mapping will not quickly replace histological methods, but rapid improvements in resolution and contrast have already established a secure place for MRI technologies, at least in rodent brain mapping (Paxinos et al., 2015). In addition, MRI, unlike histology, can be performed non-invasively and is used in human clinical diagnostics.

Since the publication of the first (MRI) atlas of the live rat in 1987 (Johnson et al., 1987), a series of important technical improvements have enabled the production of high-quality image series, such as those presented herein. MR images have progressively acquired higher spatial resolution, 3D acquisitions with isotropic voxels (Suddarth and Johnson, 1991), and a valuable set of contrast mechanisms. The major developments since 1987 include the following:

1. An increase in magnetic field strength from 1.5 T to 7.1 T, increasing the MRI signal.
2. An increased number of sequences to emphasize different contrasts in the tissue (T1—emphasizing difference in spin lattice relaxation time; T2—emphasizing difference in spin relaxation time; PD—based on differences in proton density; and diffusion tensor images (DTI)—providing a suite of contrast parameters based on tissue-specific diffusion of water).
3. The development of isotropic imaging through the use of three-dimensional (3D) imaging sequences with large arrays (Suddarth and Johnson, 1991).
4. Active staining of tissues with gadolinium contrast agents to reduce T1 and allow shorter acquisition (Johnson *et al.,* 2002).
5. The development of stronger gradient coils to increase the spatial encoding.
6. The development of an extended dynamic range (Johnson *et al.,* 2007).

Source of tissue

The data for the construction of this atlas came from the Duke Center of In Vivo Microscopy, Duke University. The images are derived from a postmortem human brain (brainstem and thalamus) of a 65-yr-old male with no known history of neurologic or psychiatric illness. The resolution is 50 μm for GRE and 200 μm for the DTI, making our GRE data 8000 x higher resolution than a typical human structural scan (0.05 x 0.05 x 0.05 mm versus 1 x 1 x 1 mm), and DTI 1000 x higher resolution than a typical human DTI scan (0.2 x 0.2 x 0.2 mm versus 2 x 2 x 2 mm; see below for more details). All the work is based on one brain, but no feature was identified unless it was confirmed on both hemispheres.

Postmortem MR images were acquired using a 7.0T MRI system (Agilent Direct Drive) at the Duke Center for In Vivo Microscopy. The system has been configured explicitly for magnetic resonance histology, with the addition of a high-strength gradient insert (Resonance Research), providing peak gradients of 770 mT/m that are linear over an 80 (diameter) x100 mm field of view. A large specimen quadrature coil (M2M imaging) with a 65x100 mm homogenous volume.

The specimen was perfused with saline ex vivo, immersion fixed in 10% formalin, and soaked in neural phosphate buffered saline doped with 2.5 mM gadoteridol (0.5 % ProHance) for 3 weeks. Structural T2*-weighted GRE images were acquired with a gradient echo pulse sequence (repetition time = 50 ms, echo time = 10 ms, flip angle = 60°, receiver bandwidth = 78 Hz/pixel) with a spatial resolution of 50 μm isotropic. GRE acquisition time was 14 hours. High-angular resolution DTI images were acquired with 120 angles at b=4000 mm^2/s, and 11 evenly spaced b=0 volumes, using a standard spin echo pulse sequence (repetition time = 100 ms, echo time = 33.6 ms, receiver bandwidth = 278 Hz/pixel) with a spatial resolution of 200 μm isotropic. DTI scan time was 208 hours. Web-based tools and environmental monitoring systems have been implemented to monitor and control the scanner over the long course of the scan to ensure consistency. The raw data for each of the 120 constitutive images was approximately 1 GB yielding a total dataset size of 120 GB. A custom computer infrastructure (acquisition, reconstruction, post processing, archiving) was designed to accommodate these very large MRI data arrays. Post-processing pipelines were developed and implemented in a computing cluster environment to facilitate data reduction (diffusion tensor estimation and generating of quantitative parametric diffusion maps).

Two specific MRI contrasts were used for this atlas: the T2*-weighted GRE, which served as the higher resolution "structural" contrast, and the directionally colorized FA image from the DTI acquisition, which served as the "functional" contrast by highlighting the directional organization of brain structures, such as white matter tracts.

The following is a more detailed description of methods from Calabrese et al., (2015):

Postmortem MR imaging was performed in a 7 Tesla small animal MRI system controlled with an Agilent console (Agilent Technologies, Santa Clara, CA). RF transmission and reception were achieved with a 65 mm inner-diameter quadrature RF coil (M2M Imaging, Cleveland, OH).

Anatomic images were acquired using a 3D gradient echo pulse sequence with repetition time (TR) = 50 ms, echo time (TE) = 10 ms, flip angle (α) = 60°, and bandwidth (BW) = 78 Hz/pixel. The field of view (FOV) was 80 × 55 × 45 mm, and the acquisition matrix was 1600 × 1100 × 900 resulting in a 50 μm isotropic voxel size. Total acquisition time was 14 h.

Diffusion data were acquired using a simple diffusion-weighted spin echo pulse sequence (TR = 100 ms, TE = 33.6 ms, BW = 278 Hz/pixel). Diffusion preparation was achieved with a pair of unipolar, half sine diffusion gradient waveforms of width (δ) = 4.7 ms, separation (Δ) = 26 ms, and gradient amplitude (G) = 50.1 G/cm. Single-shell high angular resolution diffusion imaging (HARDI) data were acquired with 120 unique diffusion directions at b = 4,000 s/mm^2 and 11 b = 0 s/mm^2 (b0) volumes dispersed evenly throughout the acquisition. The FOV was 90 × 55 × 45 mm, and the acquisition matrix was 450 × 275 × 225 resulting in a 200 μm isotropic voxel size. Total acquisition time was 208 h.

Orientation of the MR Dataset to the Paxinos et al., (2019) Histological Atlas Plane

The stereotaxic coordinates of this MRI atlas compare with those of the histological atlas of Paxinos et al., (2019).

In this MRI/DTI atlas, the stereotaxic anterior-posterior and dorso-ventral zero is at the obex, similar to our histological atlas (Paxinos et al., 2019). Researchers who section the brain on a plane perpendicular to the ventricular axis will obtain a different section from ours because our plane of section was perpendicular to the long axis of the medulla and pons (transverse plane).

The pages for all 86 diagrams in this atlas are formatted with a blue square on the bottom right of the page containing 3 lines of text: a figure number, a distance relative to the obex number, and a reference number, indicating a sections position in its respective plane (i.e., there are 60 transverse and 26 sagittal sections in total). Each diagram is also fitted with an x-y scale axis.

For the 60 transverse diagrams, the distance relative to the obex number indicates the rostro-caudal distance of the section from the vertical transverse plane passing through the obex. The ordinate (y-axis) coordinates show the plates distance from the plane through the obex and contains the longitudinal axis of the brainstem and is orthogonal to the midline plane and passes through the obex. The abscissa (x-axis) coordinates show the distance of structures from the midsagittal plane. For the 26 sagittal drawings, the distance relative to the obex number indicates the sections distance from the midline. The ordinate (y-axis) coordinates indicate the distance from the obex, and the abscissa (x-axis) coordinates show the distance from the transverse plane passing through obex.

Comparatively, the present MRI and histological (Paxinos et al., 2019) atlases are well aligned at the level of the 7th nerve (approx. +15 mm from the obex). In this MRI atlas, the posterior commissure is found at obex + 49 mm (Fig. 57b), and in the histological atlas at + 47 mm anterior to the obex (Fig. 127b).

Diagrams and Representative Images

For each diagram, image sections were printed and covered with transparency paper (0.2 mm thick), which was used to define the borders of structures using pencil. The pencil drawings were scanned and digitized using Adobe Illustrator. In general, fiber tracts are outlined by solid lines and nuclei by dashed lines. The ventricles are filled in solid blue. Additionally, several structures on sagittal drawings are color-filled corresponding to their general appearance in FAC: i.e., the medial lemniscus – blue, the transverse fibers of the pons – pink, and nerves – yellow.

Magnification

To enhance resolution, we sacrificed consistency of magnification in the transverse plane, changing it in three locations (i.e., magnification 1: Figs 1-9, magnification 2: Figs 10-21, and magnification 3: Figs 22-60). Therefore, the individual scale bars should be consulted when comparison of sizes are made across plates or figures. The representative images (i.e., GRE, DWI, and FAC) are presented as lower in magnification than their corresponding diagrams allowing all the relative content for a single section to be viewed in an open page. All sagittal diagrams and corresponding GRE, FAC, and DWI images are presented at one magnification and, therefore, are fitted with the same x-y scale axis as the diagram, in place of a scale bar.

General comments on brainstem organization

We have used the nomenclature and abbreviations introduced by Paxinos and Watson (1982). We have modified this nomenclature to be consistent with the modern brain ontology based on gene expression during development (Watson et al., 2017a).

The new ontology has important implications for the understanding of the organization of the brainstem and the boundaries of its subdivisions. Firstly, it redefines the caudal and rostral boundaries of the midbrain. At the caudal boundary of the midbrain, the isthmus is now recognized as the most rostral segment of the hindbrain. The characteristic isthmic structures include the trochlear nucleus, the dorsal raphe, the parabigeminal nucleus, the dorsal nucleus of the lateral lemniscus, the rostral interpeduncular nucleus, and the decussation of the superior cerebellar peduncle (Watson et al., 2017b). All of these structures were previously thought to belong to the midbrain. At the rostral boundary of the midbrain are a number of structures which are part of the pretectal segment of the diencephalon, but which have been mistakenly assumed to belong to the midbrain; they include the posterior commissure and associated nuclei, the nucleus of Darkschewitsch, the interstitial nucleus of Cajal, and the rostral (parvicellular) part of the red nucleus (Puelles et al., 2012a,b,c). Developmental gene expression has forced a major revision in the use of the term pons as a regional descriptor.

Pons. Traditionally the human hindbrain has been divided into two segments, the pons and the medulla, based on the appearance of their ventral surface. In most mammals, the basilar pontine enlargement is restricted to rhombomeres 3 and 4. The problem is that in the human, and to a lesser extend in other primates, the fiber systems associated with the basilar pontine nuclei form a mushroom-like expansion so large that it appears to extend from the midbrain to the seventh rhombomere. This oddity can no longer be allowed to determine the subdivision of the hindbrain. Studies of gene expression in mammalian rhombomeres show that the pontine nuclei arise in the rhombic lip in rhombomeres 6-7 and migrate to their final position in the ventral part of r3 and r4 (Farago et al., 2006). The mammalian hindbrain (including the human hindbrain) can be divided into a prepontine portion (isthmus, r1, and r2), a pontine portion (r3 and r4) and a post-pontine portion (r5 to r11). The terms pons and medulla should no longer be used as descriptive subdivisions of the hindbrain

Gene expression reveals the segmental organization of the brainstem

The segmental organization of the brainstem was first observed by embryologists in the late nineteenth century; they described a series of outpouchings in the developing vertebrate brainstem. Since the advent of developmental gene expression and molecularly defined regional progeny studies, the outpouchings have been recognized as evidence for the fundamental segmental organization of the brainstem. It is now clear that the brainstem is made up of a rostrocaudal series of segments called neuromeres that arise in early development and impose an anatomical and functional organization that persists in the adult brain. In the midbrain there are two such segments, called mesomeres (Puelles et al., 2012b), whereas the hindbrain is divided into 12 neuromeres––the isthmus and 11 rhombomeres (Puelles et al., 2013; Tomas-Roca et al., 2016; Watson et al., 2017b).

The first attempt to illustrate the boundaries and contents of the segmental elements of the brainstem was presented in the chick brain atlas of Puelles et al., (2007, 2019). More recently, the boundaries of the brainstem segments in the mammalian brain have been illustrated in diagrams of sagittal brain sections in the rat brain atlas of Paxinos and Watson (2014). Many of the segments in the brainstem can be confidently identified by the presence of one or more signature nuclei; examples are the trochlear nucleus in the isthmus, the abducens nucleus in r5, and the facial nucleus in r6. A list showing the segmental position of selected structures in the brainstem and adjacent diencephalon and spinal cord is shown here:

Diencephalon	
Prosomere 1	posterior commissure
	Darkschewitsch nucleus
	interstitial nucleus of Cajal
	parvicellular red nucleus
Midbrain	
Mesomere 1	superior colliculus
	oculomotor nucleus
	magnocellular red nucleus
Mesomere 2	sagulum nucleus
	retrorubral field (DA8)
	subbrachial nucleus
Rostral hindbrain	
Isthmus	trochlear nucleus
	parabigeminal nucleus
	microcellular tegmental nucleus
	rostral interpeduncular nucleus
Rhombomere 1	locus coeruleus
	caudal interpeduncular nucleus
	parabrachial nuclei
Rhombomere 2	rostral motor trigeminal nucleus
Rhombomere 3	caudal motor trigeminal nucleus
Rhombomere 4	emerging facial nerve
	rostral pontine nuclei
Rhombomere 5	caudal pontine nuclei
	abducens nucleus
	superior olive
Rhombomere 6	facial nucleus
Caudal hindbrain	
Rhombomere 8	compact ambiguus nucleus
	rostral inferior olive
Rhombomere 9	semicompact ambiguus nucleus
	middle inferior olive
Rhombomere 10	loose ambiguus nucleus
	caudal inferior olive
Rhombomere 11	retroambiguus nucleus
Rostral spinal cord	
C1 segment	pyramidal decussation

Nomenclature and abbreviations

The Paxinos/Watson nomenclature and abbreviation system was originally published in the rat brain atlas in 1982, and it has been retained and expanded in six subsequent editions of the rat brain atlas up to 2014 (Paxinos and Watson, 2014). It is currently used in major atlases of the mouse brain (Paxinos and Franklin, 2013, 2019), the marmoset brain (Paxinos et al., 2012b), the rhesus monkey brain (Paxinos et al., 2009, 2023 in press), the human brain (Mai et al., 2008), the developing rat brain (Paxinos and Ashwell, 2018), the developing mouse brain (Paxinos et al., 2007), the chick brain (Puelles et al., 2019) and the spinal cord (Watson et al., 2009; 2013). In NeuroNames, Bowden et al., (2012) used these abbreviations, as did Foster (1998) and Morin and Wood (2001).

In considering the merit of a particular name over synonyms, Paxinos and Watson chose terms that have been ratified by modern usage. Anglicized versions of terms have replaced older Latinized forms wherever possible. The Paxinos/Watson system avoids the use of eponyms in all but a handful of cases.

In the digital era, it is increasingly important that scientists use consistent systems of abbreviation. Unfortunately, many neuroscientists ignore this principle and construct new abbreviations without reference to standard schemes. For example, SO sometimes stands for supraoptic nucleus and sometimes for superior olive; SC may stand for either for suprachiasmatic nucleus and superior colliculus; IC may stand for either inferior colliculus or internal capsule. Worse still, “accumbens nucleus” has been abbreviated in more than 20 ways - Acb, ACB, acb, NAS, nas, A, a, Ac, ac, NA, na, AN, an, NAC, nac, ACN, acn, ACU, acu, ACC, and Acc. We recommend neuroscientists consider the Paxinos/Watson system because it is logical, coherent, comprehensive, and widely used.

The construction of abbreviations in the Paxinos/Watson nomenclature

The principles used in the construction of abbreviations are the similar to those used to derive the abbreviations for the elements of the periodic table of elements. The main rules we have followed are listed here:

1. It is now explicit that DA11 refers to a dopamine group, and NA7 declares a noradrenaline group.

2. Capital letters represent nuclei and cortical areas, and lowercase letters represent fiber tracts and fissures. Because of this, it has not been necessary to use the letter 'N' to denote nuclei or to use the letter 't' to denote fiber tracts.

3. Abbreviations representing nuclei and cortical regions are shown in uppercase characters. Fiber tracts and fissures are abbreviated in lowercase characters.

4. The general principle used in the abbreviations of the names of elements in the periodic table was followed: the capital letter representing the first letter of a word in a nucleus is followed by the lowercase letter most characteristic of that word (not necessarily the second letter; e.g., Mg = magnesium; Rt = reticular thalamic nucleus).

5. Compound names of nuclei have a capital letter for each part of the compound that will represented in the abbreviation (e.g., LPGi = lateral paragigantocellular nucleus).

6. If a word occurs in the names of a number of structures, it is usually given the same abbreviation (e.g., Rt = reticular thalamic nucleus; RtTg = reticulotegmental nucleus of the pons). Exceptions to this rule are made for well-established abbreviations, such as VTA (which we would otherwise have rendered as VTgA).

7. Abbreviations of brain regions are omitted where the identity of the region in question is clear from its position (CMn = centromedian thalamic nucleus, not CMTh).

8. Arabic numerals are used instead of Roman numerals in identifying (a) cranial nerves and nuclei (as in the Berman, 1968, atlas), (b) Rexed's laminae of the spinal cord, (c) layers of the cerebral cortex, and (d) cerebellar folia. While the spoken meaning is the same, the detection threshold for Arabic numerals is lower, and ambiguity is reduced. They are also easier to position in small spaces available on diagrams. We have added the suffix “Cb” to numbered cerebellar folia and the suffix “Sp” to Rexed's laminae of the spinal cord. For the cerebral cortical layers, numbers stand by themselves.

9. We have used the prefix A to identify numbered areas of the cerebral cortex.

10. We have modified the abbreviations for monoamine cell groups. Because we allocated the capital letter A to cortical areas, we were forced to modify the alphanumeric nomenclature for the catecholamine cell groups. For A1 we used NA1 (noradrenalin group 1), for C1 we used Ad1 (adrenalin group 1), for A8 we used DA8 (dopamine group 8), for A11 we used DA11 (dopamine group 11). Whereas this change in nomenclature was introduced to avoid duplication, it has the benefit of specifying the different monoamine groups. In the original scheme, A1 stood for noradrenalin group 1, something counterintuitive, since “A” for the naïve reader stands for adrenalin. Adrenalin on the other hand was designated as C1 originally. Dahlström and Fuxe (1964) named these monoamine groups according to the order in which they were discovered without knowing the identity of the catecholamine in each case (A was used for the first one identified).

The basis of delineation of structures

The human brainstem contains a complex amalgam of neuronal groups, some compact and some dispersed. This discussion deals first with structures related to the motor and sensory functions of the cranial nerves. However, many brainstem structures are not obviously related to a particular cranial nerve function. These include the reticular nuclei, the precerebellar nuclei, and the monoamine cell groups.

As pointed out by Herrick (1922), the motor and sensory nuclei of the cranial nerves are organized in a series of columns in the hindbrain. From ventromedial to dorsolateral they are the general somatic efferent column (hypoglossal and abducens nuclei), the special visceral (branchial) efferent column (motor trigeminal, facial and ambiguus nuclei), the general visceral efferent column (dorsal motor nucleus of the vagus, superior and inferior salivatory nuclei), the visceral afferent column (nucleus of the solitary tract), the general somatic afferent column (sensory trigeminal nuclei), and the special somatic afferent column (cochlear and vestibular nuclei).

Most brainstem nuclei were originally identified using cytoarchitecture, myeloarchitecture, and connectivity. In recent years, researchers have made increasing use of developmental, functional, and chemoarchitectonic criteria to define neuronal groups (Mehler, 1974; Paxinos et al., 2009; Paxinos and Watson, 2014). Using mainly chemoarchitectonic comparisons with the rat, Paxinos et al., (1990) and Paxinos and Huang (1995) identified 61 homologous nuclei in the human brainstem, and to these Paxinos et al., (2019) added four.

Because the AChE content of homologous nuclei is reasonably stable across mammalian species, our recent histological atlas relied mainly on AChE distribution to illustrate brainstem homologies. Some connectivity data were obtained from the study of therapeutic cordotomies (Mehler, 1974). Our guide to segmenting the present MRI/DTI atlas was our previous histological work, augmented with works of authors in The Human Nervous System (Mai and Paxinos, 2012), The Rat Nervous System (Paxinos, 2014) and The Mouse Nervous System (Watson et al., 2012). The following is a summary of our approach to the delineation of particular structures. Note that this section is organized according to the developmental ontology described by Puelles at al. (2013).

We had available five series of transverse images for the segmentation containing an evenly spaced subset of the original MR images spanning the rostrocaudal extent of the brainstem. Of the five series listed below, only the GRE and FAC were helpful for delineations:

GRE - gradient recalled echo

DWI - isotropic diffusion weighted spin echo image

ADC - apparent diffusion coefficient

RD - radial diffusivity

FAC - directionally colored fractional anisotropy image

We began delineating the GRE image series using pencil on a transparent paper overlay. First, we outlined the obvious features of the GRE image. Next, we overlaid the transparency on the FAC image to outline the major fiber tracts. FAC was occasionally helpful in the delineation of nuclear groups, memorably, for example, the superior olive was slightly green. The DWI, RD and ADC series nearly always provided no more information than the GRE.

Finally, we compared each pencil diagram with the diagrams on either side of it to harmonize the boundaries in adjacent diagrams. We continually referred to the diagrams in the Paxinos et al., (2019) Human Brainstem atlas during this stage of editing of the drawings.

Wherever possible, we relied on the actual signature of the structure on MRI. E.g., the red nucleus is all too evident because it is black in GRE. However, when we had confirmation of the presence of a structure, we could guess the location of its neighbors, much as you can safely guess the location of Italy, France, Austria and Germany once you know the borders of Switzerland.

GRE images

In the brainstem, overall, the GRE images display major fiber tracts as generally darker while nuclear groups are often pale or lighter in appearance. Because GRE images represent fiber bundles so well, they are the best choice for primary identification of brain structures.

FAC images

The color convention for FAC images displays longitudinal tracts in green (running across the rostrocaudal axis), tracts crossing the midline horizontally in red (running laterally), and vertical oriented tracts in blue (running along the dorsoventral axis), as seen in the MRI/DTI Atlas of the Rat Brain by Paxinos et al., (2015). In accordance with the orientation of the human brainstem, tracts along the dorsoventral axis are generally seen as green, running rostrocaudally as blue, and tracts crossing the midline horizontally displaying as red. Fibers with an oblique angle orientation are represented by a mix of these initial colors (i.e., tracts 45° to the dorsoventral (green) and rostrocaudal (blue) orientation are seen as cyan).

Descriptive features from GRE/FAC

Following the descriptions given below for GRE signal intensity (i.e., high - bright/white, intermediate - light/grey, low - dark/black), and FA color orientation, the reader should be able to identify each of our delineations in the same manner used to construct this atlas.

Efferent and afferent nuclei of the cranial nerves

Somatic efferent nuclei

The somatic efferent nuclei of the brainstem are the hypoglossal and the abducens. The oculomotor and trochlear nuclei were previously classified as somatic efferent, but they have been found to express Phox2b during development and, therefore, must be classed as branchial (special visceral) efferent.

Special visceral (brachial) efferent nuclei – ambiguus nucleus, facial nucleus, motor trigeminal nucleus, trochlear nucleus, and oculomotor nucleus.

The branchial motor nuclei supply the gill arch muscles of early vertebrates. The nuclei are clearly defined in early development by the expression of Phox2b. Using this criterion, we must consider the trochlear and oculomotor nuclei to be derived from ancestral gill arch motor nuclei since they both express Phox2b.

General visceral efferent nuclei

The general visceral efferent (parasympathetic) nuclei of the brainstem include the Edinger-Westphal nucleus (see oculomotor nucleus above), the superior and inferior salivatory nuclei and the dorsal motor nucleus of the vagus.

General visceral afferent nuclei and associated nuclei – the nucleus of the solitary tract, the parabrachial nuclei, and periaqueductal gray

Nucleus of the solitary tract The solitary tract (sol) is a heavily myelinated fiber bundle that extends from the spinal cord to the level of facial nucleus (shown in Figs 27-67 and 136-141 of the histological atlas). The nucleus of the solitary tract (here abbreviated to 'solitary nucleus'; Sol) occupies a large region of the dorsal tegmentum, mostly medial to the solitary tract.

General somatic afferent nuclei

The general somatic afferent (somatosensory) nuclei of the brainstem include the gracile and cuneate nuclei and the sensory trigeminal nuclei.

Special somatic afferent nuclei – vestibular and cochlear nuclei

The special somatic afferent nuclei include the auditory and vestibular nuclei. We have also included in this section a discussion of other nuclei associated with the auditory and visual pathways.

Features in GRE/FAC

The **hypoglossal nucleus** (12N) displays a higher signal intensity than the subjacent reticular formation, appearing lighter in GRE in the transverse plane (Fig 7). In contrast, the 12N shows a lower signal intensity than both the **intercalated nucleus** (In) and **the dorsal motor nucleus of vagus** (10N), appearing darker than both of which it overlies in the sagittal plane. The 12N is also darker than the **prepositus nucleus** (Pr) which lies rostral to it (Fig. 63).

In the transverse plane, **the facial nerve nucleus** (7N) is slightly brighter than its surrounds making its boundary distinguishable from the **intermediate reticular formation** (IRt) which it abuts and is silhouetted against the descending exiting root of the **facial nerve** (7th nerve), which is very dark in appearance (Fig. 24). In FAC, while its boundaries are slightly difficult to distinguish, the 7N is nonetheless detectable between the exiting **abducens nerve** (6th nerve) and 7th nerve and is seen as a blend of green and sky blue (Figs. 23 - 24). Another landmark in FAC at this level is the **central tegmental tract** (ctg) which, at this level, has an appearance of sky blue.

Continuing in the transverse plane, the **abducens nerve nucleus** (6N) is seen to have a similar signal intensity to that of 7N, also making this nucleus evident in GRE in comparison to the subjacent reticular formation. However, in FAC, the boundaries of 6N are slightly more easily distinguishable than that of 7N, appearing as a mélange of faint green and mauve-blue hues (Fig 24). In contrast to the nuclei, both the 6th and 7th nerves are very conspicuous in FAC. The exiting 6th nerve is fully green, as it parallels with the descending part of the 7th nerve; both of which are well displayed in Figs 22-25. The exiting 7th nerve is captured particularly well in Fig. 24, as it hooks around 6N and proceeds to exit through the **middle cerebellar peduncle** (mcp), clearly visible as magenta-purple in its horizontal mediolateral course and appears as pale-yellow merging into bright green along its descending course (Fig. 24).

In accordance with the transverse, 6N in the sagittal plane is brighter than the subjacent reticular formation in GRE. In FAC, 6N is principally a dull green under the purple-red crossing fibers (Fig. 65), while at more lateral levels, appears blue with patches of green (Fig. 67). The 6th nerve fibers are seen to sweep across the sagittal plane, appearing dark in GRE and green in FAC green (displayed well at Fig. 65). The exiting 7th nerve is seen as a pale yellow and purple where it takes multiple directions; mediolateral, dorsoventral, and anteroposterior (Fig. 72).

The **motor trigeminal nucleus** (5N) is quite evident and detectable in GRE by its lighter appearance and the perforation of the dark exiting motor fibers in the transverse plane (Fig. 28). In FAC, 5N is dark navy laterally, but a lighter navy blue medially, while the extending **sensory root** (s5) and **motor root** (m5) fibers appear as a blend of bright yellow-green, and the exiting **trigeminal nerve** (5th) itself is seen as solid bright yellow as it exits (Fig. 28). In the sagittal plane, 5N is lighter in GRE compared to the darker 7th nerve which it abuts while in FAC, 5N appears as dark purple and blue, with patches of green from the fiber bundles it issues (Fig. 75). The m5 and s5 trigeminal nerves are yellow, and the exiting fibers are green, as can be seen at Fig. 81.

The **trochlear nerve nucleus** (4N) in the transverse plane has a light appearance in GRE and is evident against the darker grey of the **medial longitudinal fasciculus** (mlf). In FAC, the nucleus appears as a dull green with hints of dull purple (Fig. 47). The **trochlear nerve** (4th) is very dark in GRE, making it the most obvious structure in this section as it crosses the midline. In FAC, it can be seen as a distinct orange-red and purple coloring in FAC as it proceeds caudally and is poised to cross to the other side (Fig. 41). In the sagittal plane, 4N can be seen as a green and pink structure, and the 4th nerve is visible as burnt orange-red in color (Fig. 63).

In the transverse plane, the **oculomotor nucleus** (3N) in GRE can be seen as lighter in appearance than the mlf that partly engulfs it, as well as the **dorsal raphe** (DR), and in FAC appears to be a split of blue and green (Fig. 50). The **oculomotor nerve** (3rd) fibers are very distinct, appearing quite dark in GRE compared to the brighter **ventral tegmental area** (VTA) they protrude through, while in FAC they are seen exiting as a dull green (Fig. 52). In the sagittal plane, 3N is, again, lighter in GRE compared to the darker mlf, and can be seen at an advantage at Fig. 62. In FAC, the nucleus appears blue and green (Fig. 62), becoming predominantly a deeper blue with slight purple hues more caudally at Fig. 63. The 3rd nerve exiting fibers are very dark in GRE, corresponding to the green exiting fibers visible in FAC at the same level (Figs. 66-67).

The mlf can be seen throughout its tract to have a dark appearance in GRE (at the level of the 4N; Fig. 47), and appears light blue at this level in FAC, the nerve fibers themselves being reddish in appearance. Caudally, the mlf appears a lilac blue against the deeper blue of the **interfascicular dorsal raphe** (DRI; Fig. 40), while more rostrally, the mlf appears a sky blue (Fig. 48). In the sagittal plane, the mlf is panoramically visible in GRE at Fig. 62, and is equally prominent in FAC, where we can see its coloration change from deep blue caudally, to a sky blue more rostrally at the level of the **decussation of the superior cerebellar peduncle** (xscp). The mlf is also prominent at Figs. 63-64, becoming attenuated at Fig. 65.

In the transverse plane, the **solitary tract** (sol) has an intermediate signal intensity in GRE, seen as dark grey making it easily identified against the lighter grey background of the **solitary nucleus** (Sol), and 10N (Figs. 13 - 16), The tract is also conspicuous in FAC, seen as bright purple (Fig. 13). Sol itself is quite visible in GRE along with the associated 10N, as they are brighter than the ventrally lying **parvicellular reticular formation** (PCRt) and, at times, brighter than the overlying vestibular nuclei (e.g., Figs. 15-16). In FAC, Sol has olive green elements with slightly faint red hues (Fig. 15).

The **medial parabrachial nucleus** (MPB) has a brighter intensity signal in GRE compared to surrounding structures and is excellently displayed in the transverse plane at Fig. 33, where it borders the much darker **superior cerebellar peduncle** (scp). This arrangement is also well displayed in the sagittal plane in GRE (e.g., Fig 68). The MPB is a brilliant reddish purple in FAC in the transverse plane at Fig. 33. In the sagittal plane, MPB displays similar coloring in accordance with its transverse plane presentation, seen as a bright pink-purple structure at Figs. 75-74, becoming a blueish purple more rostrally at Figs. 73-71.

In the transverse plane, the **periaqueductal gray** (PAG) is rather uniformly bright in GRE, with its intensification of brightness appearing as a tight doughnut surrounding the aqueduct (Fig. 47). In FAC, the PAG appears predominantly as deep blueish purple, however patches of pinkish red are also visible in the **ventrolateral PAG**, and green present in the **dorsomedial PAG** (Fig. 47). In the sagittal plane, the PAG from both below and above the aqueduct is very bright at the most medial level in GRE (Fig. 61), with the brightness attenuating at more lateral locations, as seen in Fig. 65. In FAC, the PAG has a rather blue appearance at the medial locations, while at Fig. 65 is a darker, multicolor mélange.

Nestled within the periaqueductal gray, and overhanging the 3N, are the **parvicellular oculomotor nucleus** (PC3), **supraoculomotor periaqueductal gray** (Su3), and **supraoculomotor cap** (Su3C), all of which appear slightly lighter than the subjacent ctg and other surrounding regions outside the PAG (Fig. 52). At rostral levels of the transverse plane, the ctg is quite visible in FAC, appearing as an oval of violet surrounded by a donut of blue or red hues (Fig. 49). However, in GRE the tract is only marginally detectable from the remainder of the **mesencephalic reticular formation** (mRt; Fig. 49). In FAC, at more caudal levels the tract still retains some of the violet hue, though is progressively seen to become more blueish before it descends to expire in the inferior olive (Fig. 29).

The dorsal column nuclei (the **gracile nucleus** (Gr) and **cuneate nucleus** (Cu)) are light grey in appearance and feature against the darker **gracile fasciculus** (gr) and **cuneate fasciculus** (cu) that surround them, especially dorsally in the GRE (e.g., Fig. 4). The FAC also displays these structures well, with the nuclei themselves being green and the fiber bundles of gr and cu being various colors, predominantly blue and pink-violet (Fig. 4). The **external cuneate nucleus** (ECu) in GRE appears similarly bright in appearance to the dorsal column nuclei (e.g., transverse Fig. 12), and is silhouetted well against the darker **inferior cerebellar peduncle** (icp). In FAC, the ECu is predominantly green with a slight area blended with purple-red, presumably due to the orientation of fibers within the nucleus (Fig. 12). In the sagittal plane, the Gr appears brighter than its surrounding fiber bundle in GRE and is shown as light green in FAC (Fig. 61). This arrangement persists for the extent of the nucleus, seen as far as laterally as Fig. 65, subsequently becoming characterized by patches of red, green, deep blue. The Cu can also be seen as principally green, with some red regions in FAC, and is brighter than the cu in GRE at Fig. 68.

In the transverse plane, the **medial lemniscus** (ml), at its origin in the dorsal column nuclei, is called the **internal arcuate fibers** (ia) and in FAC can be seen as green, consistent with their principally dorsoventral direction (Fig. 5). When entering its decussation, ml is a bright lime green, subsequently becoming blue as its direction becomes caudorostral in more rostral regions, such as Fig. 13. The ml is interrupted by fibers from the inferior olive as they rush through the midline to the opposite cerebellum via the icp, with the level of their crossing attenuating the blue color of the ml with a reddish tinge (e.g., Fig. 14). The midline at several levels features a very strong green coloration which is associated with the **medullary stria** (meds; e.g., Fig. 8-9). The ml is extremely clear in FAC at Fig 33, distinguishable from the mediolaterally directed fibers of the pons, as a blueish lilac color. Lateral to the crossing of the scp, at the same level, we see a very bright green appearance which corresponds to part of the nucleus of the **retrorubral field** (RRF) and surrounding space, the green making a distinction between the nuclei, but it fills in the space between the lateral ml and the substantia nigra compacta (Fig. 48).

In the sagittal plane, the ml can be seen at an advantage in Figs. 65-69, appearing darker than all bordering structures. The ml distinguishes itself in FAC against the red **transverse fibers of the pons** (tfp), appearing lilac in color that extends into a sky blue. More rostrally, the ml features shades of green and pink, like unmixed paints. More laterally in the sagittal plane, the ml can be seen circumnavigating the pontine bubble featuring components of slight pink, blue, and sky blue (e.g., Figs. 69-71), increasingly becoming a reddish-purple at Figs. 73-74, and purple at Figs. 75-77.

The **spinal trigeminal nucleus, caudal part** (Sp5C) is very bright in GRE in **laminae 1 & 2** (Sp5C1/2), and less bright in **laminae 3 & 4** (Sp5C3/4) in the transverse plane (Fig. 1). In FAC, the nucleus itself is red and green across different areas while, external to this, the **spinal tract of 5** (sp5) appears as blue (Fig. 1). A similar arrangement is observed in Fig. 4, where laminae 1 & 2 are very bright in GRE, and the remaining caudal part of Sp5 is also quite bright. In GRE, sp5 itself is white in both Figs.1 and 4, which is unusual given that other tracts visible at the same level, such as the **pyramidal tract** (py) and cu, both appear darker. Therefore, the sp5 can be distinguished from the cu due to its brightness. In the sagittal plane, the sp5 can be seen as a blue structure in FAC at Fig. 77, with the crossing cerebellar fibers visible both above and below it, appearing red in their medial lateral course to join up the inferior cerebellar peduncle, which is seen as green caudally and blue more rostrally.

The **principle sensory trigeminal nucleus** (Pr5) can be seen at an advantage in GRE in the transverse plane at Figs. 27-28, with the nucleus's boundaries evident appearing slightly brighter than the mcp and s5 and, in FAC, is visible as a blend of both light and dark green, located medial to the mcp and s5 (Fig. 27). The **spinal trigeminal oral subnucleus** (Sp5O) can be seen in GRE as brighter than its tract in which it nestles, as seen at an advantage at Fig. 22.
The **medial vestibular nucleus** (MVe) is evident in the sagittal plane in GRE showing a brighter signal intensity than other bordering structures, and in FAC appears principally as a dull green interspersed with dark blue-purple (Fig. 68). The **superior olive** (SOl) is not distinctive in GRE in the transverse plane, though in some regions it is seen to be cradled by the darker fibers of the **lateral lemniscus** (ll; Fig. 28). In FAC, SOl can be seen as a dark green, appropriate for its name (Fig. 26). In the sagittal plane SOl can be seen as patches of teal and olive green in FAC (Fig. 74). The **ventral lateral lemniscus** (VLL) appears lighter in GRE against the darker ll in the transverse plane (Fig. 35), while in FAC we can see VLL being a denser blue than the sky blue of the ll itself (Fig. 35).

In the transverse plane, the **central nucleus of the inferior colliculus** (CIC) in GRE is has a lower signal intensity, seen to be dark in appearance while, in contrast, the **dorsal cortex of the inferior colliculus** (DCIC) is brighter, as is the dorsolateral aspect of the **external cortex of the colliculus** (ECIC). Conversely, the ventral and medial aspect of the external cortex of inferior colliculus, where the lateral lemniscus invades the colliculus, is darker (Fig. 46). In FAC, the CIC is purple and red in appearance, with the red portion extending into parts of the DCIC. The remaining DCIC is blue. The ECIC shows both slightly red and light green color, with the invading lm visible in blue. The **commissure of the inferior colliculus** (cic) shows dark in GRE at Fig. 47, and as expected, red in FAC. In the sagittal plane, the inferior colliculus can be seen at an advantage at Figs. 66 and 67. In GRE, the CIC is shown to be darker than both the DCIC and ECIC, and in FAC, the ECIC is greener in color than the CIC, which does not look homogeneous but is rather constituted by red, purple, and blue patches (Fig. 67). The **brachium of the inferior colliculus** (bic) in GRE has a lower signal intensity, seen as dark in appearance and in FAC, shows as a dull dark green in the transverse plane (Fig. 50).

In the transverse plane, the **medial geniculate** (MG) is optimally seen in Fig. 55, and while it is hard to distinguish its subnuclei, the **ventral** (MGV) and **dorsal** (MGD) nuclei do appear slightly darker than the **suprageniculate nucleus** (SG) which abuts them dorsally and medially, as seen Fig. 56. In FAC, the MGV subnucleus shows a distinctly brighter red than the other subnuclei, which appear as a mélange of colors (Fig. 56).

Reticular and tegmental nuclei of the brainstem

Many nuclei in the brainstem that are not directly associated with the cranial nerves, or the cerebellum have been labelled as reticular or tegmental nuclei. In the past, the reticular nuclei were considered to form a functional group which was named the reticular formation. This collective concept has been abandoned in favor of a separate consideration of the individually named reticular nuclei. Some nuclei that are named 'reticular' belong to entirely different functional entities. These include the reticulotegmental and lateral reticular nuclei (both precerebellar nuclei).

In the hindbrain the large celled (gigantocellular) reticular nuclei are medially placed, and the small celled (parvicellular) nuclei are laterally placed (Figs. 8-24, and Figs. 61-72). The narrow region between these two large nuclei was named, by Paxinos and Watson (1986), the intermediate reticular nucleus.

Features in GRE/FAC

In the transverse plane, the reticular formation at the caudal level shows various colors in FAC, with reddish and green regions featured dorsally and blueish reddish regions more ventrally (Fig. 8). Rostral to the caudal nucleus, at the levels of the **interpositus** (IPo), the PCRt still has an olive-green tinge while the **gigantocellular reticular nucleus** (Gi) has a blueish tinge (Fig. 15). In the sagittal plane, the **intermediate reticular nucleus** (IRt) is displayed with the warmer colors red-purple and is bracketed by the PCRt which is seen as a deep royal blue, and the **lateral paragigantocellular nucleus** (LPGi), which is also blue (Fig 72). This arrangement persists at Fig. 73.

Monoaminergic nuclei of the brainstem

Four distinct populations of monoamine nuclei are present in the brainstem. These are the dopamine nuclei (substantia nigra, ventral tegmental area, and retrorubral field), the noradrenaline nuclei (locus coeruleus, NA7, NA5, NA2 and NA1), the adrenaline nuclei (Ad1 and Ad2), and the serotonin nuclei of the raphe.

Features in GRE/FAC

In the transverse plane, the **substantia nigra** (SN) is visible in GRE as a very dark structure in its reticular part and has a slightly lighter appearance in its compact part. At times, it becomes difficult to distinguish the two as there are instances where some of the compacta invades or is found wholly within the reticulata (Figs. 53-54). In FAC, the SN is seen as blue and green, with blue dominating in the reticulata and the green dominating in the compacta. The green areas are at times blended with green corresponding to the invading compacta. Medial to the SN, and below the crossings of the scp, there is the **parabrachial pigmented nucleus** (PBP) and the ml. This area in FAC is bright green, indistinctly spreading from structure to structure (Fig. 54).

In the sagittal view, the PBP is the brightest structure in GRE, between the SN and ml. It is brighter than the compact part of the reticular formation, with the reticulata nicely shown to be blueish in color in FAC, and the compacta seen as green (Fig. 75). The SN is seen at an advantage at Fig. 78, where it is a gravel of pebbles that are sky blue, yellow, and green. Here, the reticulata is of a duller color than the PBP, which is seen as a brighter green (Fig. 78).

In the transverse plane, the **locus coeruleus** (LC) is brighter in its alpha part (Fig. 31), which is below the fibers separating the central grey from the reticular formation. In the FAC, these fibers show some purple-red coloration, which distributes and attenuates dorsally where the main LC lies (Fig. 31).

The **median raphe nucleus** (MnR) in GRE is light grey followed dorsally by the bright white of the **dorsal raphe nucleus, caudal part** (DRC; Fig. 37). Here, the FAC shows the midline as a light mint-green, with the same color for the DR. At the level of the 4th nerve, the DR is bright in GRE, and in FAC it is blend of navy with patches of magenta, indistinguishable from the PAG (Fig. 47). Continuing caudally, the MnR becomes slightly brighter in GRE than its surrounds, and in the FAC it is shown to be a light green midline, surrounded by the deep blue **paramedian raphe** (PMnR; Fig. 40). In the sagittal plane, the DR is lighter than the fibers of the mlf that are below it, but darker than the region of the PAG near the midline. More laterally, the PAG and the DR are indistinguishable in GRE (Fig. 64).

Precerebellar Nuclei and Red Nucleus

The precerebellar nuclei (nuclei that project directly to the cerebellum) include the pontine nuclei, the inferior olive, the lateral reticular nucleus, and smaller hindbrain nuclear groups. A comprehensive description of the cytoarchitecture and connectivity of the precerebellar nuclei is to be found in Voogd and Ruigrok (2012).

The pontine nuclei. The pontine nuclei are interspersed between the longitudinal fasciculus of the pons and the transverse fibers of the pons. In GRE the cellular regions are lighter in color (e.g., Fig. 40). In FAC, the descending fibers are strikingly blue and the crossing fibers strikingly red, with the cellular regions often light green.

Inferior olive. All the subdivisions of the inferior olive can be seen in the myelin series of Paxinos et al., (2019; their Figs. 133-143). In GRE of the present study, the inferior olive is conspicuously bright in its cellular regions (e.g., Fig 11), while in FAC, the cellular regions look greenish and strikingly different from the hilus of the inferior olive which appears red, consistent with the mediolateral disposition of fibers that exit the olive to cross the midline and reach the contralateral cerebellum (Fig. 11).

Features in GRE/FAC

The **lateral reticular nucleus** (LRt) is hard to identify in FAC, however, guided by expected position, it has green components in FAC (Fig. 6). More rostrally, it can be seen principally as blue with some slight purple coloration inside (Fig. 9). In the transverse plane, the principal subnucleus of the inferior olive is conspicuous in GRE, being bright white in the cellular areas, and darker in the hilus and elsewhere (Fig. 16). The smaller subnuclei of the olive, such as the **inferior olive, dorsomedial cell group** (IODM), **inferior olive, medial nucleus** (IOM), **inferior olive, cap of Kooy** (IOK), and **inferior olive, beta subnucleus** (IOBe) are also quite bright regions adjacent to the darker fiber bundles of the ml and the py (Fig. 16). In the FAC, the principal subnucleus is green in its cellular regions, and the **hilus of the inferior olive** (hio) is predictably (by the fiber orientation) bright red. The red color is also present occasionally in other areas such as above the py and in some of the septa between the folds of the cellular regions; the red presumably produced by the mediolaterally directed fibers as they rush to cross the midline and reach the cerebellum of the opposite side (Fig. 16). The associated **arcuate nucleus** (Ar) which also projects to the cerebellum and is principally surrounding the pyramids is strikingly visible in GRE and it is the dark green region that is visible in the corresponding FAC (Fig. 10). In the sagittal plane, the inferior olive can be seen most prominently in Fig. 68 as a wrinkled/crumpled purse, bright in cellular regions and dark in fibrous regions in GRE. In FAC, the cellular regions are green, whereas the hilus can be seen as strong orange-red (Figs. 67-68). Remaining in the sagittal plane the ctg, when close to the olive, is excellently displayed at Fig. 67 and Fig. 68, and is seen in a panoramically in Fig. 68. In FAC it appears lilac blue, though caudally, at rostral levels above the crossing of the scp, it appears sky blue when approaching its destination, the inferior olive.

The py, once it leaves the **pontine nuclei** (Pn), is principally royal blue in color, but within it exist clouds of sky blue and even reddish-purple at its most ventral region (Fig. 19). Advancing caudally, the situation persists, and where the pyramid is preparing to enter the decussation the color changes to a purple pink, with the **pyramidal decussation** (pyx) is itself principally a mint green (e.g., Fig. 3). In the sagittal plane, the py is evident, more so in the midline where it can be seen as blue to sky blue and a brighter blue green as it crosses in Fig. 61. In Fig. 62, the blue predominates with the crossing appearing a light green. In Fig. 63 it is principally seen as deep blue or sky blue and persists in this arrangement.

The Pn hug the py as it struggles to exit and form the pyramids. In the transverse plane at these levels, the Pn are extremely bright in GRE, silhouetting the **longitudinal fasciculus of the pons** (lfp) as well as the tfp (e.g., Fig. 20). In FAC, these cell groups are principally green and dark green at the external end (Fig. 20). As we go rostrally, the lfp are embedded in the Pn and are bracketed by the crossing tfp, with pontine cell groups dispersed between. At this level, the Pn are light in color (however not as bright as seen in Fig. 20) and retain their green in appearance in FAC (Fig. 24). Internally, we see the lfp as various shades of blue (Fig. 24).

Continuing in the transverse plane, many crossing fibers can also be seen to an advantage in Fig. 36, where the tfp, which are a vibrant red, are indigitated with the lfp, appearing as a range of deep blue, violet and lighter sky-blue fibers. These crossing fibers form the likeness of a big sandwich with many layers. The vibrant red tfp fibers terminate medially to form the mcp, and the blue lfp fibers are surrounded by patches of green, followed by red, and then blue again, repeatedly, like an interdigitation of fingers (Fig. 36). The mcp and tfp are also excellently shown at Fig. 43, where they allow some cellular groups to interpose themselves between them. The tracts appear dark in GRE, but there are visible light streaks appearing between them, owing to these cell groups. Notably, in the midline of the basilar pons there is always a green appearance which may be related to the step pattern that the crossing fibers seem to form. In preparations of myelin in Human Brainstem (Paxinos et al., 2019), we can see that the crossing fibers follow a step pattern when the bundles arrive at the midline, changing direction to advance slight upwards or downwards to interdigitate with their opposites on the other side and, possibly, this change in orientation produces some of the green midline (their Figs 146-148). However, the existence of a pontine midline fiber bundle orthogonal to the lfp and tfp cannot be excluded (Fig. 43).

On the midline of the sagittal plane, we find vast displays of the green Pn in FAC, and in GRE they appear brighter than the surrounding fibers (especially at the caudal levels). At the midline, the tfp are seen to congregate in the more rostral part of the **pontine oral nucleus** (PnO). The Pn predominate at Fig. 61, however following this point the tfp and lfp become more predominant, with the green cellular Pn regions filling the matrix between the red tfp and blue lfp (FAC Fig. 63). One of the most striking images of this in the sagittal plane can be seen in Fig. 66. The lfp fibers are also seen to become continuous with the py at Figs 66-68 in GRE and, in FAC, together display a brilliant fibrous projection that is purple, cornflower blue, and sky blue in color.

In the transverse plane, the **cerebral peduncle** (cp) looks like a psychedelic painting in FAC containing sky blue, royal blue, pink and purple hues all together

(Fig. 52). In GRE, the cp appears as solid grey, however it is not as dark as the SNR which it cradles (Fig. 52). At this level, the rostral region of the cp appears dark, however not as dark as the underlying SNR or SN (Fig. 57). In FAC, there are many purple-red fibers and patches that can be seen to be intermingled with the green and light blue (Fig. 57).

In the sagittal plane, we meet the **internal capsule** (ic) at Fig. 81 as it metamorphoses into the cp at the level of the **subthalamic nucleus** (STh). The ic is principally blue, becoming a blend of blue, pink, and purple as it goes into the cp (Fig. 81). Continuing in the sagittal plane, the Ar, which can be viewed as a continuation of the Pn, is visible in Figs. 61-62, and is seen as green in appearance at more lateral levels such as Fig. 63, with the color shifting to deep blue at Fig. 64.

The **reticular tegmental nucleus** (RtTg) is seen as lighter than the surrounding ml or pontine reticular formation in GRE, and in FAC one can see patches of green and blue (Fig. 27). In the sagittal plane, the RtRg can be seen throughout its extent in Fig. 63 and is lighter than the subjacent fibrous zone of the pons in GRE. In FAC, it appears as light blue, lilac and green, best seen in Fig. 63.

The **red nucleus** (R) is a mélange of colors in Fig 55, while occasionally, at more rostral levels, the green predominates (Fig 58). At Fig. 59, the **fasciculus retroflexus** (fr) is a bright yellow-green and is shown to be advancing through green patches of the medial region of the R. The R loses green coloring more caudally (Fig. 52) appearing blue medially, and purple in the lateral half. The arrangement persists until the medial region becomes greener and the lateral region displays purple hues (Fig. 55). Even more rostrally, the green takes over until there is no more purple, as in Fig. 59, although some blue elements in the center remain. In the sagittal plane the R, when close to the midline (the most medial edge on the sagittal plane being at Fig. 63), can be seen to feature green and blue in a sea of principally mint green fibers of the scp. Yet more laterally, it looks mosaic with green, blue, and magenta patches, and is also seen as an intensification of brighter green in FAC (Fig. 65). At Fig. 67 we see the R as being patchwork of green and pink with white parts surrounded by the scp. The red nucleus at more lateral levels is seen to be a conglomerate of green, deep blue and magenta (Fig. 74).

The scp is very dark in GRE as it exits the cerebellum and is shown as an intense sky blue in FAC (Fig. 34). In contrast, the cerebellum itself appears extremely bright in GRE and a forest green in FAC (Fig. 30). In Fig. 34 the arrangement persists, where the scp is within the brainstem, still appearing as sky blue in FAC (Fig. 37). The scp indeed becomes dark again in GRE at Fig. 43. As it then proceeds to reach its crossing, the lateral elements that have not yet crossed appear a paler blue, becoming white at the elbow, and subsequently red as they cross the midline (Fig. 41). At the decussation itself, the lateral parts are light blue, moving into white, finally becoming principally red as they cross the midline (Fig. 44). In GRE, we can see a striation of the fiber bundles as they cross the midline. At more rostral levels, in GRE the scp is still as dark where the nigra is about to expire. It has strips of redness interdigitated with a navy blue and lilac in FAC (Fig. 48).

In the sagittal plane, the scp as it leaves the cerebellum is a sky blue through its extent (Fig. 71). At its crossing, the scp is evident as principally green (Fig. 61), though interdigitated with strips of red and yellow. This persists at Figs. 63-65, with the red now predominating, and the yellow lost at Fig. 67.

Nuclei Related to the Visual System

In the midbrain, the superior colliculus and tectal gray are important visual centers.

The **dorsal lateral geniculate nucleus** (DLG) in GRE is slightly bright and distinguishable amongst the fibers of the **optic tract** (opt) that terminate in it (Fig. 59). In FAC, it is seen as a dull green surrounded by the brighter green of the tract, with the tract itself becoming yellow as it follows a rostral and medial and ventral course girthing the cp (Fig. 59). The overlying **pulvinar** (Pul) appears a red-dull orange, as can be seen in Fig. 59, and in GRE stripes of fibers that traverse it appear darker than the cellular regions. The **brachium superior colliculus** (bsc) is dark in color and can be seen to surround the **anterior pretectal nucleus** (APT) which is light in color in GRE (Fig. 56). The APT in Fig. 56 is predominantly a dull teal green with purple-red components in FAC. In the sagittal plane, the opt is seen as bright yellow at Figs. 82-83.

In the transverse plane, the **superior colliculus** (SC) in GRE features a bright **zona layer** (Zo) and **superficial gray layer** (SuG), and appears rather uniformly grey in the remainder of the colliculus, except for part of the **optic nerve layer** (Op), which is also partly bright (Fig. 52). The **deep white layer** (DpWh) features a dense fibrous component. In FAC, the SC shows a green color in its Zo, SuG, and **intermediate grey layers** (InG), with this green color attenuated in the **intermediate white layer** (InW). The **deep grey** (DpG) and **deep white** (DpWh) layers are green in their medial edge but are a pink-purple color more laterally. Close to the midline, the layers above the DpG and DpWh show a slightly navy color (Fig. 52). The green from the SC layers continues with a green component of the PAG, ending halfway between the dorsal part of the PAG and the aqueduct. The **posterior commissure** (pc) is visible as bright red in its crossing at Fig. 58.

In the sagittal plane, a striking feature of the SC is the red color of the DpWh, seen to be continuous with the red of the pc at Fig. 61. This layer is dark in GRE, outlined nicely against the subjacent PAG or the overlying DpG. The SC can be seen at an advantage in Fig. 65, again with the DpWh being obvious in GRE and showing purple and red in the FAC. The **pineal gland** (Pi) can be seen as green in FAC and very white in GRE. The SC has a substantial amount of green close to the midline except for its rostral level which is lilac (Fig. 65). This can also be seen at Fig. 67 with the DpWh appearing as magenta, and the remaining layers being principally green.

Other regions

In the transverse plane, the regions surrounding the **central canal** (CC) are bright compared to the surrounding fibrous regions such as the subjacent pyx in GRE (Fig. 3), while in FAC, the region is royal to navy blue in color (Fig 3).

The **interpeduncular nucleus** (IP) is shown as bright in GRE (Fig. 48), while in FAC is seen as extremely dark navy blue (Fig. 48). In the sagittal plane, the IP is shown as very bright in contrast to the adjacent xscp in GRE and appears a deep blue in FAC, becoming a sky blue as it reaches towards the MnR (Figs. 62-61).

The crossing of the **retromammillary decussation** (rmx) in GRE is seen as a slightly darker grey than the surrounding cellular regions of the hypothalamus. The **lateral hypothalamus** (LH), **ventromedial hypothalamus** (VMH) and surrounding area have a high signal intensity and are seen as are extremely bright in the transverse plane (Fig 59). In FAC, the rmx is predominantly deep red in coloration containing some central green fibers/patches (Fig 59). The **mammillary body** (MB) in cellular areas has an intermediate signal intensity, silhouetting the principle **mammillary tract** (pm) and **fornix** (f) at this level, both of which are darker in GRE (Fig 59). Remaining in the hypothalamus, the STh is very evident in GRE, appearing as a dark structure just dorsal and anterior to the SN, while the **zona inserta** (ZI) is slightly lighter in appearance in GRE, distinct from the STh at the most caudal level (Fig 59). In FAC, the STh demonstrates a mosaic of colors, with yellow, green, and dark green dominating, while the ZI looks as though its ready for the Carnival - displaying orange, pink and yellow mélange (Fig. 60).

Moving into the sagittal plane, the ZI is brighter than the ventrally lying STh in GRE, and in FAC is seen as flame red and orange appearance (Fig. 86). The f is significantly darker than the surrounding hypothalamic region, which appears very bright (in accordance with the transverse plane), while in FAC it has a mélange of colors as it prepares to invade the MB, showing components of lilac, deep blue, and sky blue (Fig 66). The **mammillothalamic tract** (mt) is predominantly green at this level, consistent with the dorsal ventral orientation at Fig. 66. The APT is distinct from the SC and subjacent reticular formation in FAC, predominantly appearing as a dark teal green (Fig. 75).

Comparison of this Atlas with Prior Work Using the Same Data

The MRI data used to create this atlas has also been used in several previously published works including Calabrese et al., (2015), Sitek et al., (2019), Rushmore et al., (2020), and Adil et al., (2021). The present atlas differs substantially from prior work in terms of purpose, scope, and level of detail. Two of these studies did not attempt any in depth anatomic delineation, but rather were focused on DTI fiber tracking for deep brain stimulator electrode targeting (Calabrese et al., 2015) or for mapping subcortical auditory pathways (Sitek et al., 2019). The two remaining studies from Rushmore et al., (2020) and Adil et al., (2021) each provided substantial anatomic delineations, but these did not approach the level of detail provided in the present work. For example, Rushmore et al., (2020) identified 47 unique structures, and Adil et al., (2021) identified 48, while the present work identified 363 structures. In addition, Rushmore et al., (2020) used only the GRE images, thus limiting their ability to identify certain structures. The format of anatomic delineations also differs substantially with Rushmore et al., (2020) providing only annotated axial images (no diagrams) and Adil et al., (2021) providing only 3-dimensional voxel-based labels. Finally, the present atlas is the first and only such atlas to provide fully annotated sagittal plates and diagrams, giving a unique perspective on brainstem organization.

References

Adil, S. M., Calabrese, E., Charalambous, L. T., Cook, J. J., Rahimpour, S., Atik, A. F., Cofer, G.P., Parente, B.A., Johnson, G.A., Lad, S.P., White, L. E. (2021). A high-resolution interactive atlas of the human brainstem using magnetic resonance imaging. NeuroImage, 237, 118-135.

Berman AL. (1968). The Brainstem of the Cat: A Cytoarchitectonic Atlas with Stereotaxic Coordinates. University of Wisconsin Press, Madison.

Bowden, D. M., et al. (2012). NeuroNames: an ontology for the BrainInfo portal to neuroscience on the web. Neuroinformatics 10 (1): 97-114.

Calabrese, E, Hickey, P, Hulette, C, Zhang, J, Parente, B, Lad, SP, Johnson, GA. Postmortem diffusion MRI of the human brainstem and thalamus for deep brain stimulator electrode localization. Human brain mapping 36 (8), 3167-3178.

Capecchi, M. R. (1989). The new mouse genetics: altering the genome by gene targeting. Trends Genet 5 (3), 70-76.

Dahlström, A., and Fuxe, K. (1964) Localization of monoamines in the lower brainstem. Experimentia 20 (7), 398-399

Farago AF, Awatramani RB, Dymecki SM. (2006). Assembly of the brainstem cochlear nuclear complex is revealed by intersectional and subtractive genetic fate maps. Neuron 50 (2), 205–218.

Foster, G. (1998), Chemical Neuroanatomy of the Prenatal Rat Brain. A Developmental Atlas: 416. Oxford University Press. New York.

Herrick, C. J. (1922). "An Introduction to Neurology," 3rd Ed. W. B. Saunders, Philadelphia.

Johnson GA, Ali-Sharief A, Badea A, Brandenburg J, Cofer G, Fubara B, Gewalt S, Hedlund L, Upchurch L. (2007). High-throughput morphologic phenotyping of the mouse brain with magnetic resonance histology. Neuroimage 37 (1), 82–89.

Johnson GA, Cofer GP, Gewalt SL, Hedlund LW. (2002). Morphologic phenotyping with MR microscopy: the visible mouse. Radiology 222 (3), 789–793.

Johnson GA, Thompson MB, Drayer BP. (1987). Three-dimensional MR microscopy of the normal rat brain. Magn Reson Med 4 (4), 351–365.

Joyner, A. L. and M. Zervas (2006). Genetic inducible fate mapping in mouse: establishing genetic lineages and defining genetic neuroanatomy in the nervous system. Dev Dyn 235(9), 2376-2385.

Mai JK, Paxinos G, Voss T. (2008). Atlas of the Human Brain. (3rd edition). Elsevier Academic Press, San Diego.

Mai, J., and Paxinos, G. (2012). "The Human Nervous System". 3rd ed. Academic Press, San Diego

Mehler, W. R., (1974) Central Pain and the Spinothalamic Tract. Adv. Neurol. 4, 127 – 146.

Morin LP, Wood RI. (2001). A Stereotaxic Atlas of the Golden Hamster Brain. Academic Press, San Diego. fine

Olszewski, J. and D. Baxter (1954). Cytoarchitecture of the Human Brain Stem, Lippincott.

Paxinos, G, The Rat Nervous System, 4th Edition, Academic Press, San Diego, 2014.

Paxinos, G, Ashwell, KWS and Atlas of the Developing Rat Nervous System, 4rd Ed, Academic Press, San Diego, 2018

Paxinos G, Carrive P, Wang H, Wang P-Y. (1999). Chemoarchitecture of the Rat Brainstem. Academic Press, San Diego.

Paxinos G, Kassem, MS, Kirkcaldie, M and Carrive P, Chemoarchitectonic Atlas of the Rat Brain, Third Edition, Academic Press, San Diego, 2022.

Paxinos G, Franklin K. (2013). Paxinos and Franklin's, The Mouse Brain in Stereotaxic Coordinates, 4th ed. Elsevier Academic Press, San Diego.

Paxinos G, Halliday G, Watson C, Koutcherov Y, Wang HQ (2007) Atlas of the Developing Mouse Brain at E17.5, P0, and P6. Elsevier Academic Press, San Diego.

Paxinos G, Huang X-F, Sengul G, Watson C. (2012a). Organization of human brainstem. In Mai JK and Paxinos G (Eds.), The Human Nervous System. (3rd edition). Elsevier Academic Press, San Diego.

Paxinos G, Kus L, Ashwell K, Watson C. (1999a). Chemoarchitecture of the Rat Forebrain. Academic Press, San Diego.

Paxinos G, Watson C, Carrive P, Kirkcaldie M, Ashwell K. (2009b). Chemoarchitectonic Atlas of the Rat Brain. 2nd edition. Elsevier Academic Press. San Diego.

Paxinos G, Watson C. (1986). The Rat Brain in Stereotaxic Coordinates", 2nd Ed.

Paxinos G, Watson, C., (2005). The Rat Brain in Stereotaxic Coordinates, 5th Ed. Elsevier Academic Press. San Diego.

Paxinos G, Watson, C., (2007). The Rat Brain in Stereotaxic Coordinates. 6th Ed. Elsevier Academic Press. San Diego.

Paxinos G. (Ed.) (2004). The Rat Nervous System, 3rd ed., Elsevier Academic Press, San Diego.

Paxinos, G. (1983). Evidence for the existence of a parvicellular nucleus medial to the parabigeminal (PBg) nucleus. Neurosci. Lett., Suppl. 14, S276.

Paxinos, G. (1985). The microcellular tegmental nucleus (Mi). Neurosci. Lett., Suppl. 1-9, 589.

Paxinos, G., and Franklin, K.B.J, (2019) Paxinos and Franklin's the Mouse Brain in Stereotaxic Coordinates, Compact 5th Edition, Academic Press, San Diego,

Paxinos G, Furlong T and Watson, C, Human Brainstem: Cytoarchitecture, Chemoarchitecture, Myeloarchitecture, Academic Press Elsevier, San Diego, 2019.

Paxinos, G., and Huang, X. F. (1995). Atlas of the Human Brainstem. Academic Press, San Diego.

Paxinos, G and Watson, C. The Rat Brain in Stereotaxic Coordinates, Academic Press, Sydney, 1982.

Paxinos, G and Watson, C, Paxinos and Watson's The Rat Brain in Stereotaxic Coordinates, 7th Edition, Academic Press, San Diego, 2014.

Paxinos, G, Watson, C, Calabrese, E, Badea, A and Johnson, GA, MRI/DTI Atlas of the Rat Brain, Academic Press/Elsevier, San Diego, 2015

Paxinos, G., Huang, X. F., Petrides, M. and Toga, A. W. (2009). The Rhesus Monkey Brain in Stereotaxic Coordinates, 2nd ed., Academic Press, San Diego.

Paxinos, G., Tork, I., Halliday, G. and Mehler, W.R. (1990). Human homologues to brainstem nuclei identified in other animals as revealed by acetylcholinesterase activity, In G. Paxinos, The Human Nervous System, Academic Press, San Diego, 149-202.

Paxinos, G., Watson. C., Petrides, M., Rosa, M.G., and Tokuno H. (2012b). "The Marmoset Brain in Stereotaxic Coordinates,". Academic Press, San Diego.

Puelles L, Martinez-de-la-Torre M, Bardet S, Rubenstein JLR. (2012b). Hypothalamus. In Watson C, Paxinos G, and Puelles L (Eds.), The Mouse Nervous System. Elsevier Academic Press, San Diego.

Puelles L, Martinez-de-la-Torre M, Ferran J-L, Watson C. (2012c). Diencephalon. In Watson C, Paxinos G, Puelles L (Eds.), The Mouse Nervous System. Elsevier Academic Press, San Diego.

Puelles L, Martinez-de-la-Torre M, Paxinos G, Watson, C, Martinez S., (2007). The Chick Brain in Stereotaxic Coordinates. An Atlas Featuring Neuromeres and Mammalian Homologies. Elsevier Academic Press. San Diego.

Puelles L, Martinez-de-la-Torre M, Martinez S, Watson C, Paxinos G, The Chick Brain in Stereotaxic Coordinates and Alternate Stains, 2nd Edition, Academic Press, San Diego, 2019.

Rushmore, Richard Jarrett, et al. (2020). 3D exploration of the brainstem in 50-micron resolution MRI. Frontiers in neuroanatomy 14, 40.

Suddarth SA, Johnson GA. (1991). Three-dimensional MR microscopy with large arrays. Magn. Reson. Med. 18 (1), 132–141.

Tomás-Roca, L., Corral-San-Miguel, R., Aroca, P., Puelles, L., Marin, F., (2016) Crypto-rhombomeres of the mouse medulla oblongata, defined by molecular and morphological features. Brain Structure and Function. 221 (2), 815 - 838.

Toulmin, S. (1953). The philosophy of science: An introduction. London: Hutchinson's University Library.

Voogd, J., and Ruigrok, T. (2012) Cerebellum and Precerebellar Nuclei. The Human Nervous System, J. Mai and G. Paxinos Eds, 471-545. Academic Press, San Diego.

Watson C, Paxinos G, Puelles L. (Eds.). (2012). The Mouse Nervous System. Elsevier Academic Press, San Diego.

Watson, C, Paxinos, G, Kayalioglu, and Heise, C. (2009). Atlas of the mouse spinal cord. Academic Press.

Watson, C, Mitchelle, A, and Puelles L. (2017a). A New Mammalian Brain Ontology Based on Developmental Gene Expression. Evolution of the Nervous System 53-75.

Watson, C., Shimogori, T., and Puelles L. (2017b). Mouse Fgf8-Cre-LacZ lineage analysis defines the territory of the postnatal mammalian isthmus. J Comp Neurol 525 (12), 2782-2799.

List of Structures

3rd ventricle 3V
4th ventricle 4V

A
abducens nerve 6n
abducens nucleus 6N
acoustic radiation ar
Ad1 adrenalin cells Ad1
Ad1 adrenalin cells/A1 noradrenalin cells Ad1/NA1
ambiguus nucleus, compact part AmbC
ambiguus nucleus, loose part AmbL
ambiguus nucleus, semicompact part AmbSC
amiculum of the olive ami
anterior commissure ac
anterior pretectal nucleus APT
aqueduct Aq
arcuate hypothalamic nucleus Arc
arcuate nucleus of the caudal hindbrain Ar
area postrema AP
ascending fibers of the facial nerve asc7
austral nucleus Au

B
B9 serotonin cells B9
Barrington's nucleus Bar
brachium of the inferior colliculus bic
brachium of the superior colliculus bsc

C
caudal dorsal paramedian nucleus CDPMn
caudal linear nucleus of the raphe CLi
caudal linear nucleus of the raphe, azygos part CLiAz
caudal linear nucleus of the raphe, zygos part CLiZ
central canal CC
central cervical nucleus of the spinal cord CeCv
central gray CG
central gray of the pons CGPn
central gray, beta part CGB
central nucleus of the inferior colliculus CIC
central tegmental tract ctg
Centromedian thalamic nucleus CMn
cerebellum Cb
cerebral peduncle cp
cochlear root of the vestibulocochlear nerve 8cn
commissure of the inferior colliculus cic
commissure of the superior colliculus csc
conterminal nucleus Ct
corticobulbar tract cbu
corticospinal tract csp
cribiform nucleus Crb
cuneate fasciculus cu
cuneate nucleus Cu
cuneiform nucleus CnF

D
decussation of the medial lemniscus xml
decussation of the superior cerebellar peduncle xscp
decussation of the superior cerebellar peduncle/caudal linear nucleus of the raphe, azygos part xscp/CliAz
decussation of the superior cerebellar peduncle/caudal linear nucleus of the raphe, zygos part xscp/CLiz
decussation of the trochlear nerve x4n
deep gray layer of the superior colliculus DpG
deep white layer of the superior colliculus DpWh
dorsal cortex of the inferior colliculus DCIC
dorsal lateral geniculate nucleus DLG
dorsal motor nucleus of vagus 10N
dorsal nucleus of the lateral lemniscus DLL
dorsal paragigantocellular nucleus DPGi
dorsal paramedian sulcus dpms
dorsal periolivary region DPO
dorsal raphe nucleus DR
dorsal raphe nucleus, caudal part DRC
dorsal raphe, interfascicular part DRI
dorsal spinocerebellar tract dsc
dorsal tegmental nucleus DTg
dorsolateral periaqueductal gray DLPAG
dorsomedial periaqueductal gray DMPAG
dorsomedial tegmental area DMTg
dorsomedian sulcus dms

E
Edinger-Westphal nucleus EW
endolemniscal nucleus EL
endorestiform nucleus EnR
ependyma and subependymal layer E
epicoeruleus nucleus EC
external cortex of the inferior colliculus ECIC
external cuneate nucleus ECu

F
facial nerve 7n
facial nucleus 7N
fasciculus retroflexus fr
fornix f
frontopontine fibres frpn

G
gamma pontine nucleus Gam
genu of the facial nerve g7
gigantocellular reticular nucleus Gi
gigantocellular reticular nucleus, alpha part GiA
gigantocellular reticular nucleus, ventral part GiV
glossopharyngeal nerve 9n
gracile fasciculus gr
gracile nucleus Gr

H
habenular commissure hbc
habenular nuclei Hb
hilus of the inferior olive hio
hypoglossal nerve 12n
hypoglossal nucleus 12N
hypoglossal nucleus, lateral part 12L
hypoglossal nucleus, medial part 12M
hypoglossal nucleus, ventral part 12V
hypothalamus hypothal

I
inferior cerebellar peduncle icp
inferior olive, beta subnucleus IOBe
inferior olive, cap of Kooy of the medial nucleus IOK
inferior olive, dorsal nucleus IOD
inferior olive, dorsomedial cell group IODM
inferior olive, medial nucleus IOM
inferior olive, principal nucleus IOPr
inferior olive, subnucleus A of medial nucleus IOA
inferior olive, subnucleus B of medial nucleus IOB
inferior olive, subnucleus C of medial nucleus IOC
inferior salivatory nucleus IS
intercalated nucleus In
interfascicular nucleus IF
interfascicular trigeminal nucleus IF5
intermediate gray layer of the superior colliculus InG
intermediate nucleus of the lateral lemniscus ILL
intermediate reticular nucleus IRt
intermediate white layer of the superior colliculus InWh
internal arcuate fibers ia
internal capsule ic
interoculomotor nucleus I3
interpeduncular fossa ipf
interpeduncular nucleus IP
interpositus IPo
interstitial nucleus of Cajal InC
interstitial nucleus of Cajal/mammillotegmental tract InC/mtg
interstitial nucleus of the decussation of the superior cerebellar peduncle ID

isthmic reticular formation isRt

J

juxtaolivary nucleus JxO

K

Kölliker-Fuse nucleus KF

L

lamina 10 of the spinal gray 10Sp
lamina 9 of the spinal gray 9Sp
lateral corticospinal tract lcs
lateral hypothalamic area LH
lateral lemniscus ll
lateral mammillary nucleus LM
lateral parabrachial nucleus LPB
lateral parabrachial nucleus, dorsal part LPBD
lateral parabrachial nucleus, external part LPBE
lateral parabrachial nucleus, superior part LPBS
lateral paragigantocellular nucleus LPGi
lateral periaqueductal gray LPAG
lateral pericuneate nucleus LPCu
lateral reticular nucleus LRt
lateral reticular nucleus, subtrigeminal part LRtS5
lateral terminal nucleus of the accessory optic tract LT
lateral vestibular nucleus LVe
lateral vestibulospinal tract lvs
laterodorsal tegmental nucleus LDTg
laterodorsal tegmental nucleus, ventral part LDTgV
linear nucleus of the medulla Li
locus coeruleus LC
longitudinal fasciculus of the pons lfp
longitudinal fasciculus of the pons/pontine nuclei lfp/Pn

M

magnocellular nucleus of the posterior commissure MCPC
mammillary body MB
mammillotegmental tract mtg
mammillotegmental tract/central tegmental tract mtg/ctg
mammillothalamic tract mt
matrix region of the medulla Mx
medial accessory oculomotor nucleus MA3
medial geniculate nucleus MG
medial geniculate nucleus, dorsal part MGD
medial geniculate nucleus, medial part MGM
medial geniculate nucleus, ventral part MGV
medial lemniscus ml
medial longitudinal fasciculus mlf
medial longitudinal fasciculus/mammillotegmental tract mlf/mtg
medial mammillary nucleus, lateral part ML
medial mammillary nucleus, lateral part/interstitial nucleus of Cajal mlf/InC
medial mammillary nucleus, medial part MM
medial parabrachial nucleus external part MPBE
medial parabrachial nucleus MPB
medial pericuneate nucleus MPCu
medial tuberal nucleus MTu
medial vestibular nucleus MVe
median raphe nucleus MnR
medullary reticular nucleus, dorsal part MdD
medullary reticular nucleus, ventral part MdV
medullary stria meds
mesencephalic reticular formation mRt
mesencephalic trigeminal tract me5
microcellular tegmental nucleus MiTg
middle cerebellar peduncle mcp
motor root of the trigeminal nerve m5
motor trigeminal nucleus 5N

N

noradrenaline cells A1 NA1
noradrenaline cells A2 NA2
noto cuneate nucleus Nt
nucleus of Darkschewitsch Dk
nucleus of origin of the efferent fibers of the vestibular nerve EVe
nucleus of Roller Ro
nucleus of the brachium of the inferior colliculus BIC
nucleus of the posterior commissure PCom
nucleus of the trapezoid body Tz
nucleus U U
nucleus X X

O

occipitopontine fibres ocpn
oculomotor nerve 3n
oculomotor nucleus 3N
olivocerebellar bundle decussation ocx
olivocerebellar tract oc
olivocerebellar tract/hilus of the inferior olive oc/hio
optic chiasm och
optic nerve layer of the superior colliculus Op
optic tract opt
oral dorsal paramedian nucleus ODPMn

P

p1 periaqueductal gray p1PAG
p1 periventricular gray p1PVG
paraabducens nucleus Pa6
parabigeminal nucleus PBG
parabrachial pigmented nucleus of the VTA PBP
parafascicular thalamic nucleus PaF
parainterfascicular nucleus of the VTA PIF
paramedian raphe nucleus PMnR
paranigral nucleus of the VTA PN
parapeduncular nucleus PaP
parasolitary nucleus PSol
paratrochlear nucleus Pa4
paravestibular nucleus PaVe
parietal pontine fibres ptpn
parvicellular oculomotor nucleus PC3
parvicellular reticular nucleus PCRt
parvicellular reticular nucleus, alpha part PCRtA
pedunculotegmental nucleus PTg
perifacial zone P7
peripeduncular nucleus PP
peritrigeminal nucleus Pe5
peritrigeminal zone P5
periventricular gray PVG
pineal gland Pi
pineal recess PiRe
pleioglia periaqueductal gray PlGl
pontine nuclei Pn
pontine nuclei/longitudinal fasciculus of the pons Pn/lfp
pontine nuclei/transverse fibers of the pons Pn/tfp
pontine raphe nucleus PnR
pontine reticular nucleus, caudal part PnC
pontine reticular nucleus, oral part PnO
pontobulbar nucleus PnB
post-olivary fissure pof
posterior commissure pc
posterior hypothalamic nucleus PH
posterior intralaminar thalamic nucleus PIL
posterior limitans thalamic nucleus PLi
posterior perforated substance PPS
posterior thalamic nuclear group, triangular part PoT
posterodorsal tegmental nucleus PDTg
pre-Bötzinger complex PrBo
pre-Bötzinger complex/intermediate reticular nucleus PrBo/IRt
pre-Edinger-Westphal nucleus PrEW
precommissural nucleus PrC
precuneiform area PrCnF
preolivary sulcus pos
prepositus nucleus Pr
principal mammillary tract pm
principal sensory trigeminal nucleus Pr5
prosomere 1 reticular formation p1Rt
pulvinar nuclei Pul
pyramidal decussation pyx
pyramidal tract py

R

raphe cap RC
raphe interpositus nucleus RIP
raphe magnus nucleus RMg
raphe obscurus nucleus ROb
raphe pallidus nucleus RPa
red nucleus R
red nucleus, parvicellular part RPC
reticulotegmental nucleus of the pons RtTg
reticulotegmental nucleus, lateral part RtTgL
retroambiguus nucleus RAmb
retrochiasmatic area RCh
retroisthmic nucleus RIs
retromammillary decussation rmx
retromammillary nucleus RM
retrorubral field RRF
rhabdoid nucleus Rbd
rostral linear nucleus (midbrain) RLi
rostral ventral respiratory group RVRG
rubrospinal tract rs

S

sagulum nucleus Sag
sensory root of the trigeminal nerve s5
solitary nucleus Sol
solitary nucleus, dorsolateral part SolDL
solitary nucleus, gelatinous part SolG
solitary nucleus, intermediate part SolIM
solitary nucleus, interstitial part SolI
solitary nucleus, medial part SolM
solitary nucleus, ventral part SolV
solitary nucleus, ventrolateral part SolVL
solitary tract sol
sphenoid nucleus Sph
spinal trigeminal nucleus, caudal part Sp5C
spinal trigeminal nucleus, caudal part, lamina 1 & 2 Sp5C1/2
spinal trigeminal nucleus, caudal part, lamina 3 & 4 Sp5C3/4
spinal trigeminal nucleus, interpolar part Sp5I
spinal trigeminal nucleus, oral part Sp5O
spinal trigeminal tract sp5
spinal vestibular nucleus SpVe
spinothalamic tract spth
spinothalamic tract/medial lemniscus spth/ml
stria medullaris of the thalamus sm
subbrachial nucleus SubB
subcoeruleus nucleus, dorsal part SubCD
subcoeruleus nucleus, ventral part SubCV
submammillothalamic nucleus SMT
subpeduncular tegmental nucleus SPTg
substantia nigra, compact part, dorsal tier SNCD
substantia nigra, compact part, medial tier SNCM
substantia nigra, compacta part, ventral tier SNCV
substantia nigra, lateral part SNL
substantia nigra, reticular part SNR
subthalamic nucleus STh
sulcus limitans sl
superficial gray layer of the superior colliculus SuG
superior cerebellar peduncle scp
superior colliculus SC
superior medullary velum smv
superior olive SOl
superior thalamic radiation str
superior vestibular nucleus SuVe
suprageniculate thalamic nucleus SG
supragenual nucleus SGe
supralemniscal nucleus SuL
supraoculomotor cap Su3C
supraoculomotor periaqueductal gray Su3
supraoptic decussation sox
supratrigeminal nucleus Su5

T

tectal gray TG
tectospinal tract ts
temporopontine fibres tepn
thalamus thal
transverse fibers of the pons tfp
transverse fibers of the pons/pontine nuclei tfp/Pn
trapezoid body tz
trigeminal nerve 5n
trigeminal transition zone 5Tr
trigeminal-solitary transition zone 5Sol
trigeminothalamic tract tth
trochlear nerve 4n
trochlear nucleus 4N
tuberomamillary nucleus TuM

V

vagus nerve 10n
ventral median fissure vmf
ventral nucleus of the lateral lemniscus VLL
ventral posterior inferior nucleus VPI
ventral posterolateral thalamic nucleus VPL
ventral posteromedial thalamic nucleus VPM
ventral spinocerebellar tract vsc
ventral tegmental area VTA
ventral tegmental area, caudal part VTAC
ventral tegmental area, rostral part VTAR
ventral tegmental nucleus VTg
ventrolateral periaqueductal gray VLPAG
ventrolateral sulcus vls
ventrolateral tegmental area VLTg
ventromedial hypothalamic nucleus VMH
vestibular root of the vestibulocochlear nerve 8vn
vestibulocochlear nerve 8n
vestibulomesencephalic tract veme

Z

zona incerta ZI
zona incerta, caudal part ZIC
zona incerta, dorsal part ZID
zona layer of the superior colliculus Zo

Index of Abbreviations

A

B

C

D

E

F

G

H

I

J

K

L

M

N

O

P

R

S

T

U

V

X

Z

FIGURES

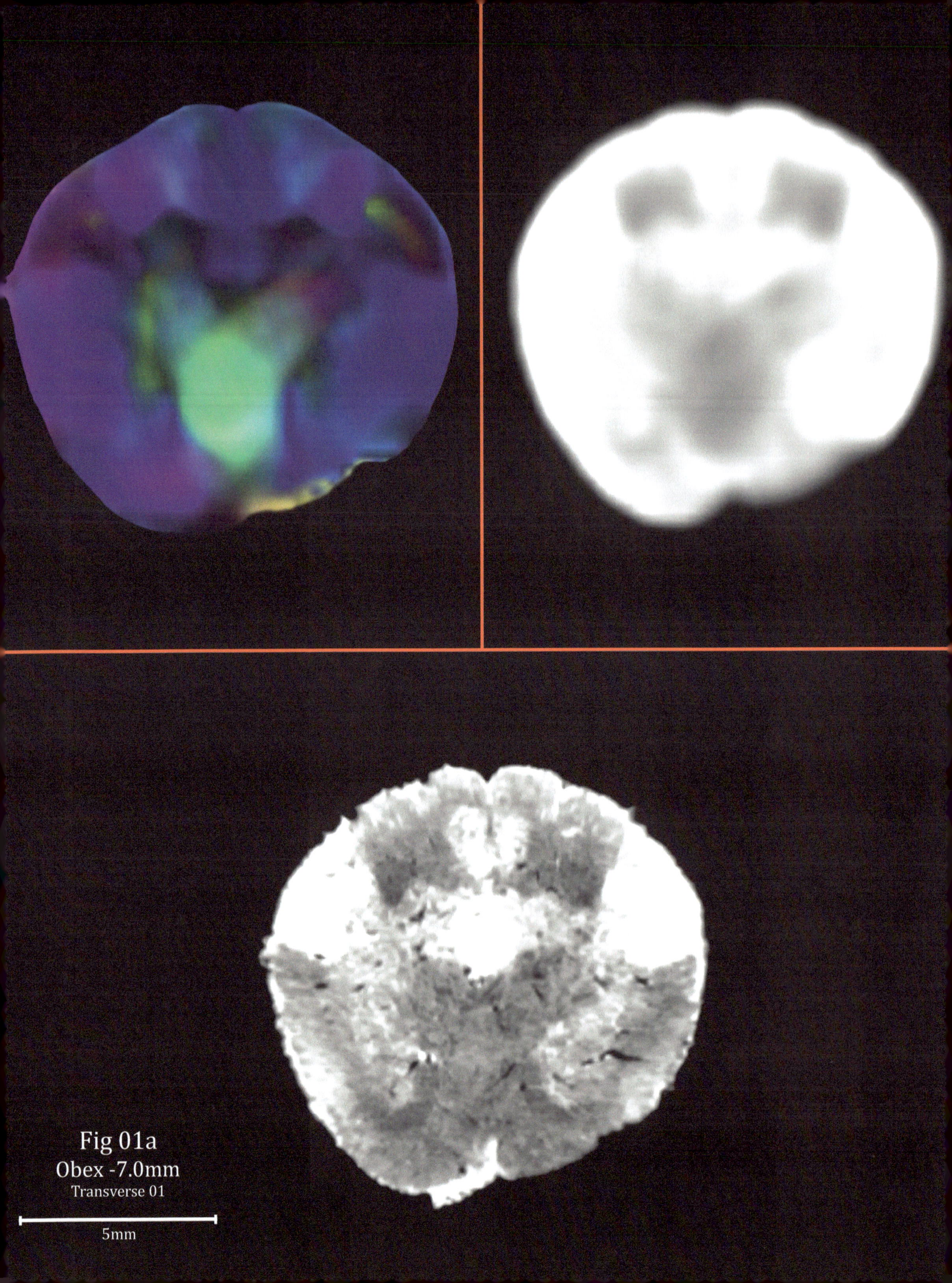

Fig 01a
Obex -7.0mm
Transverse 01

5mm

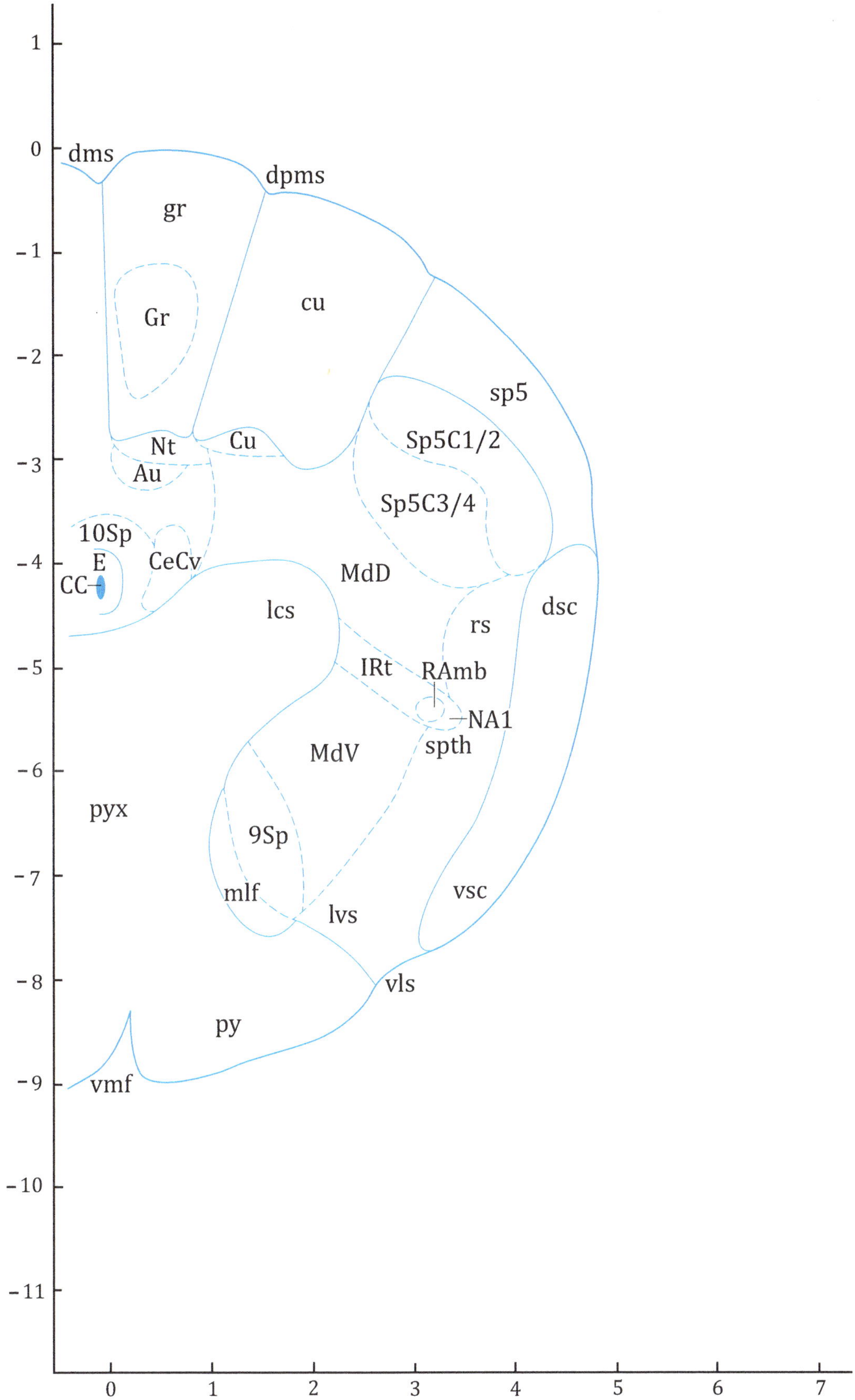

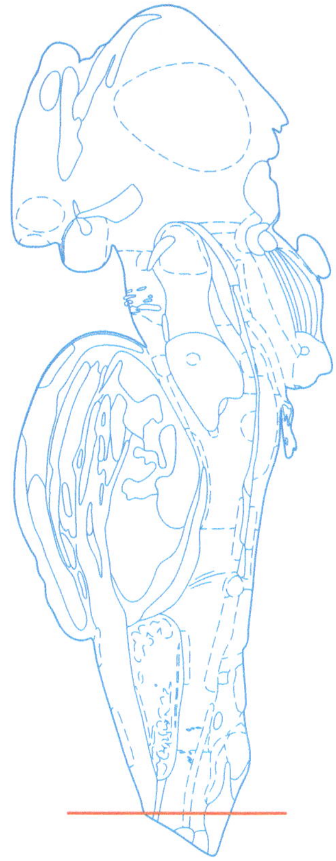

10Sp lamina 10 of the spinal gray
9Sp Rexed's lamina
Au austral nucleus
CC central canal
CeCv central cervical nu
Cu cuneate nu
cu cuneate fasciculus
dms dorsomedian sulcus
dpms dorsal paramedian sulcus
dsc dorsal spinocereb tract
E ependyma/subependymal
Gr gracile nu
gr gracile fasciculus
IRt intermed reticular nu
lcs lateral corticospinal tract
lvs lateral vestibulospinal tract
MdD medullary reticular nu, dors
MdV medullary reticular nu, vent
mlf med longitudinal fasciculus
NA1 A1 noradrenaline cells
Nt noto cuneate nucleus
py pyramidal tr
pyx pyramidal decussation
RAmb retroambiguus nu
rs rubrospinal tract
sp5 spinal trigem tract
Sp5C1/2 spinal 5 nu, caud, lam 1 & 2
Sp5C3/4 spinal 5 nu, caud, lam 3 & 4
spth spinothalamic tract
vls ventrolateral sulcus
vmf ventral median fissure
vsc ventral spinocerebellar tr

Fig 01b
Obex -7.0mm
Transverse 01

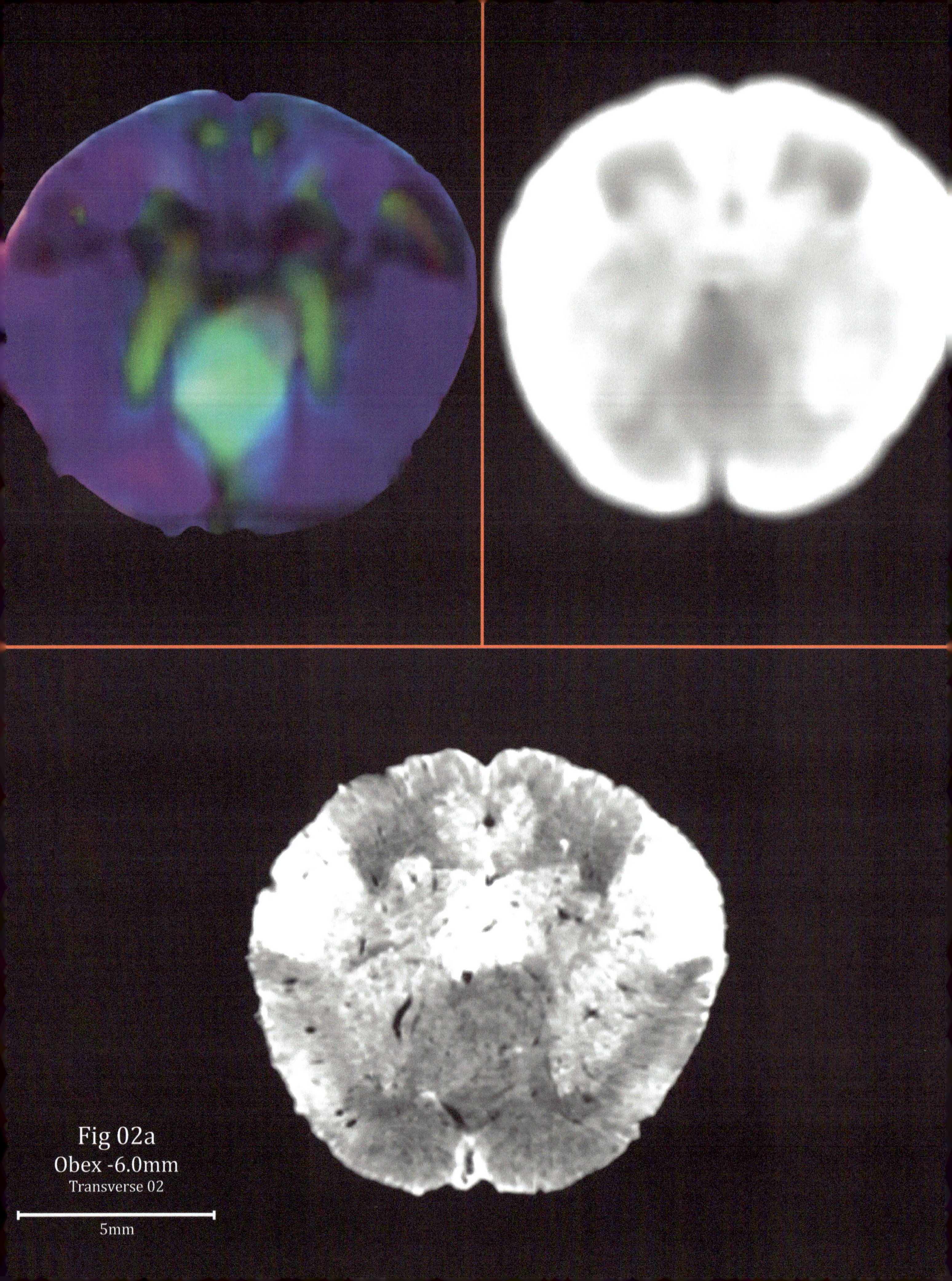

Fig 02a
Obex -6.0mm
Transverse 02

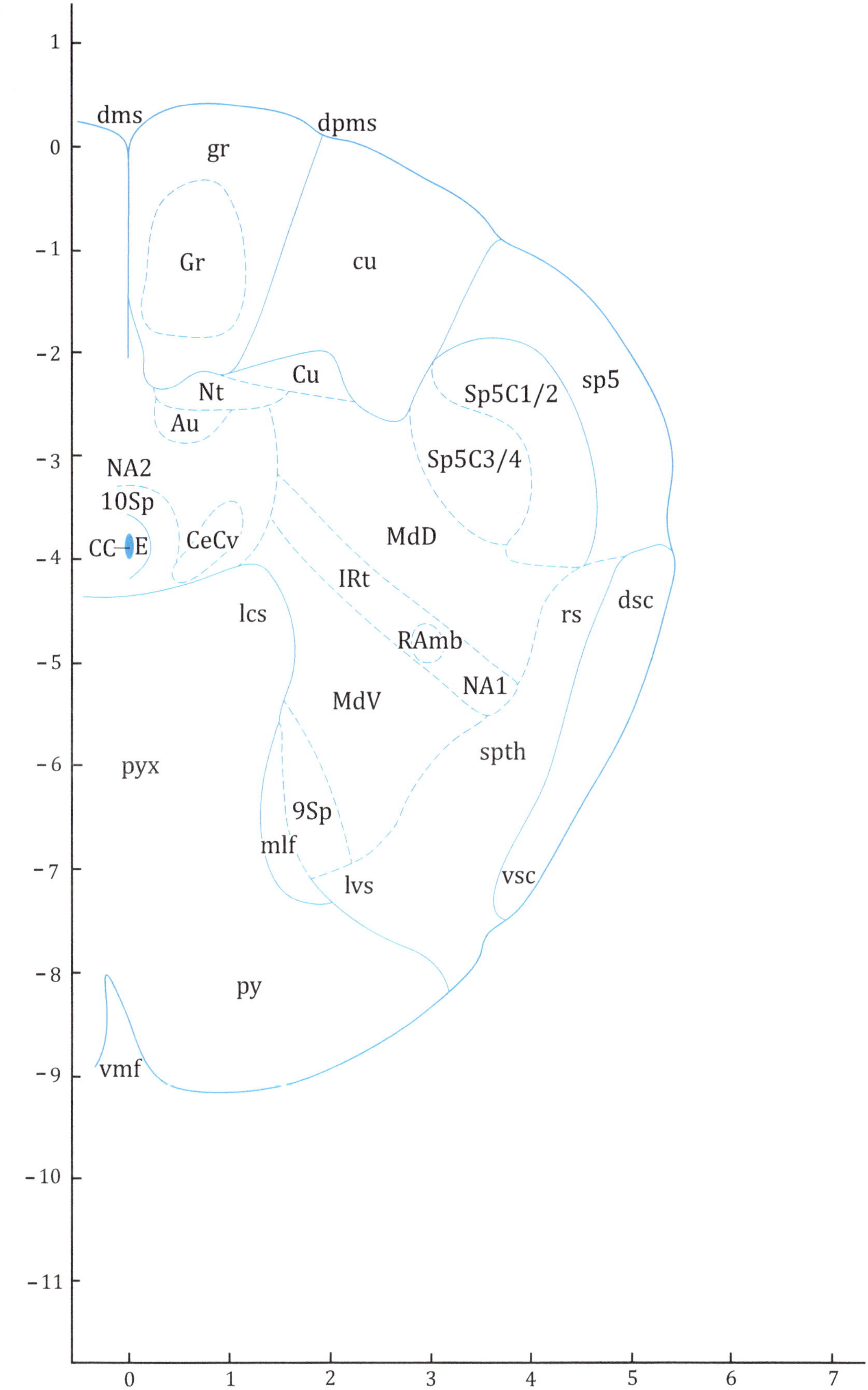

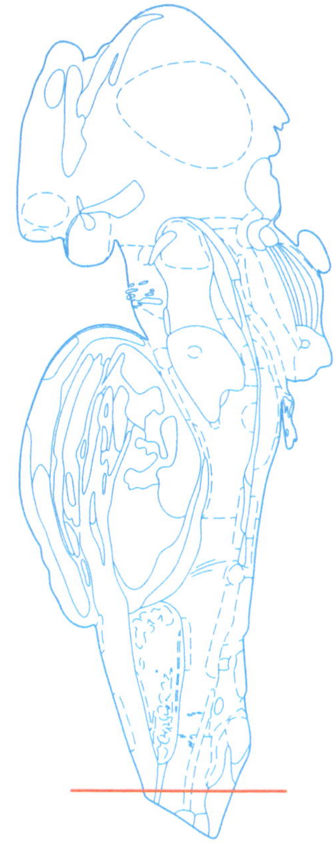

10Sp lamina 10 of the spinal gray
9Sp Rexed's lamina
Au austral nucleus
CC central canal
CeCv central cervical nu
Cu cuneate nu
cu cuneate fasciculus
dms dorsomedian sulcus
dpms dorsal paramedian sulcus
dsc dorsal spinocereb tract
E ependyma/subependymal
Gr gracile nu
gr gracile fasciculus
IRt intermed reticular nu
lcs lateral corticospinal tract
lvs lateral vestibulospinal tract
MdD medullary reticular nu, dors
MdV medullary reticular nu, vent
mlf med longitudinal fasciculus
NA1 A1 noradrenaline cells
NA2 A2 noradrenaline cells
Nt noto cuneate nucleus
py pyramidal tr
pyx pyramidal decussation
RAmb retroambiguus nu
rs rubrospinal tract
sp5 spinal trigem tract
Sp5C1/2 spinal 5 nu, caud, lam 1 & 2
Sp5C3/4 spinal 5 nu, caud, lam 3 & 4
spth spinothalamic tract
vmf ventral median fissure
vsc ventral spinocerebellar tr

Fig 02b
Obex -6.0mm
Transverse 02

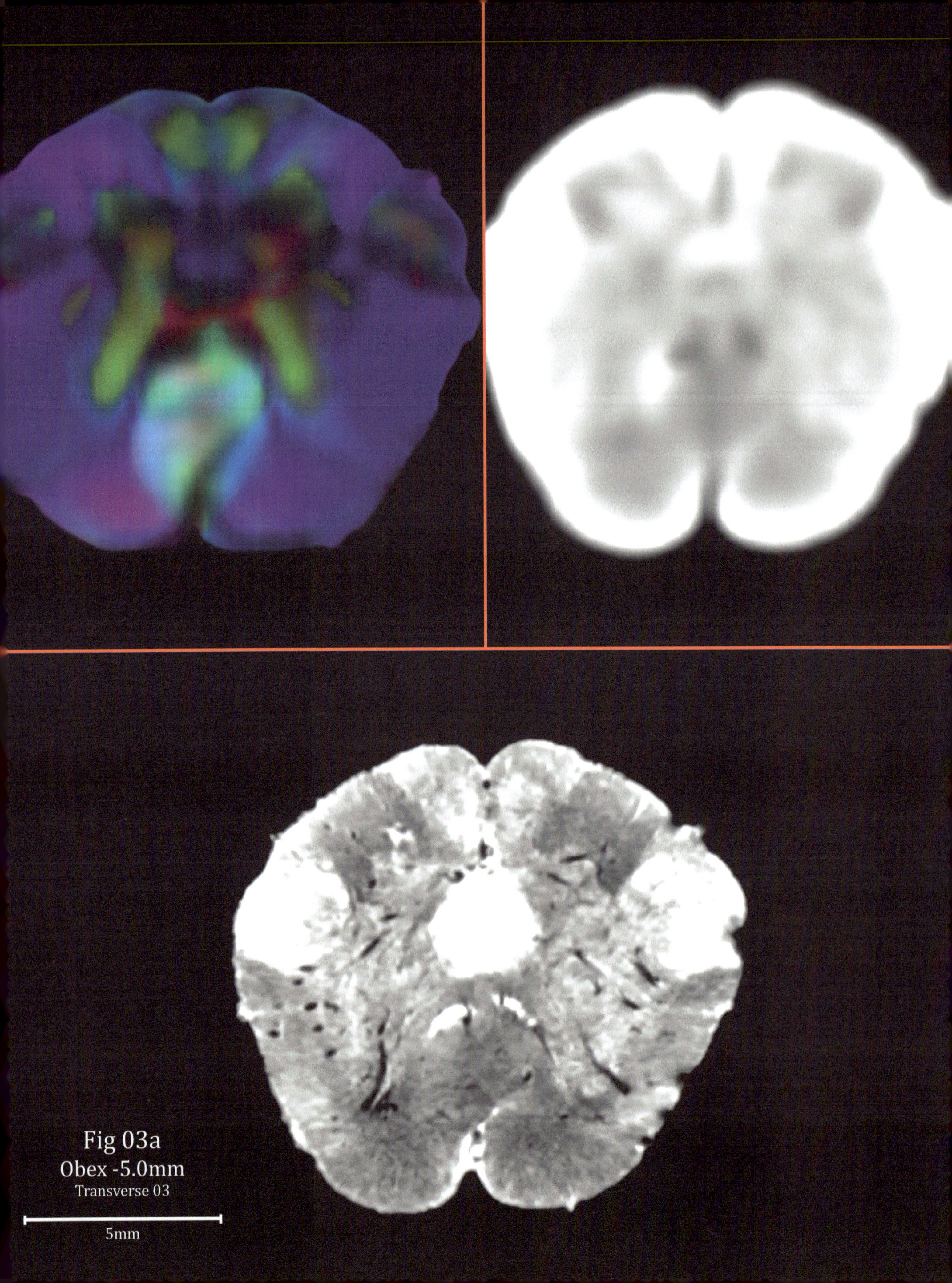

Fig 03a
Obex -5.0mm
Transverse 03

5mm

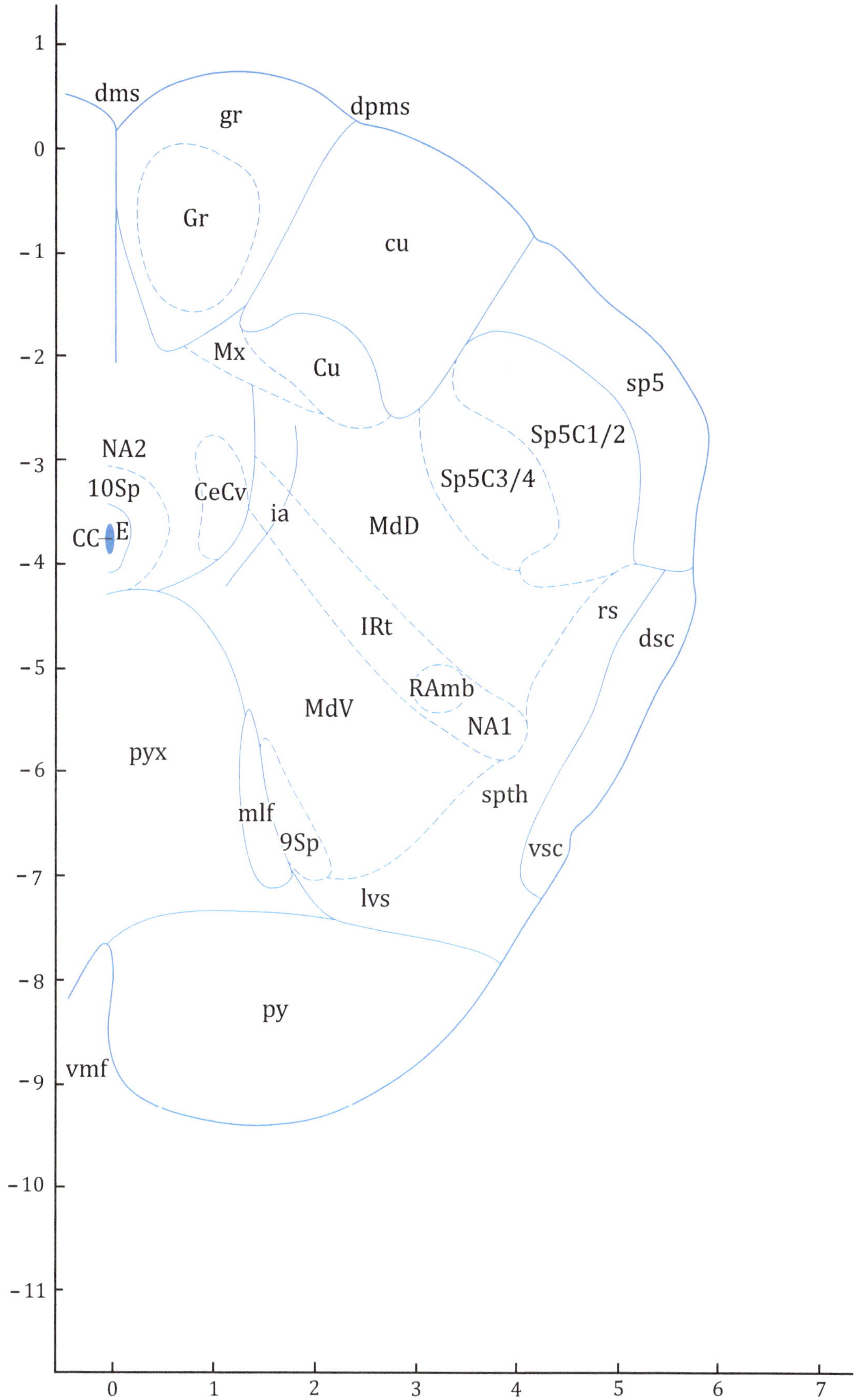

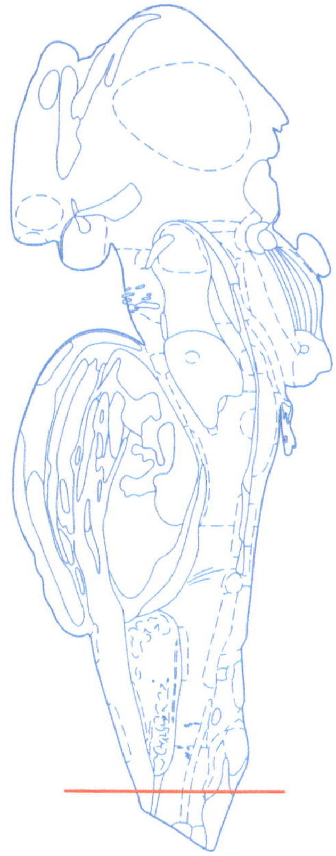

10Sp lamina 10 of the spinal gray
9Sp Rexed's lamina
CC central canal
CeCv central cervical nu
Cu cuneate nu
cu cuneate fasciculus
dms dorsomedian sulcus
dpms dorsal paramedian sulcus
dsc dorsal spinocereb tract
E ependyma/subependymal
Gr gracile nu
gr gracile fasciculus
ia internal arcuate fibers
IRt intermed reticular nu
lvs lateral vestibulospinal tract
MdD medullary reticular nu, dors
MdV medullary reticular nu, vent
mlf med longitudinal fasciculus
Mx matrix region medulla
NA1 A1 noradrenaline cells
NA2 A2 noradrenaline cells
py pyramidal tr
pyx pyramidal decussation
RAmb retroambiguus nu
rs rubrospinal tract
sp5 spinal trigem tract
Sp5C1/2 spinal 5 nu, caud, lam 1 & 2
Sp5C3/4 spinal 5 nu, caud, lam 3 & 4
spth spinothalamic tract
vmf ventral median fissure
vsc ventral spinocerebellar tr

Fig 03b
Obex -5.0mm
Transverse 03

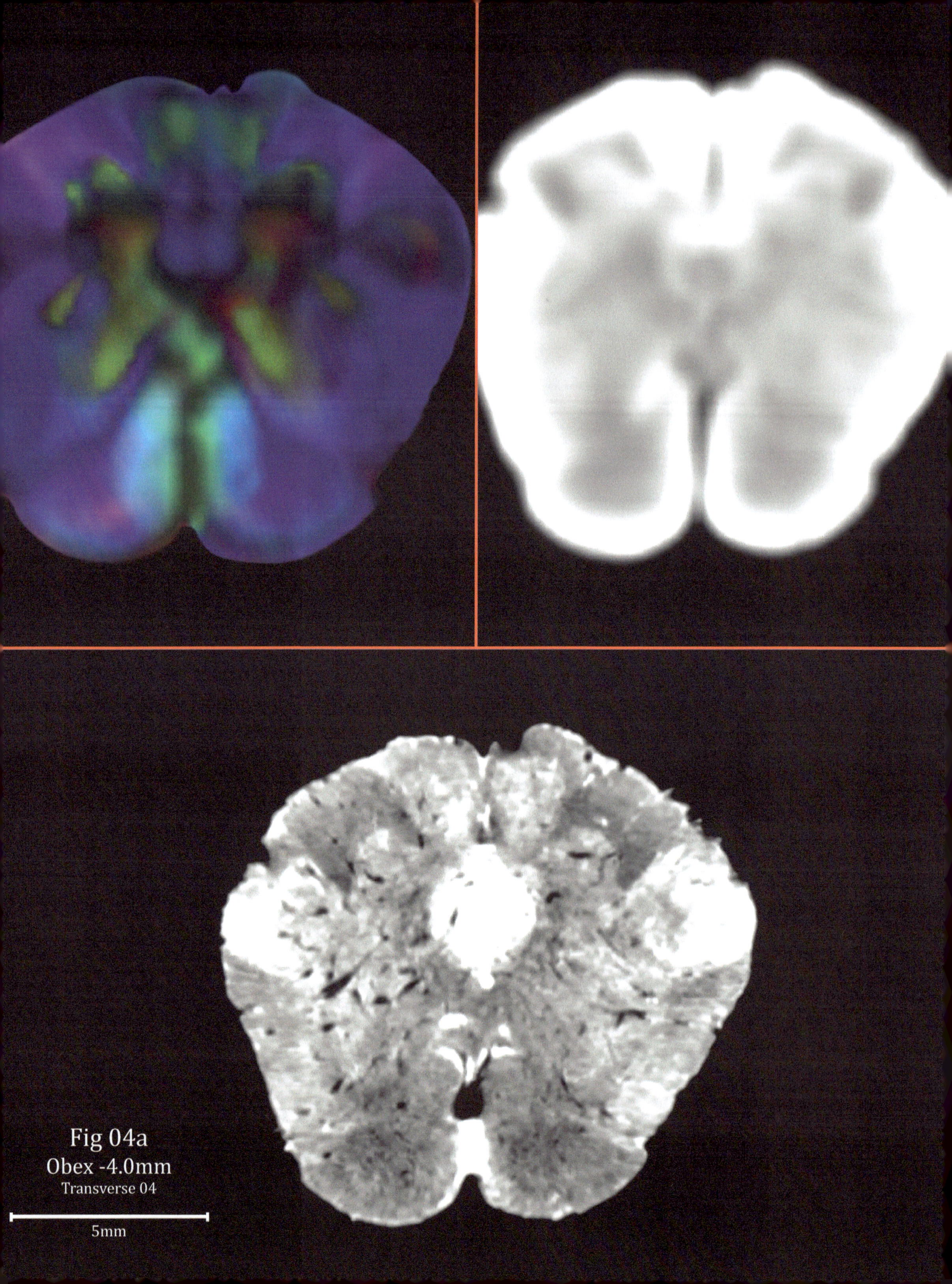

Fig 04a
Obex -4.0mm
Transverse 04

5mm

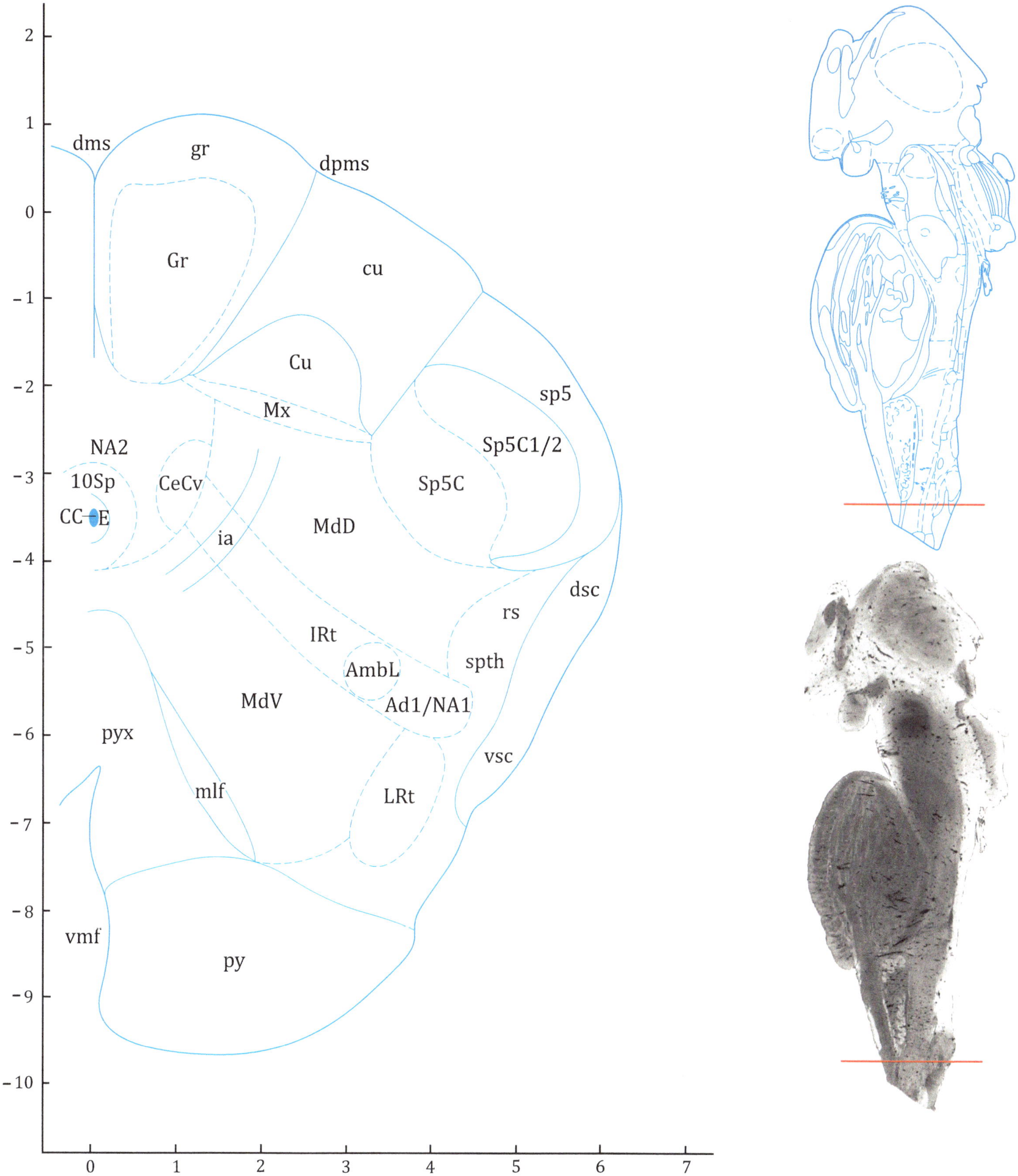

10Sp lamina 10 of the spinal gray
Ad1/NA1 Ad1 adren NA1 noradr
AmbL ambiguus nu, loose part
CC central canal
CeCv central cervical nu
cu cuneate fasciculus
Cu cuneate nu
dms dorsomedian sulcus
dpms dorsal paramedian sulcus
dsc dorsal spinocereb tract
E ependyma/subependymal
gr gracile fasciculus
Gr gracile nu
ia internal arcuate fibers
IRt intermed reticular nu
LRt lateral reticular nu
MdD medullary reticular nu, dors
MdV medullary reticular nu, vent
mlf med longitudinal fasciculus
Mx matrix region medulla
NA2 A2 noradrenaline cells
py pyramidal tr
pyx pyramidal decussation
rs rubrospinal tract
sp5 spinal trigem tract
Sp5C spinal trigem nu, caudal
Sp5C1/2 spinal 5 nu, caud, lam 1 & 2
spth spinothalamic tract
vmf ventral median fissure
vsc ventral spinocerebellar tr

Fig 04b
Obex -4.0mm
Transverse 04

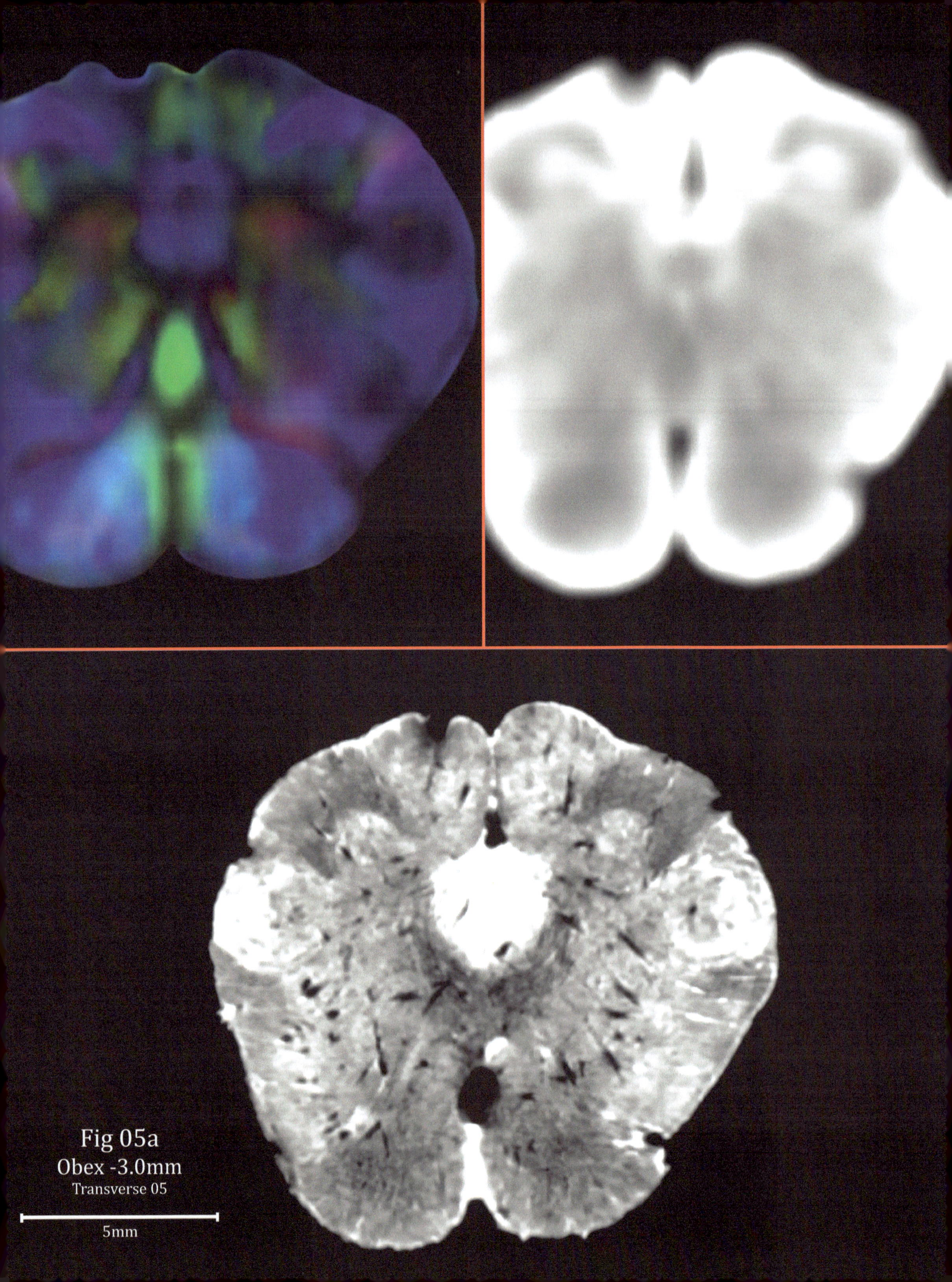

Fig 05a
Obex -3.0mm
Transverse 05

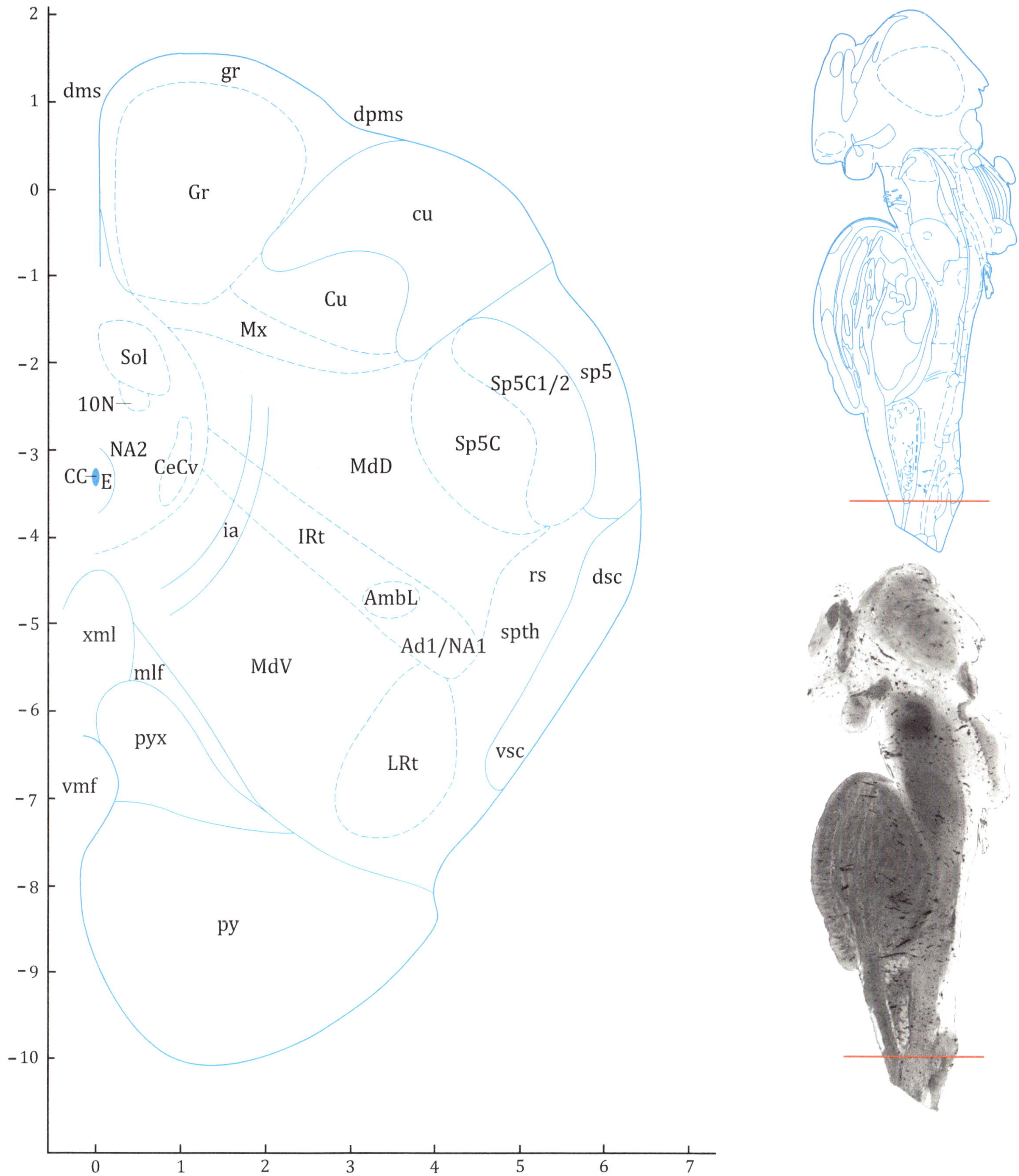

10N dorsal motor nu of vagus
Ad1/NA1 Ad1 adren NA1 noradr
AmbL ambiguus nu, loose part
CC central canal
CeCv central cervical nu
cu cuneate fasciculus
Cu cuneate nu
dms dorsomedian sulcus
dpms dorsal paramedian sulcus
dsc dorsal spinocereb tract
E ependyma/subependymal
gr gracile fasciculus
Gr gracile nu
ia internal arcuate fibers
IRt intermed reticular nu
LRt lateral reticular nu
MdD medullary reticular nu, dors
MdV medullary reticular nu, vent
mlf med longitudinal fasciculus
Mx matrix region medulla
NA2 A2 noradrenaline cells
py pyramidal tr
pyx pyramidal decussation
rs rubrospinal tract
Sol nu of solitary tract
sp5 spinal trigem tract
Sp5C spinal trigem nu, caudal
Sp5C1/2 spinal 5 nu, caud, lam 1 & 2
spth spinothalamic tract
vmf ventral median fissure
vsc ventral spinocerebellar tr
xml decussation of medial lemniscus

Fig 05b
Obex -3.0mm
Transverse 05

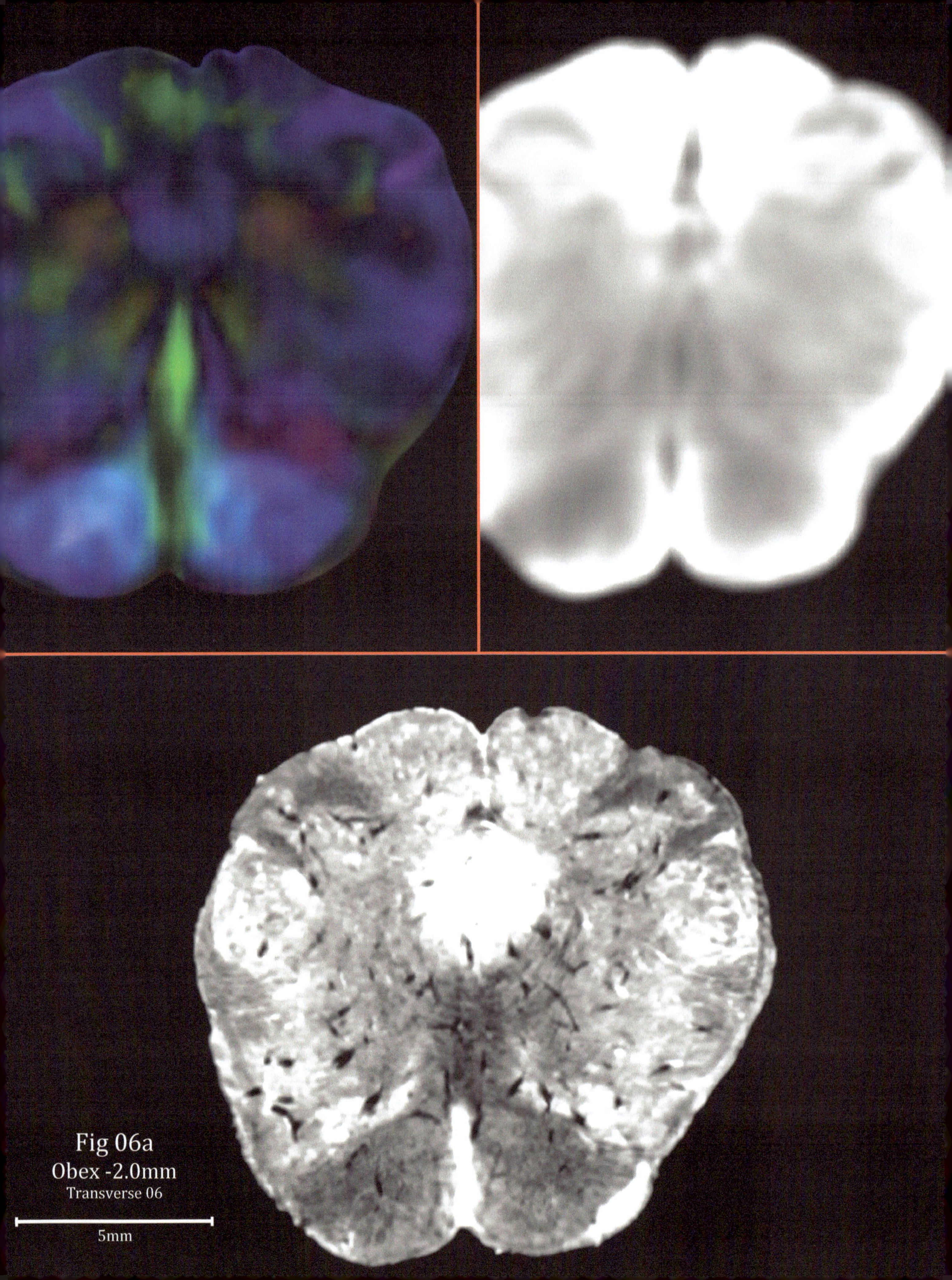

Fig 06a
Obex -2.0mm
Transverse 06

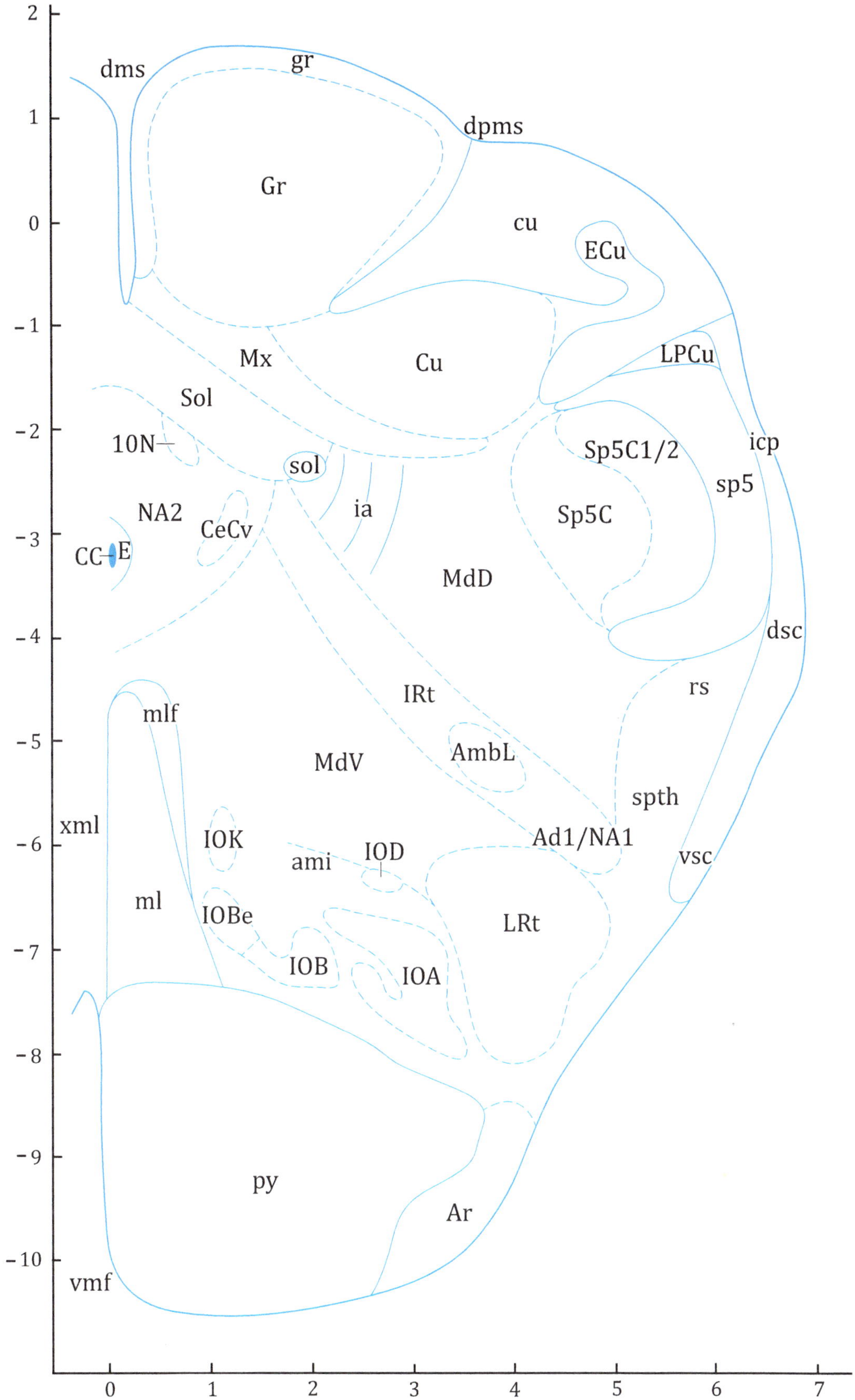

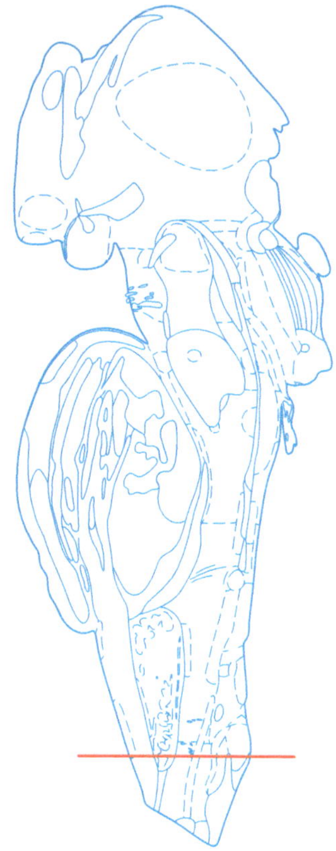

10N dorsal motor nu of vagus
Ad1/NA1 Ad1 adren NA1 noradr
AmbL ambiguus nu, loose part
ami amiculum of the olive
Ar arcuate nu of the caudal hindbrain
CC central canal
CeCv central cervical nu
Cu cuneate nu
cu cuneate fasciculus
dms dorsomedian sulcus
dpms dorsal paramedian sulcus
dsc dorsal spinocereb tract
E ependyma/subependymal

ECu external cuneate nu
gr gracile fasciculus
Gr gracile nu
ia internal arcuate fibers
icp inf cerebellar peduncle
IOA inf olive, subnu A medial nu
IOB inf olive, subnu B med nu
IOBe inf olive, beta subnu
IOD inf olive, dorsal nu
IOK inf olive, cap of Kooy med nu
IRt intermed reticular nu
LPCu lateral pericuneate nucleus
LRt lateral reticular nu

MdD medullary reticular nu, dors
MdV medullary reticular nu, vent
ml medial lemniscus
mlf med longitudinal fasciculus
Mx matrix region medulla
NA2 A2 noradrenaline cells
py pyramidal tr
rs rubrospinal tract
sol solitary tract
Sol nu of solitary tract
sp5 spinal trigem tract
Sp5C spinal trigem nu, caudal
Sp5C1/2 spinal 5 nu, caud, lam 1 & 2

spth spinothalamic tract
vmf ventral median fissure
vsc ventral spinocerebellar tr
xml decussation of medial lemniscus

Fig 06b
Obex -2.0mm
Transverse 06

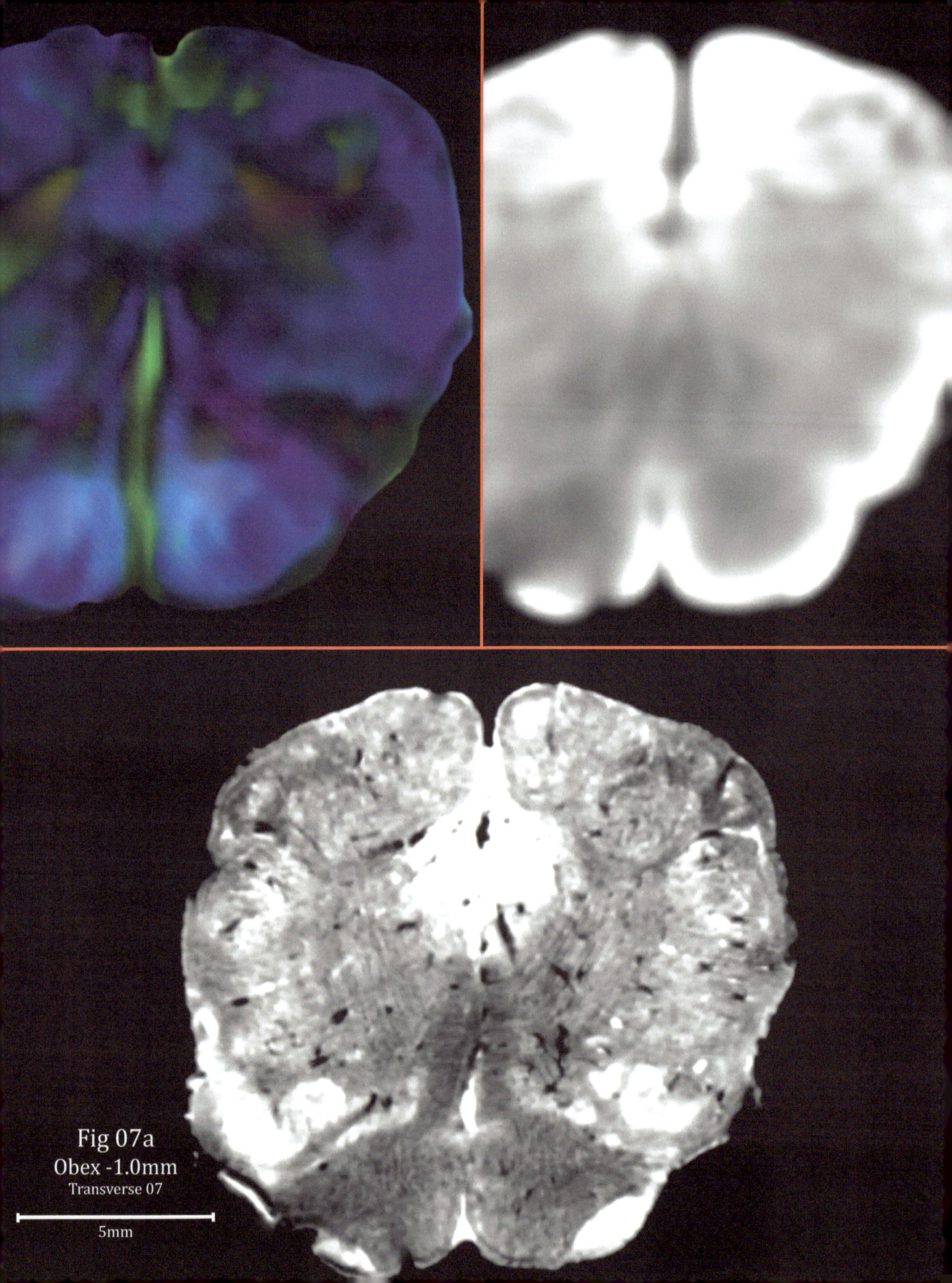

Fig 07a
Obex -1.0mm
Transverse 07

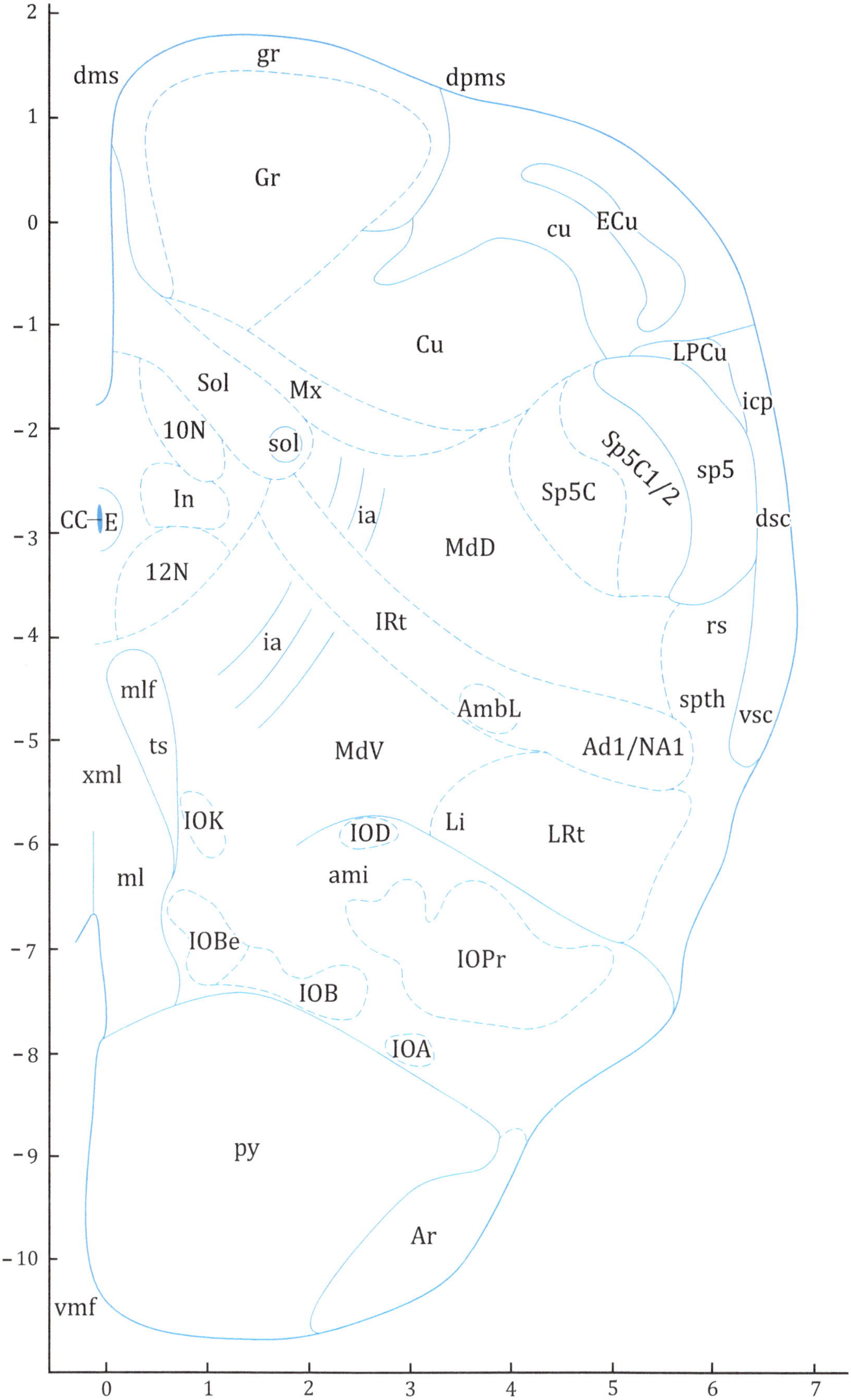

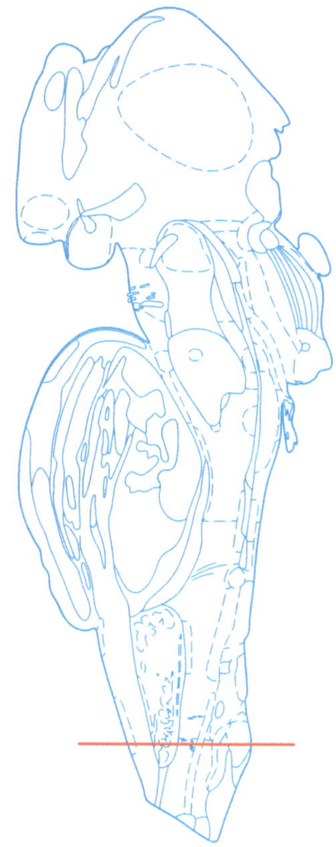

10N dorsal motor nu of vagus
12N hypoglossal nu
Ad1/NA1 Ad1 adren NA1 noradr
AmbL ambiguus nu, loose part
ami amiculum of the olive
Ar arcuate nu of the caudal hindbrain
CC central canal
cu cuneate fasciculus
Cu cuneate nu
dms dorsomedian sulcus
dpms dorsal paramedian sulcus
dsc dorsal spinocereb tract
E ependyma/subependymal
ECu external cuneate nu
Gr gracile nu
gr gracile fasciculus
ia internal arcuate fibers
icp inf cerebellar peduncle
In intercalated nu medulla
IOA inf olive, subnu A medial nu
IOB inf olive, subnu B med nu
IOBe inf olive, beta subnu
IOD inf olive, dorsal nu
IOK inf olive, cap of Kooy med nu
IOPr inf olive, principal nu
IRt intermed reticular nu
Li linear nu of the medulla
LPCu lateral pericuneate nucleus
LRt lateral reticular nu
MdD medullary reticular nu, dors
MdV medullary reticular nu, vent
ml medial lemniscus
mlf med longitudinal fasciculus
Mx matrix region medulla
py pyramidal tr
rs rubrospinal tract
sol solitary tract
Sol nu of solitary tract
sp5 spinal trigem tract
Sp5C spinal trigem nu, caudal
Sp5C1/2 spinal 5 nu, caud, lam 1 & 2
spth spinothalamic tract
ts tectospinal tr
vmf ventral median fissure
vsc ventral spinocerebellar tr
xml decussation of medial lemniscus

Fig 07b
Obex -1.0mm
Transverse 07

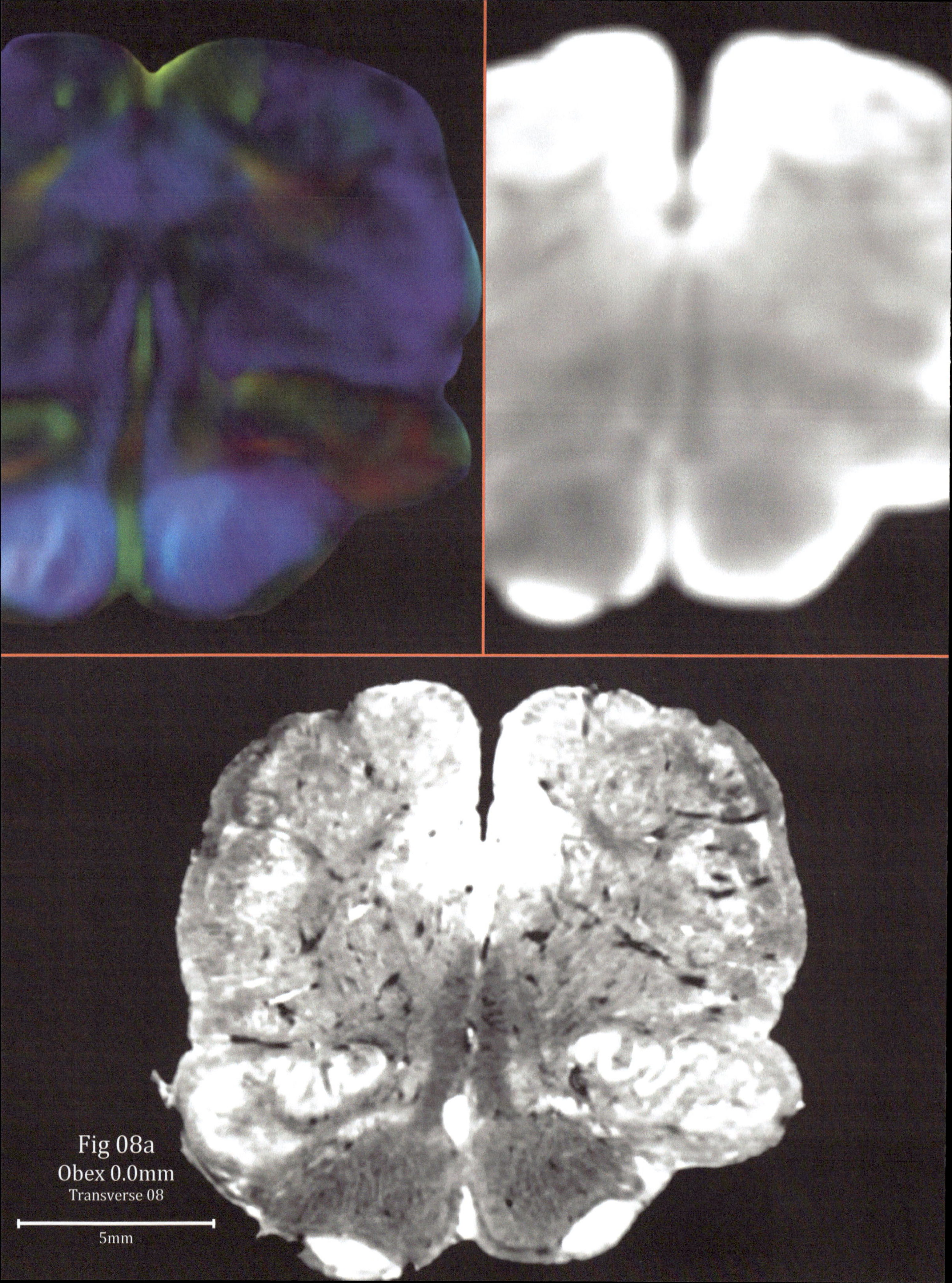

Fig 08a
Obex 0.0mm
Transverse 08

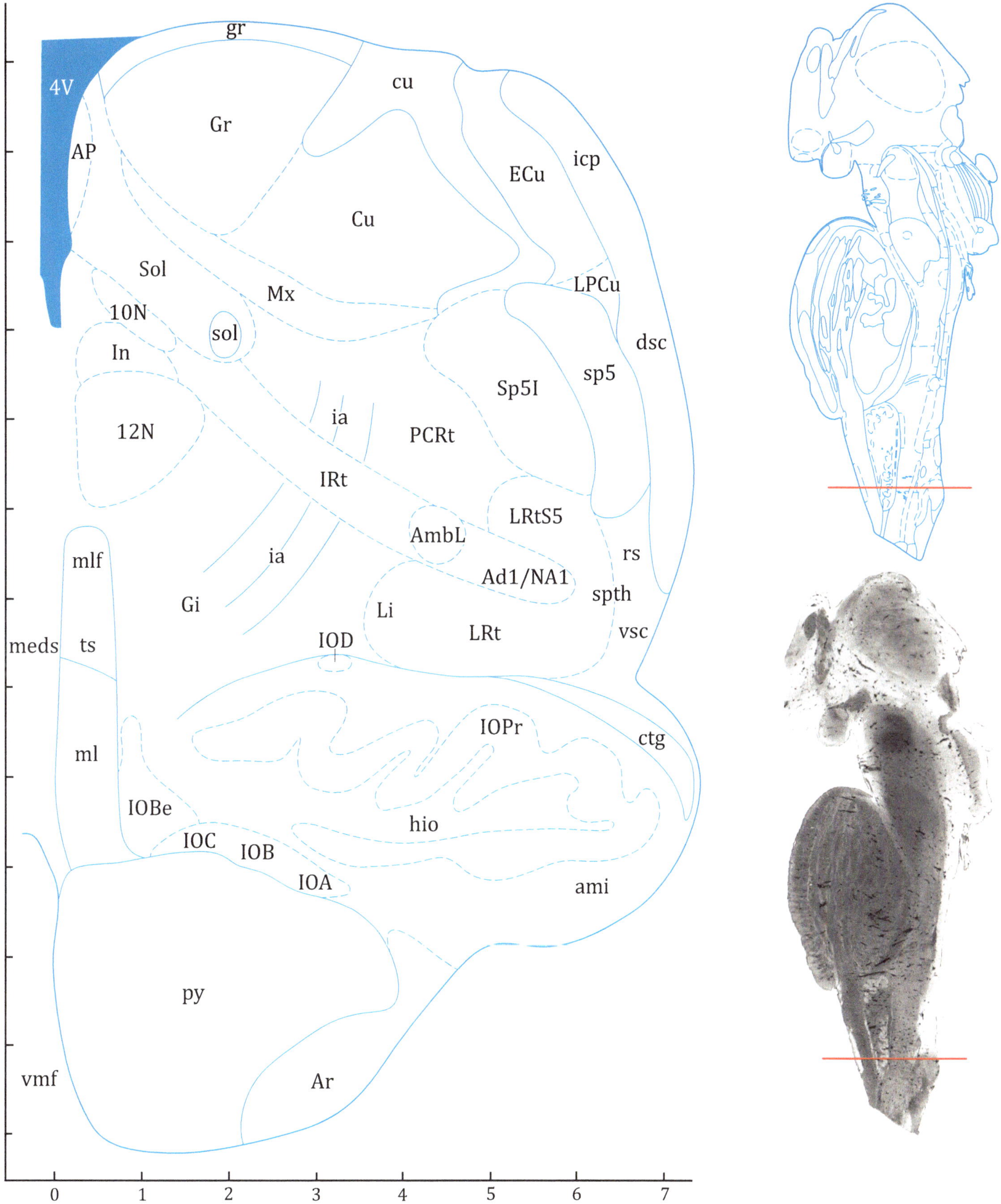

10N dorsal motor nu of vagus
12N hypoglossal nu
4V 4th ventricle
Ad1/NA1 Ad1 adren NA1 noradr
AmbL ambiguus nu, loose part
ami amiculum of the olive
AP area postrema
Ar arcuate nu of the caudal hindbrain
ctg central tegmental tract
Cu cuneate nu
cu cuneate fasciculus
dsc dorsal spinocereb tract
ECu external cuneate nu

Gi gigantocellular reticular nu
gr gracile fasciculus
Gr gracile nu
hio hilus of the inferior olive
ia internal arcuate fibers
icp inf cerebellar peduncle
In intercalated nu medulla
IOA inf olive, subnu A medial nu
IOB inf olive, subnu B med nu
IOBe inf olive, beta subnu
IOC inf olive, subnu C of med nu
IOD inf olive, dorsal nu
IOPr inf olive, principal nu

IRt intermed reticular nu
Li linear nu of the medulla
LPCu lateral pericuneate nucleus
LRt lateral reticular nu
LRtS5 lat reticular nu, subtrigeminal
meds medullary stria
ml medial lemniscus
mlf med longitudinal fasciculus
Mx matrix region medulla
PCRt parvicell reticular nu
py pyramidal tr
rs rubrospinal tract
sol solitary tract

Sol nu of solitary tract
sp5 spinal trigem tract
Sp5I spinal trigem nu, interpolar
spth spinothalamic tract
ts tectospinal tr
vmf ventral median fissure
vsc ventral spinocerebellar tr

Fig 08b
Obex +0.0mm
Transverse 08

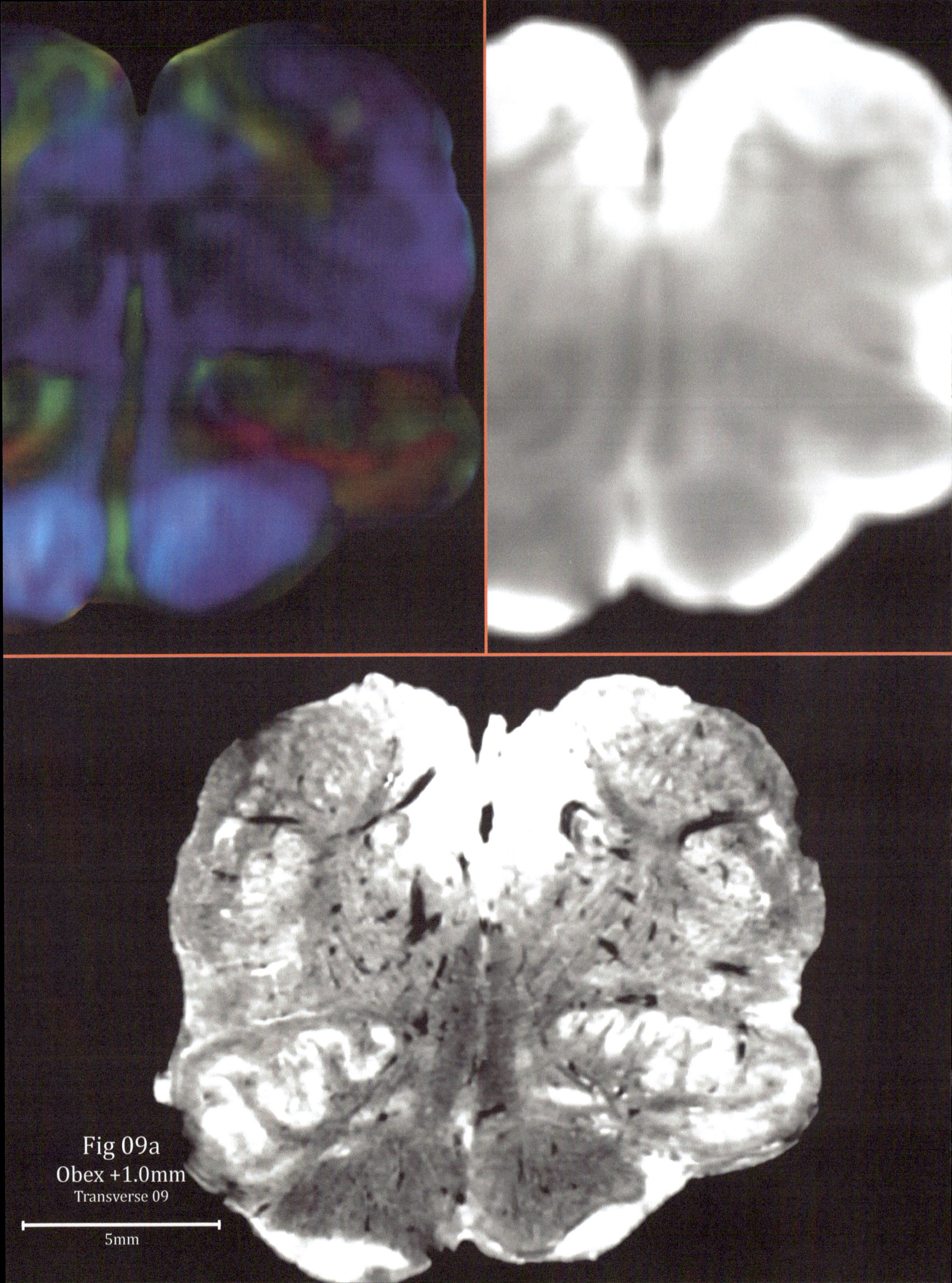

Fig 09a
Obex +1.0mm
Transverse 09

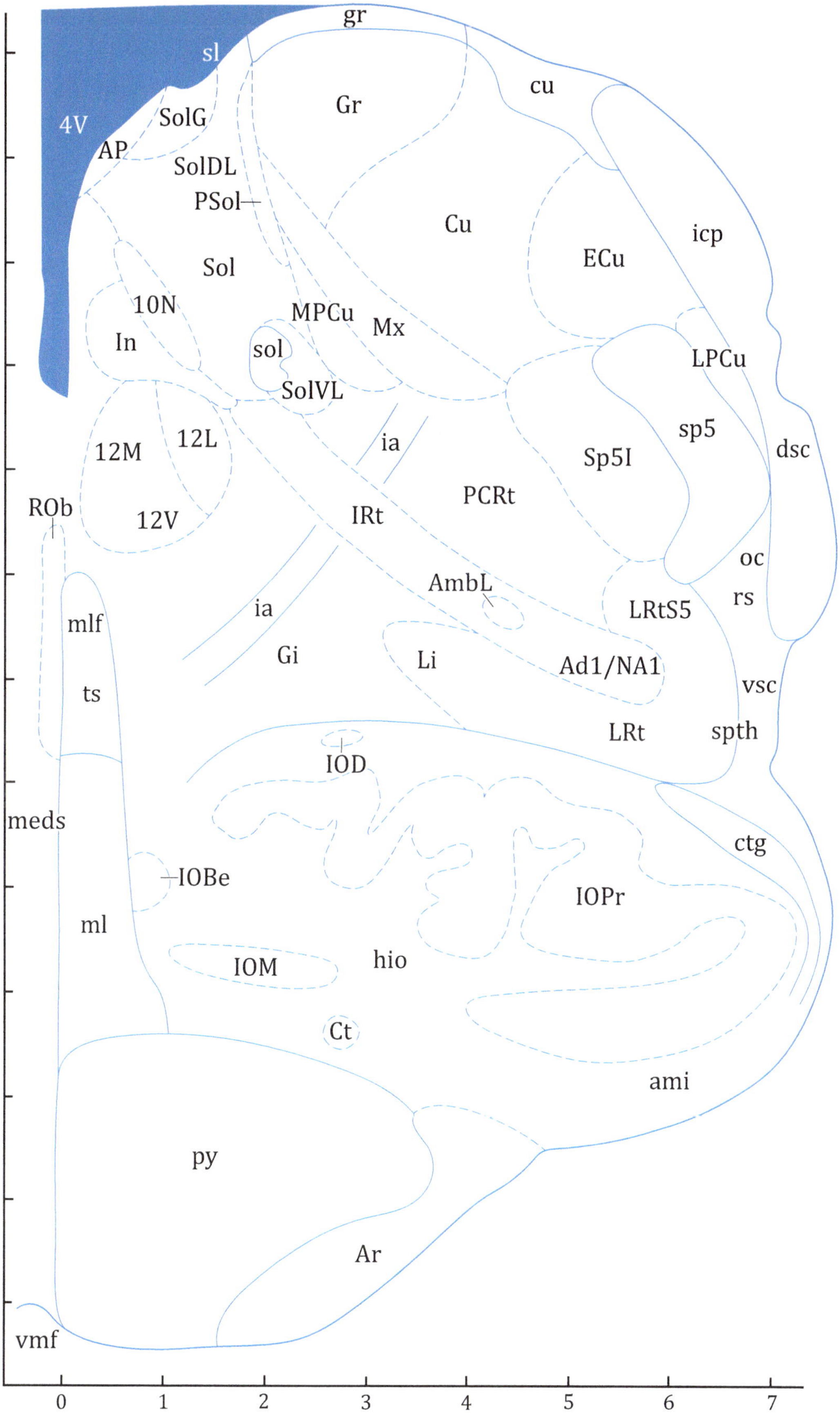

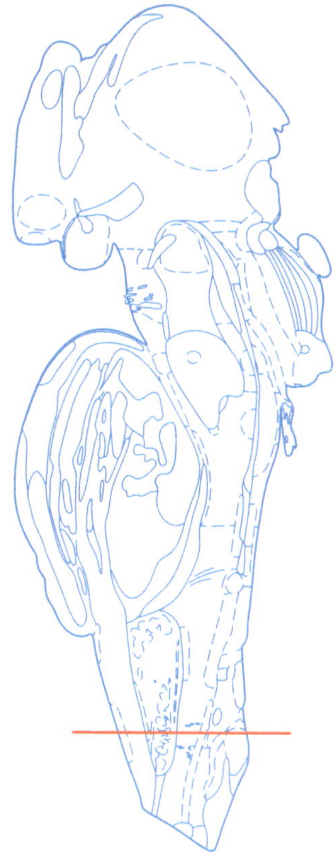

Fig 09b
Obex +1.0mm
Transverse 09

10N dorsal motor nu of vagus
12L hypoglossal nucleus, lateral part
12M hypoglossal nucleus, medial part
12V hypoglossal nucleus, ventral part
4V 4th ventricle
Ad1/NA1 Ad1 adren NA1 noradr
AmbL ambiguus nu, loose part
ami amiculum of the olive
AP area postrema
Ar arcuate nu of the caudal hindbrain
Ct conterminal nu
ctg central tegmental tract
cu cuneate fasciculus
Cu cuneate nu
dsc dorsal spinocereb tract
ECu external cuneate nu
Gi gigantocellular reticular nu
gr gracile fasciculus
Gr gracile nu
hio hilus of the inferior olive
ia internal arcuate fibers
icp inf cerebellar peduncle
In intercalated nu medulla
IOBe inf olive, beta subnu
IOD inf olive, dorsal nu
IOM inf olive, med nu
IOPr inf olive, principal nu
IRt intermed reticular nu
Li linear nu of the medulla
LPCu lateral pericuneate nucleus
LRt lateral reticular nu
LRtS5 lat reticular nu, subtrigeminal
meds medullary stria
ml medial lemniscus
mlf med longitudinal fasciculus
MPCu medial pericuneate nucleus
Mx matrix region medulla
oc olivocerebellar tract
PCRt parvicell reticular nu
PSol parasolitary nu
py pyramidal tr
ROb raphe obscurus nu
rs rubrospinal tract
sl sulcus limitans
sol solitary tract
Sol nu of solitary tract
SolDL sol nu, dorsolateral
SolG nu sol tract, gelatinous
SolVL nu sol tract, ventrolateral
sp5 spinal trigem tract
Sp5I spinal trigem nu, interpolar
spth spinothalamic tract
ts tectospinal tr
vmf ventral median fissure
vsc ventral spinocerebellar tr

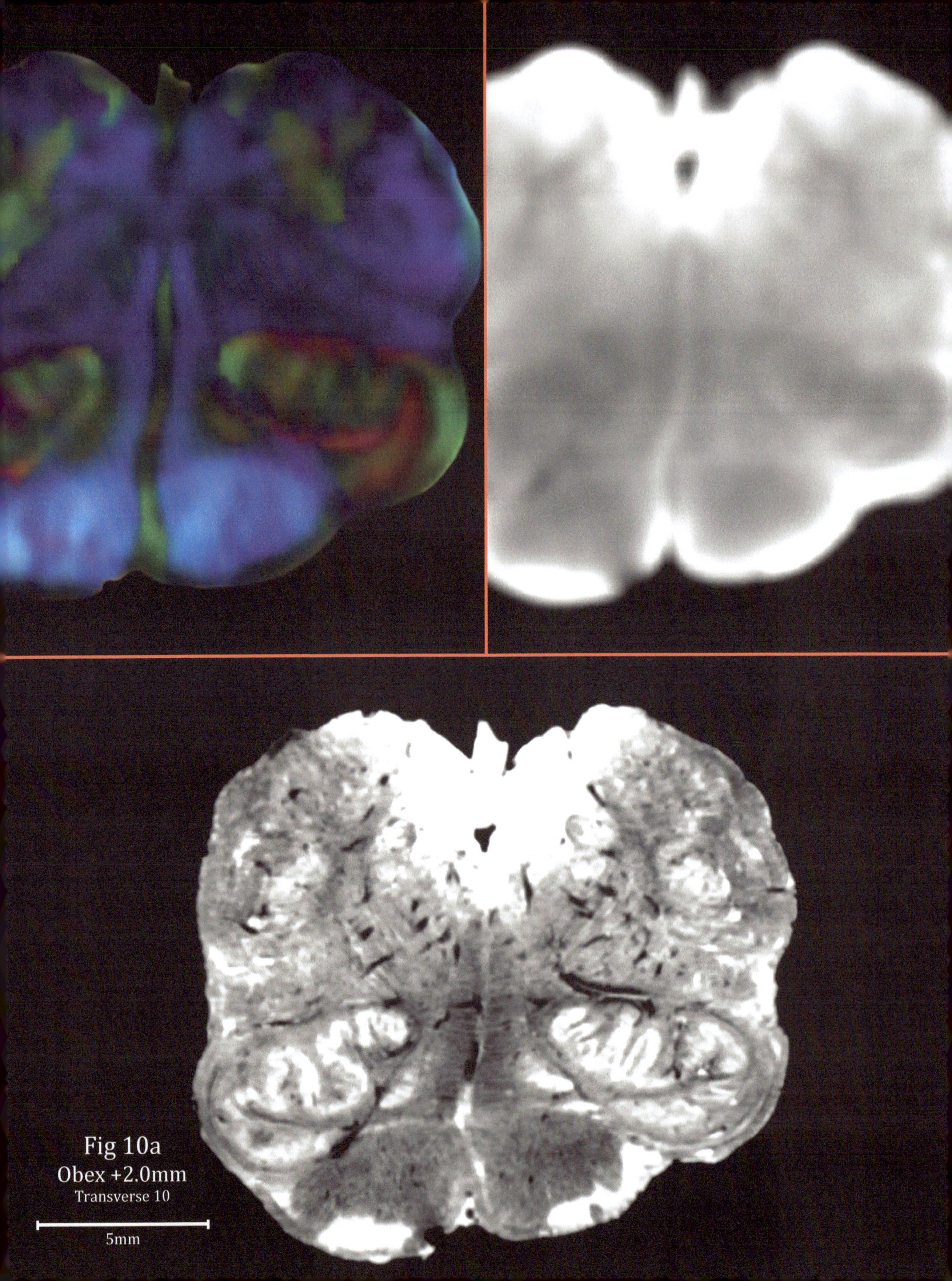

Fig 10a
Obex +2.0mm
Transverse 10

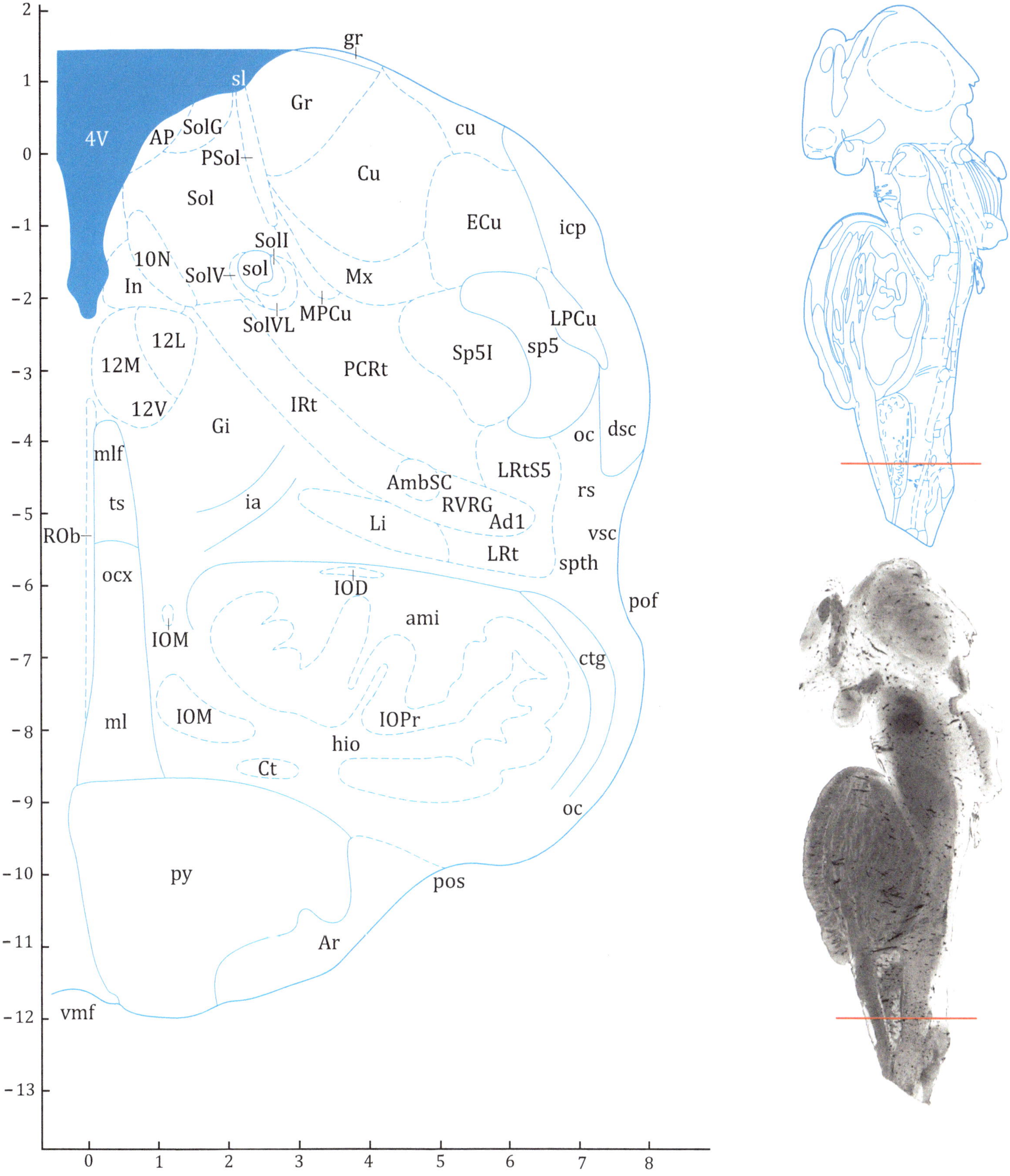

10N dorsal motor nu of vagus
12L hypoglossal nucleus, lateral part
12M hypoglossal nucleus, medial part
12V hypoglossal nucleus, ventral part
4V 4th ventricle
Ad1 Ad1 adrenalin cells
AmbSC ambiguus nu, subcompact
ami amiculum of the olive
AP area postrema
Ar arcuate nu of the caudal hindbrain
Ct conterminal nu
ctg central tegmental tract
cu cuneate fasciculus
Cu cuneate nu
dsc dorsal spinocereb tract
ECu external cuneate nu
Gi gigantocellular reticular nu
gr gracile fasciculus
Gr gracile nu
hio hilus of the inferior olive
ia internal arcuate fibers
icp inf cerebellar peduncle
In intercalated nu medulla
IOD inf olive, dorsal nu
IOM inf olive, med nu
IOPr inf olive, principal nu
IRt intermed reticular nu
Li linear nu of the medulla
LPCu lateral pericuneate nucleus
LRt lateral reticular nu
LRtS5 lat reticular nu, subtrigeminal
ml medial lemniscus
mlf med longitudinal fasciculus
MPCu medial pericuneate nucleus
Mx matrix region medulla
oc olivocerebellar tract
ocx olivocerebellar bundle decussation
PCRt parvicell reticular nu
pof post-olivary fissure
pos preolivary sulcus
PSol parasolitary nu
py pyramidal tr
ROb raphe obscurus nu
rs rubrospinal tract
RVRG rost vent respiratory
sl sulcus limitans
Sol nu of solitary tract
sol solitary tract
SolG nu sol tract, gelatinous
SolI nu sol tract, interstitial
SolV sol nu, ventral part
SolVL nu sol tract, ventrolateral

Fig 10b
Obex +2.0mm
Transverse 10

sp5 spinal trigem tract
Sp5I spinal trigem nu, interpolar
spth spinothalamic tract
See List of Structures

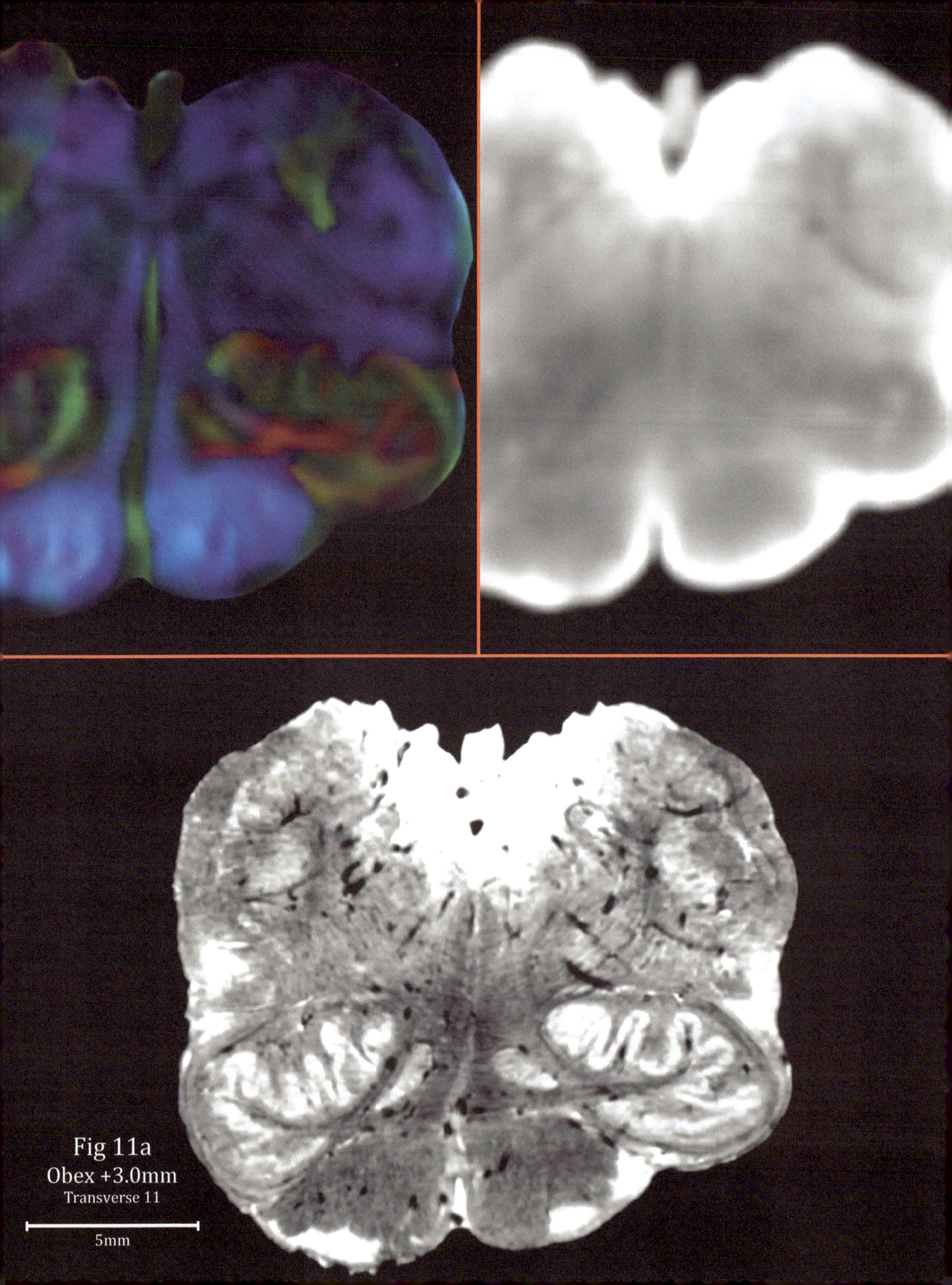

Fig 11a
Obex +3.0mm
Transverse 11

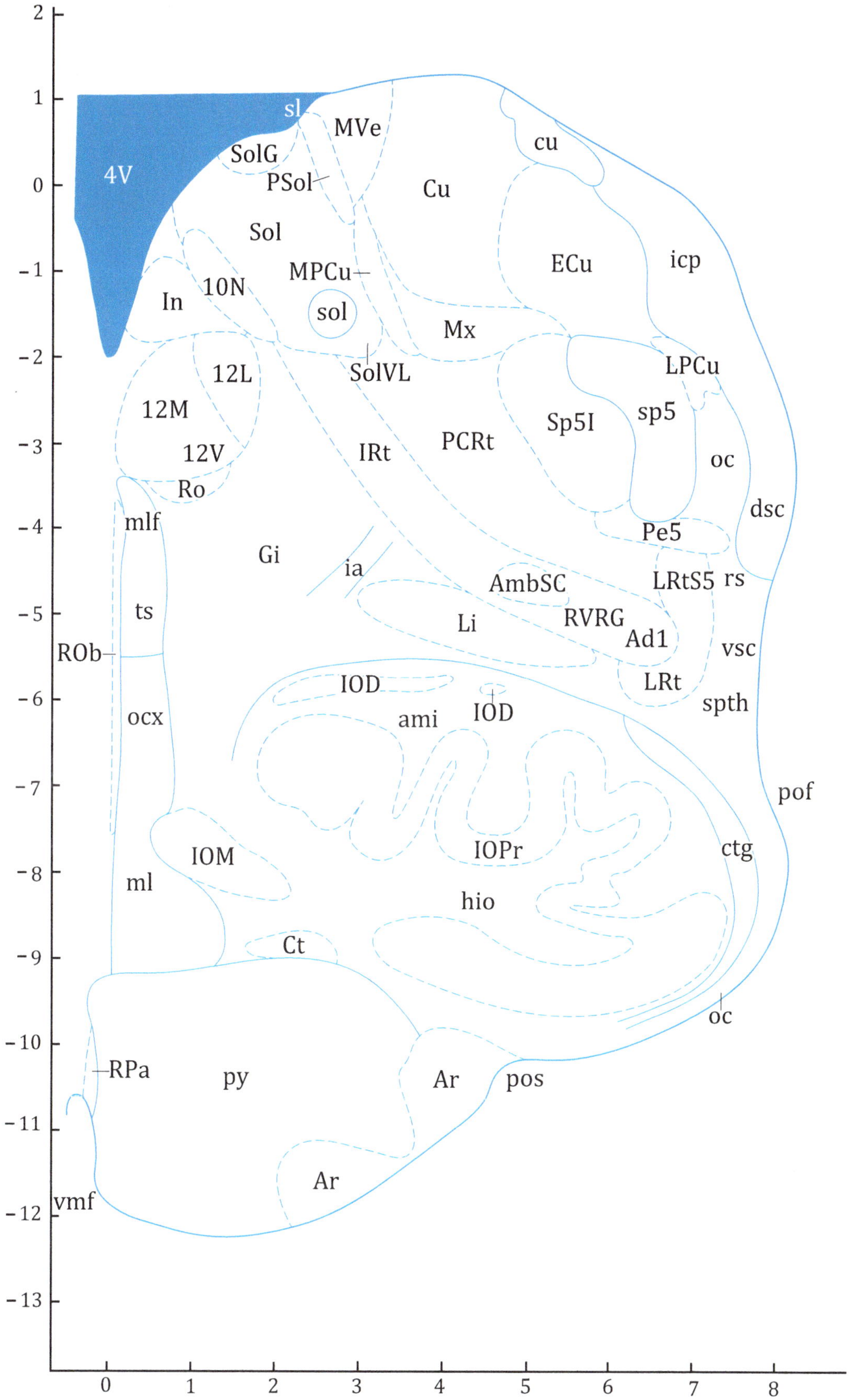

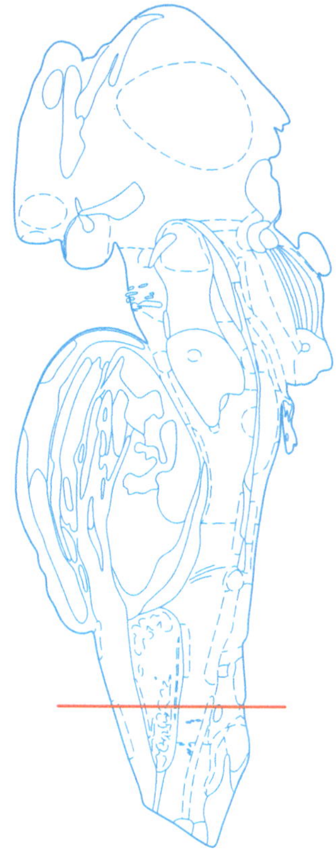

10N dorsal motor nu of vagus
12L hypoglossal nucleus, lateral part
12M hypoglossal nucleus, medial part
12V hypoglossal nucleus, ventral part
4V 4th ventricle
Ad1 Ad1 adrenalin cells
AmbSC ambiguus nu, subcompact
ami amiculum of the olive
Ar arcuate nu of the caudal hindbrain
Ct conterminal nu
ctg central tegmental tract
cu cuneate fasciculus
Cu cuneate nu
dsc dorsal spinocereb tract
ECu external cuneate nu
Gi gigantocellular reticular nu
hio hilus of the inferior olive
ia internal arcuate fibers
icp inf cerebellar peduncle
In intercalated nu medulla
IOD inf olive, dorsal nu
IOM inf olive, med nu
IOPr inf olive, principal nu
IRt intermed reticular nu
Li linear nu of the medulla
LPCu lateral pericuneate nucleus
LRt lateral reticular nu
LRtS5 lat reticular nu, subtrigeminal
ml medial lemniscus
mlf med longitudinal fasciculus
MPCu medial pericuneate nucleus
MVe medial vestibular nu
Mx matrix region medulla
oc olivocerebellar tract
ocx olivocerebellar bundle decussation
PCRt parvicell reticular nu
Pe5 peritrigeminal nucleus
pof post-olivary fissure
pos preolivary sulcus
PSol parasolitary nu
py pyramidal tr
Ro nu of Roller
ROb raphe obscurus nu
RPa raphe pallidus nu
rs rubrospinal tract
RVRG rost vent respiratory
sl sulcus limitans
Sol nu of solitary tract
sol solitary tract
SolG nu sol tract, gelatinous
SolVL nu sol tract, ventrolateral
sp5 spinal trigem tract

Fig 11b
Obex +3.0mm
Transverse 11

Sp5I spinal trigem nu, interpolar
spth spinothalamic tract
ts tectospinal tr
See List of Structures

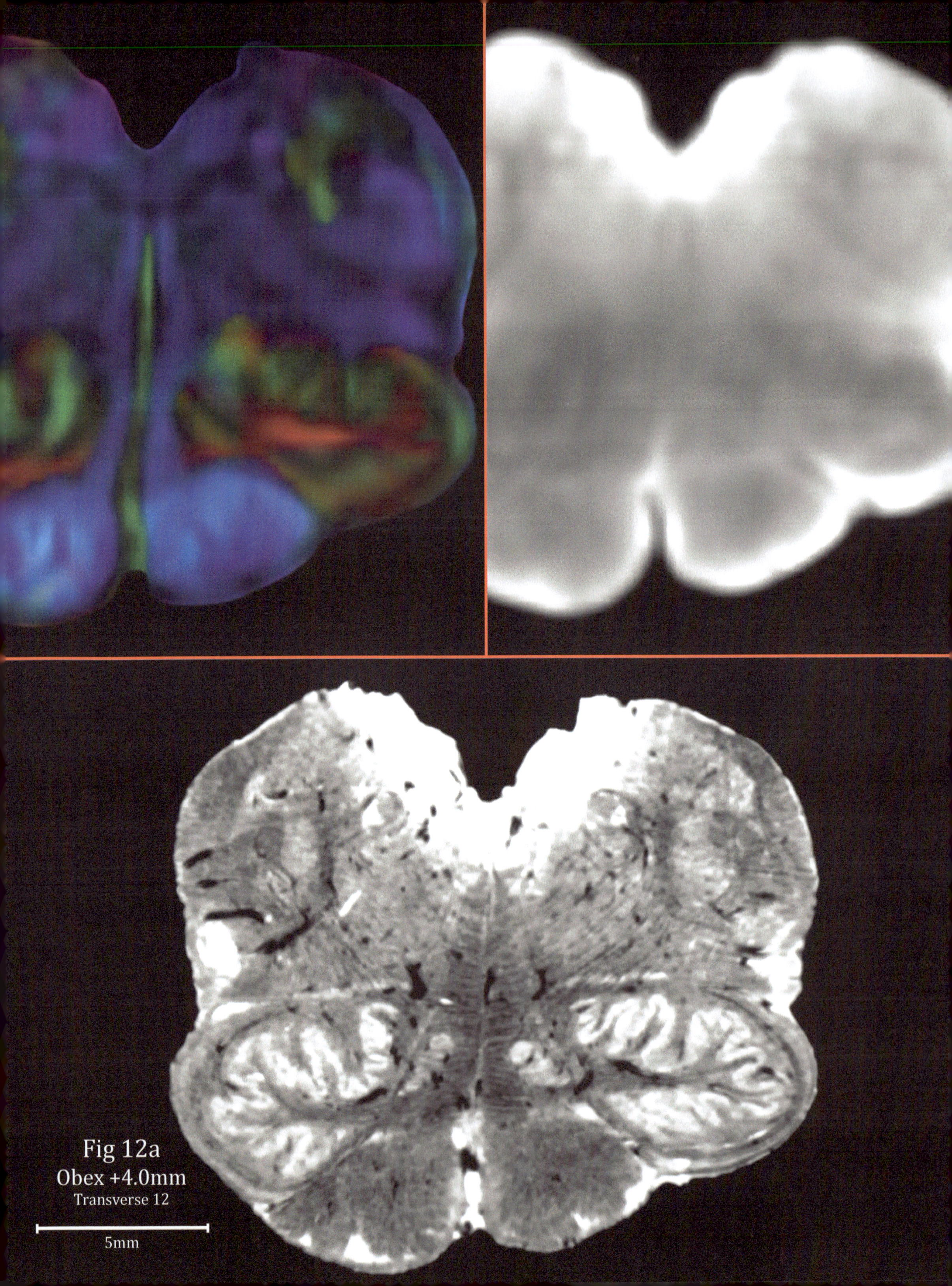

Fig 12a
Obex +4.0mm
Transverse 12

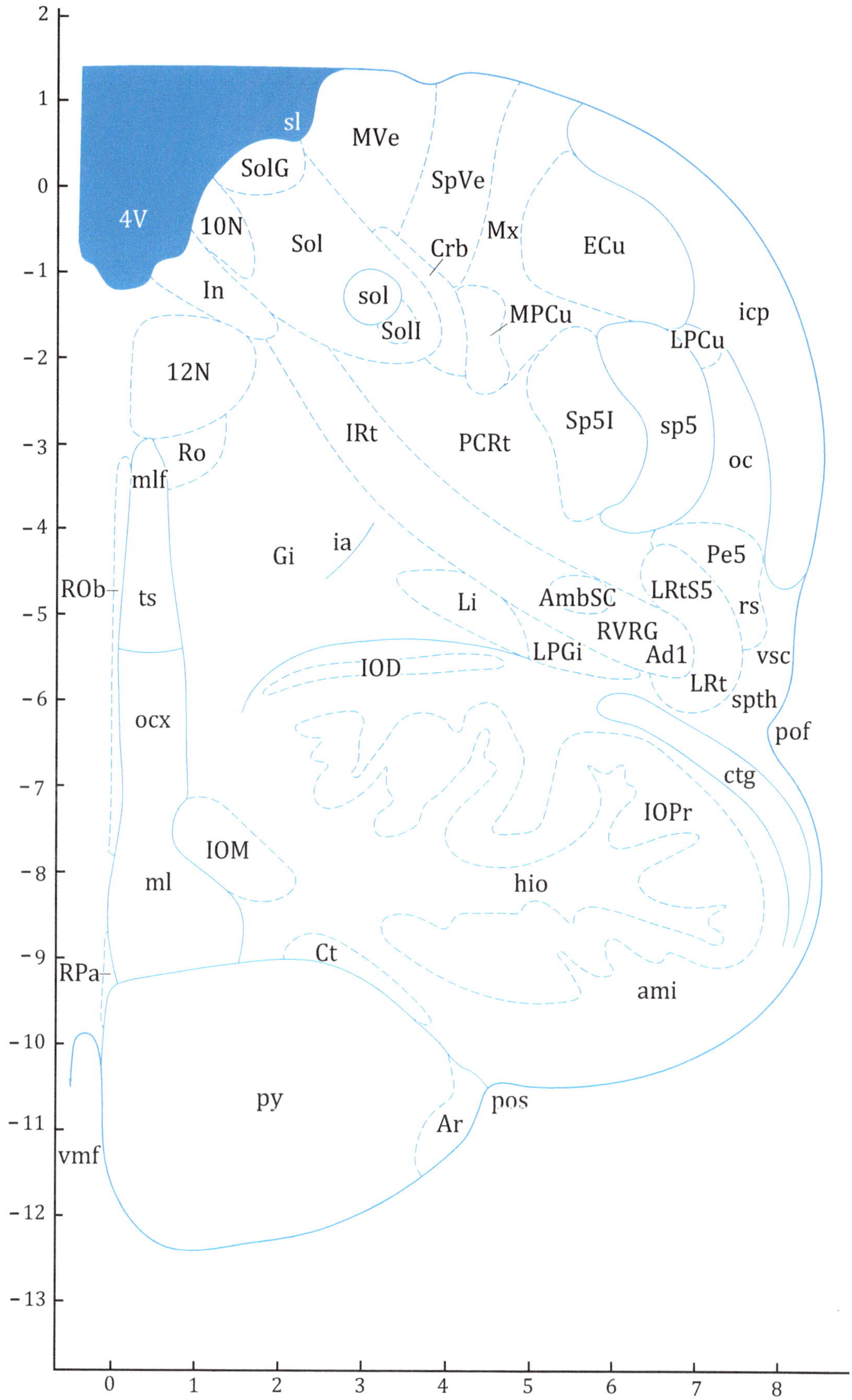

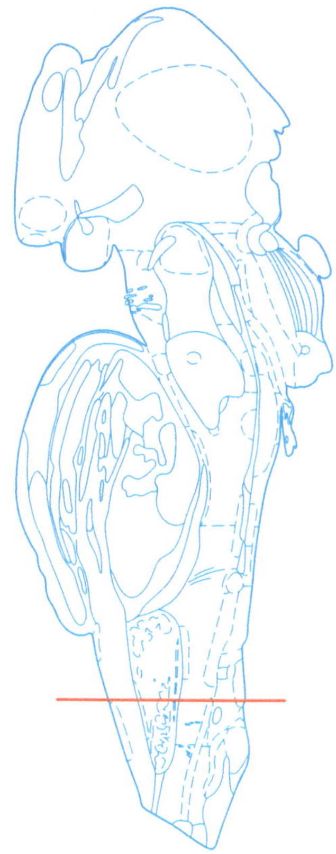

10N dorsal motor nu of vagus
12N hypoglossal nu
4V 4th ventricle
Ad1 Ad1 adrenalin cells
AmbSC ambiguus nu, subcompact
ami amiculum of the olive
Ar arcuate nu of the caudal hindbrain
Crb cribiform nucleus
Ct conterminal nu
ctg central tegmental tract
ECu external cuneate nu
Gi gigantocellular reticular nu
hio hilus of the inferior olive
ia internal arcuate fibers
icp inf cerebellar peduncle
In intercalated nu medulla
IOD inf olive, dorsal nu
IOM inf olive, med nu
IOPr inf olive, principal nu
IRt intermed reticular nu
Li linear nu of the medulla
LPCu lateral pericuneate nucleus
LPGi lat paragigantocellular nu
LRt lateral reticular nu
LRtS5 lat reticular nu, subtrigeminal
ml medial lemniscus
mlf med longitudinal fasciculus
MPCu medial pericuneate nucleus
MVe medial vestibular nu
Mx matrix region medulla
oc olivocerebellar tract
ocx olivocerebellar bundle decussation
PCRt parvicell reticular nu
Pe5 peritrigeminal nucleus
pof post-olivary fissure
pos preolivary sulcus
py pyramidal tr
Ro nu of Roller
ROb raphe obscurus nu
RPa raphe pallidus nu
rs rubrospinal tract
RVRG rost vent respiratory
sl sulcus limitans
sol solitary tract
Sol nu of solitary tract
SolG nu sol tract, gelatinous
sp5 spinal trigem tract
Sp5I spinal trigem nu, interpolar
spth spinothalamic tract
SpVe spinal vestibular nu
ts tectospinal tr
vmf ventral median fissure
vsc ventral spinocerebellar tr

Fig 12b
Obex +4.0mm
Transverse 12

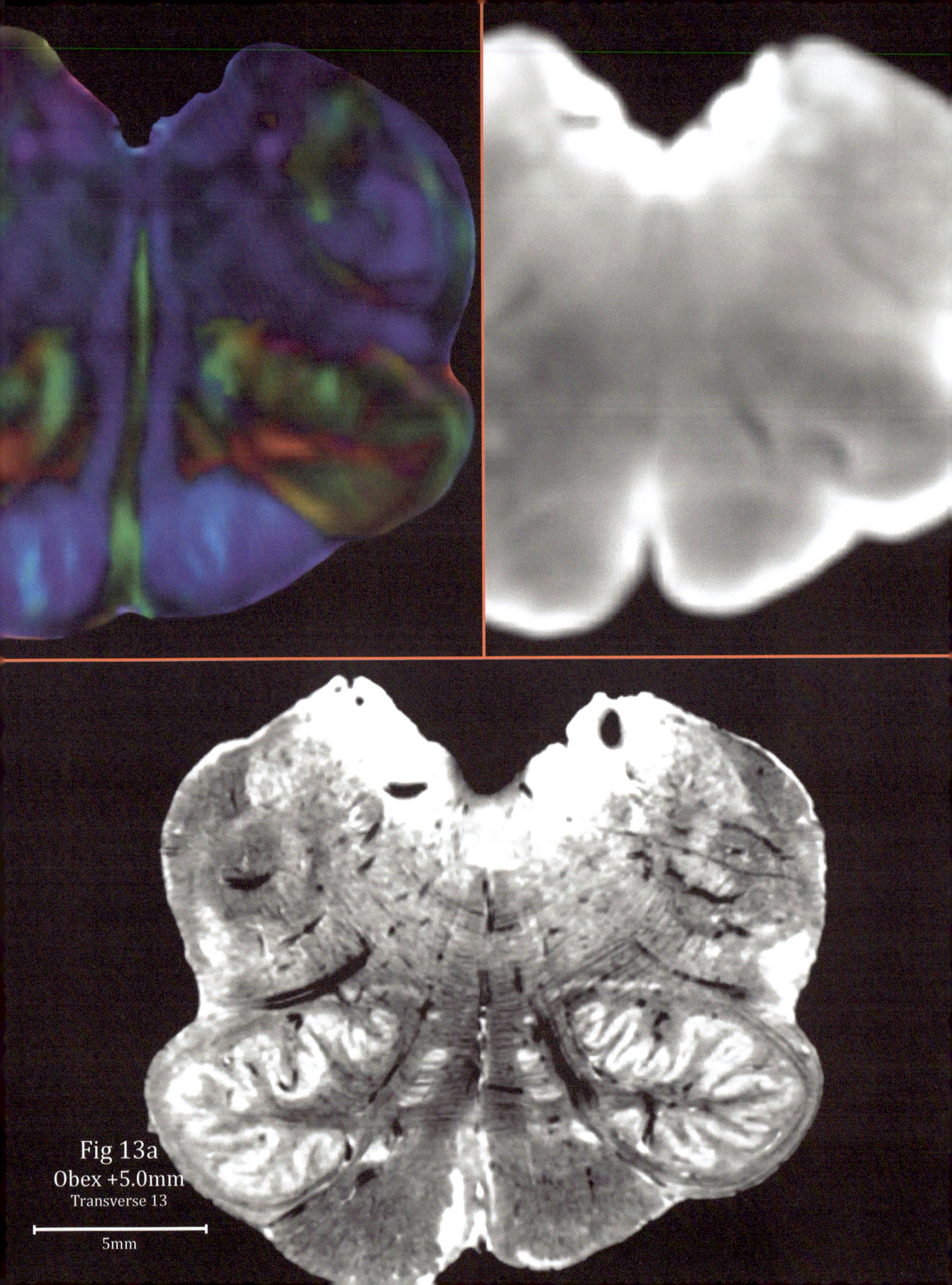

Fig 13a
Obex +5.0mm
Transverse 13

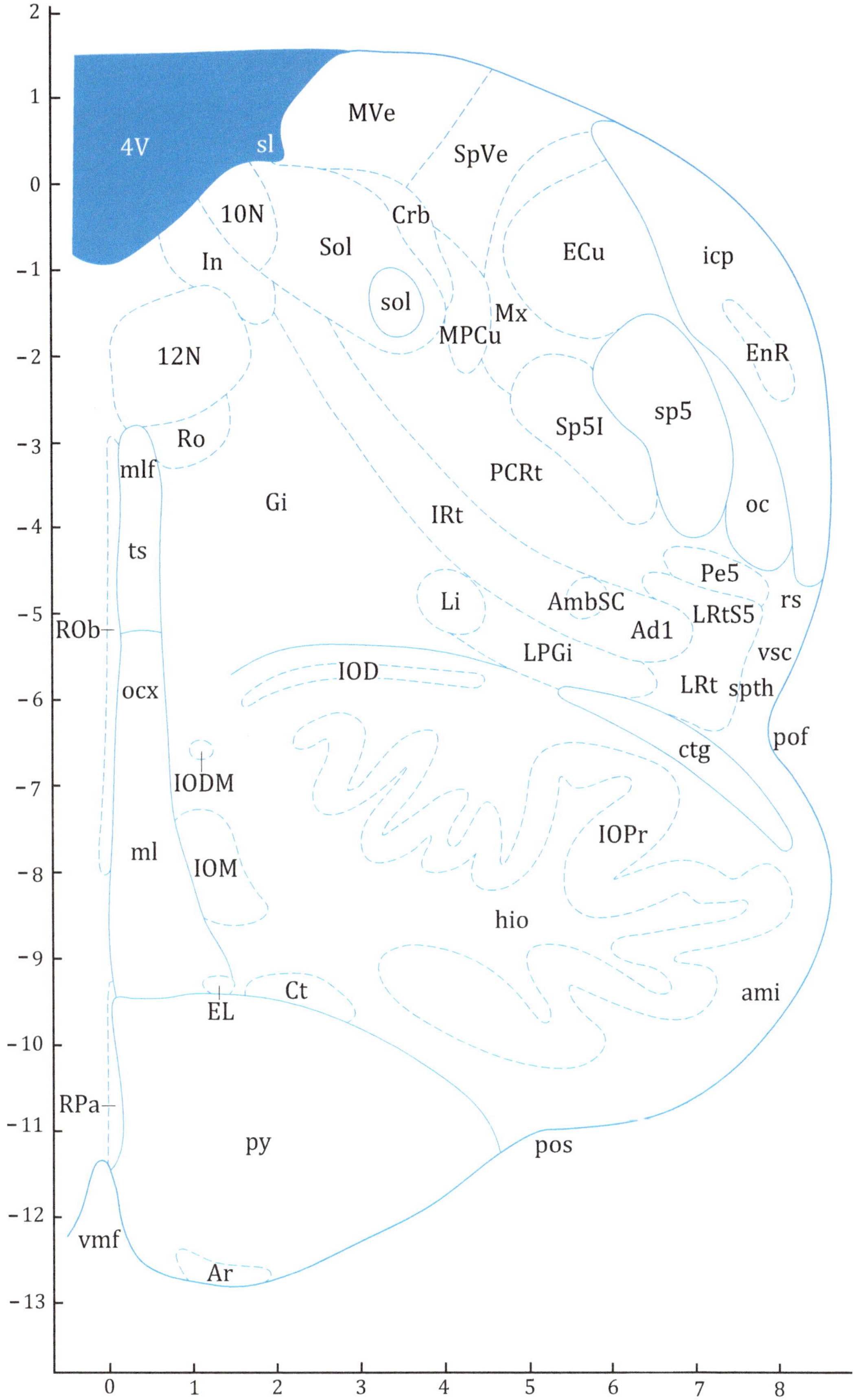

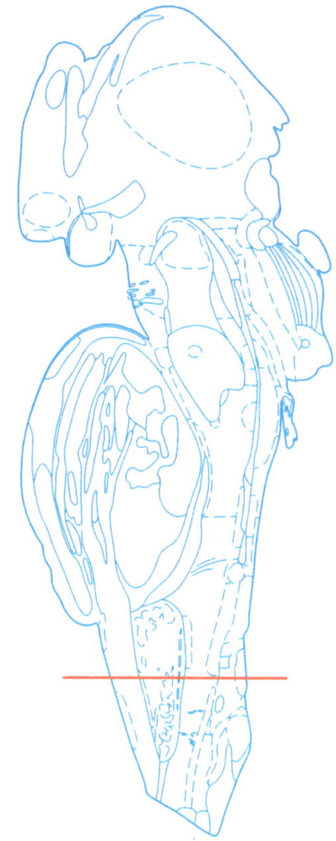

10N dorsal motor nu of vagus
12N hypoglossal nu
4V 4th ventricle
Ad1 Ad1 adrenalin cells
AmbSC ambiguus nu, subcompact
ami amiculum of the olive
Ar arcuate nu of the caudal hindbrain
Crb cribiform nucleus
Ct conterminal nu
ctg central tegmental tract
ECu external cuneate nu
EL endolemniscal nucleus
EnR endorestiform nucleus
Gi gigantocellular reticular nu
hio hilus of the inferior olive
icp inf cerebellar peduncle
In intercalated nu medulla
IOD inf olive, dorsal nu
IODM inf olive, dorsomed cell group
IOM inf olive, med nu
IOPr inf olive, principal nu
IRt intermed reticular nu
Li linear nu of the medulla
LPGi lat paragigantocellular nu
LRt lateral reticular nu
LRtS5 lat reticular nu, subtrigeminal
ml medial lemniscus
mlf med longitudinal fasciculus
MPCu medial pericuneate nucleus
MVe medial vestibular nu
Mx matrix region medulla
oc olivocerebellar tract
ocx olivocerebellar bundle decussation
PCRt parvicell reticular nu
Pe5 peritrigeminal nucleus
pof post-olivary fissure
pos preolivary sulcus
py pyramidal tr
Ro nu of Roller
ROb raphe obscurus nu
RPa raphe pallidus nu
rs rubrospinal tract
sl sulcus limitans
Sol nu of solitary tract
sol solitary tract
sp5 spinal trigem tract
Sp5I spinal trigem nu, interpolar
spth spinothalamic tract
SpVe spinal vestibular nu
ts tectospinal tr
vmf ventral median fissure
vsc ventral spinocerebellar tr

Fig 13b
Obex +5.0mm
Transverse 13

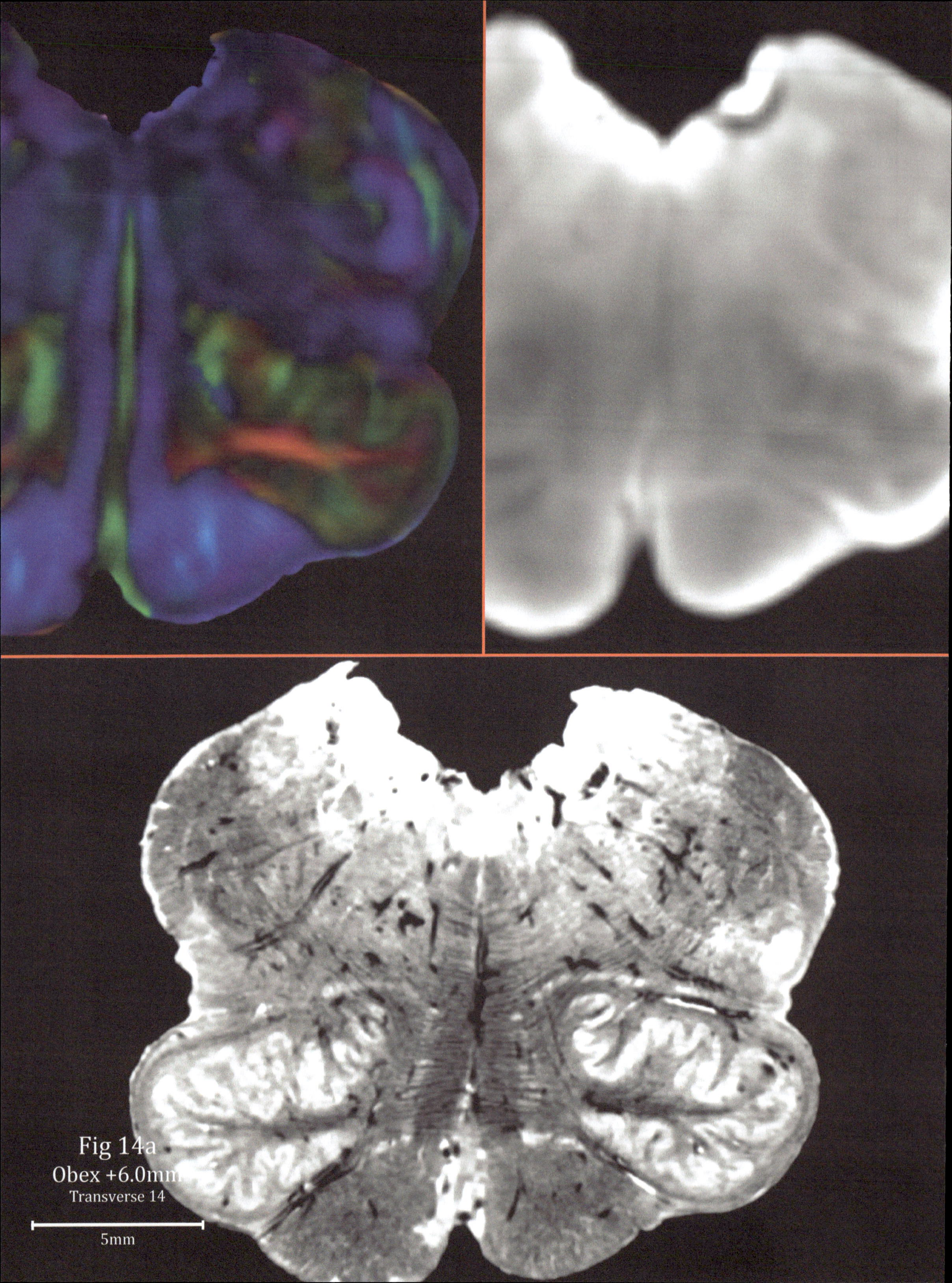
Fig 14a
Obex +6.0mm
Transverse 14
5mm

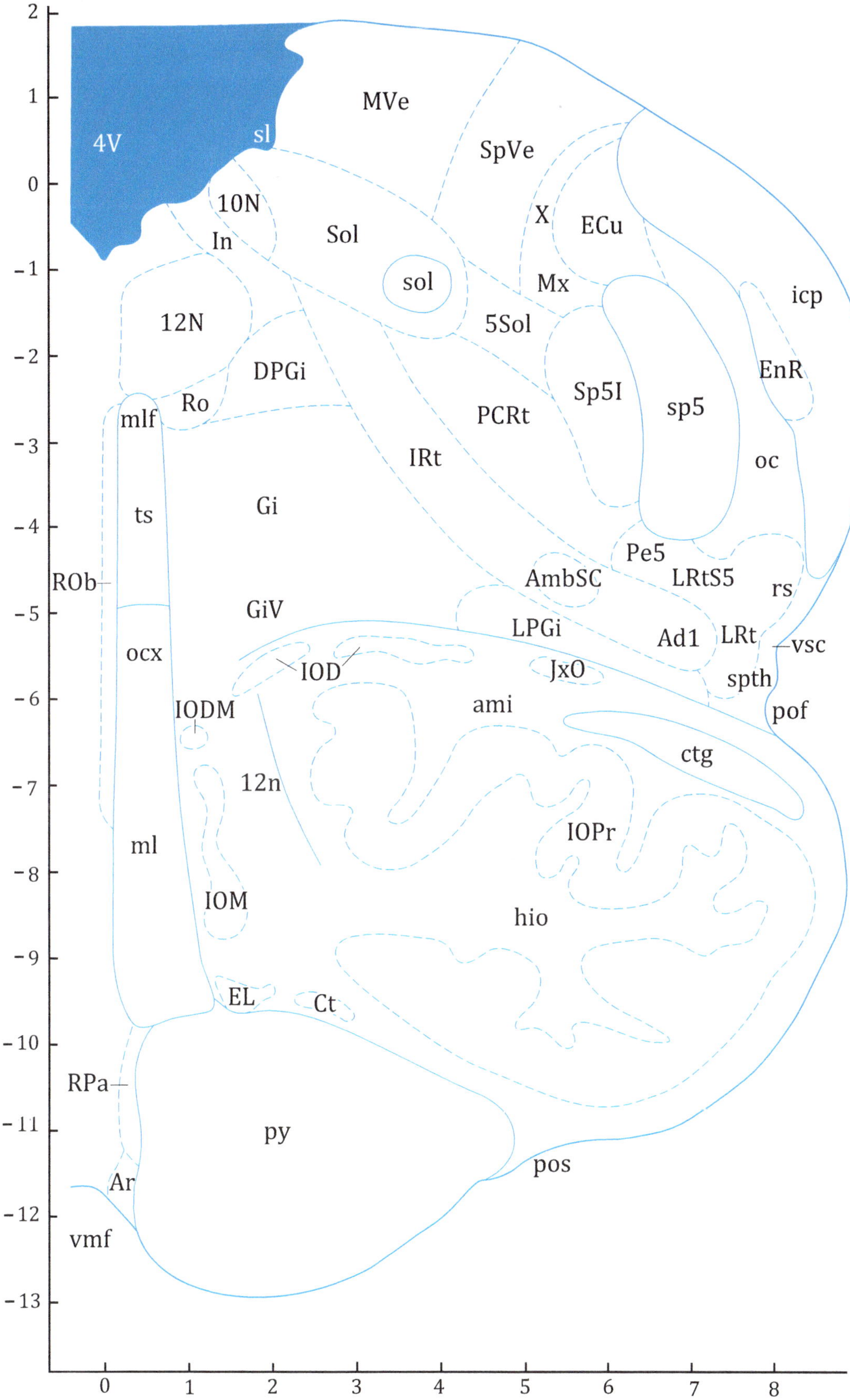

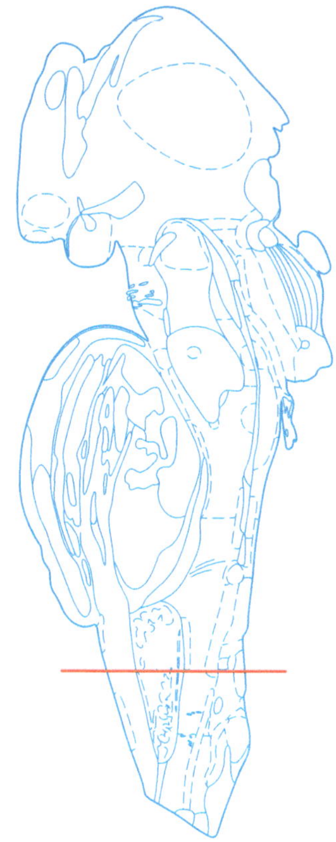

10N dorsal motor nu of vagus
12n root of hypoglossal nerve
12N hypoglossal nu
4V 4th ventricle
5Sol trigeminal-solitary trans
Ad1 Ad1 adrenalin cells
AmbSC ambiguus nu, subcompact
ami amiculum of the olive
Ar arcuate nu of the caudal hindbrain
Ct conterminal nu
ctg central tegmental tract
DPGi dors paragigantocellular nu
ECu external cuneate nu

EL endolemniscal nucleus
EnR endorestiform nucleus
Gi gigantocellular reticular nu
GiV gigantocellular retic, vent
hio hilus of the inferior olive
icp inf cerebellar peduncle
In intercalated nu medulla
IOD inf olive, dorsal nu
IODM inf olive, dorsomed cell group
IOM inf olive, med nu
IOPr inf olive, principal nu
IRt intermed reticular nu
JxO juxtaolivary nu

LPGi lat paragigantocellular nu
LRt lateral reticular nu
LRtS5 lat reticular nu, subtrigeminal
ml medial lemniscus
mlf med longitudinal fasciculus
MVe medial vestibular nu
Mx matrix region medulla
oc olivocerebellar tract
ocx olivocerebellar bundle decussation
PCRt parvicell reticular nu
Pe5 peritrigeminal nucleus
pof post-olivary fissure
pos preolivary sulcus

py pyramidal tr
Ro nu of Roller
ROb raphe obscurus nu
RPa raphe pallidus nu
rs rubrospinal tract
sl sulcus limitans
Sol nu of solitary tract
sol solitary tract
sp5 spinal trigem tract
Sp5I spinal trigem nu, interpolar
spth spinothalamic tract
SpVe spinal vestibular nu
ts tectospinal tr

Fig 14b
Obex +6.0mm
Transverse 14

vmf ventral median fissure
vsc ventral spinocerebellar tr
X nu X

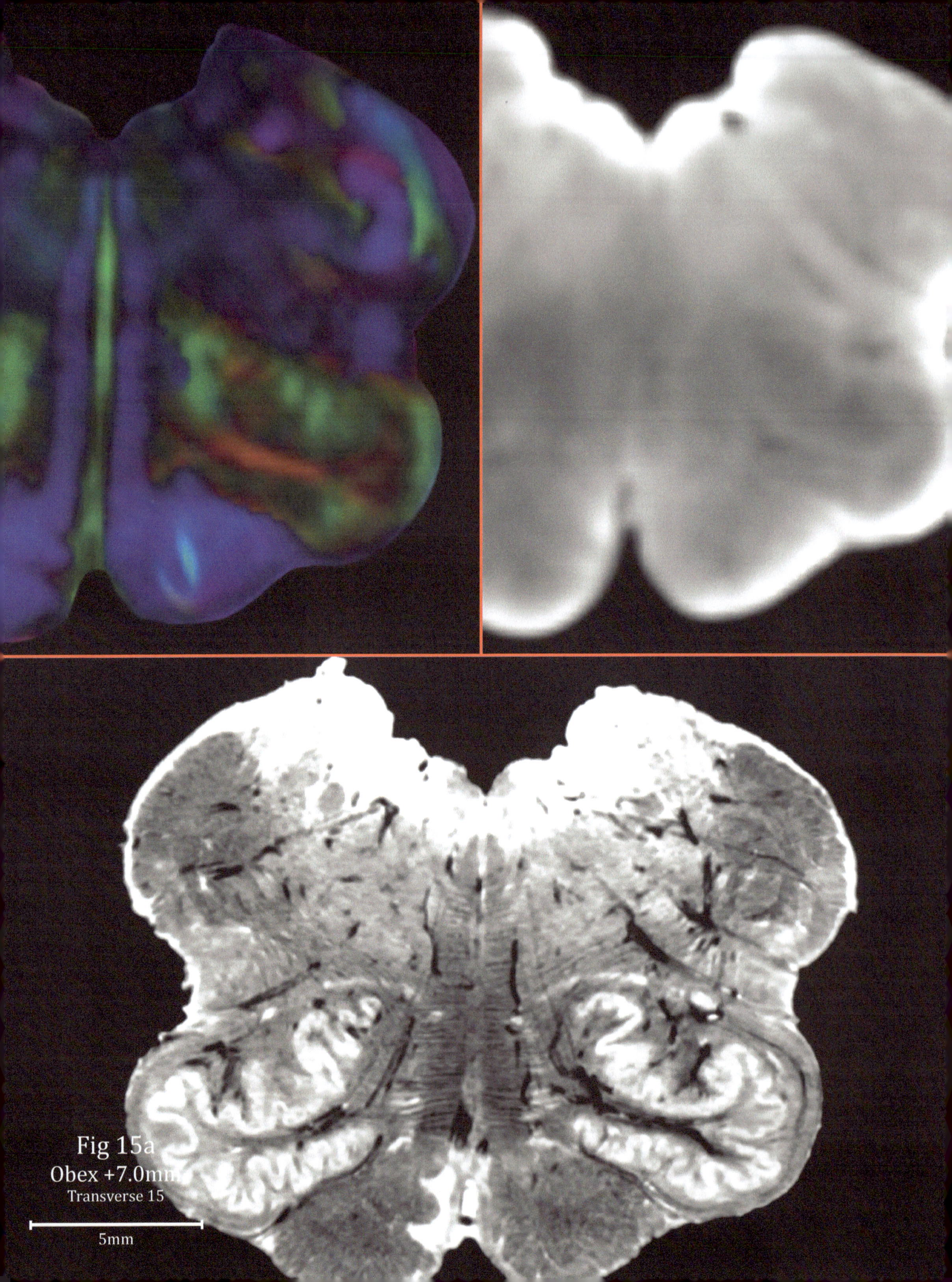
Fig 15a
Obex +7.0mm
Transverse 15
5mm

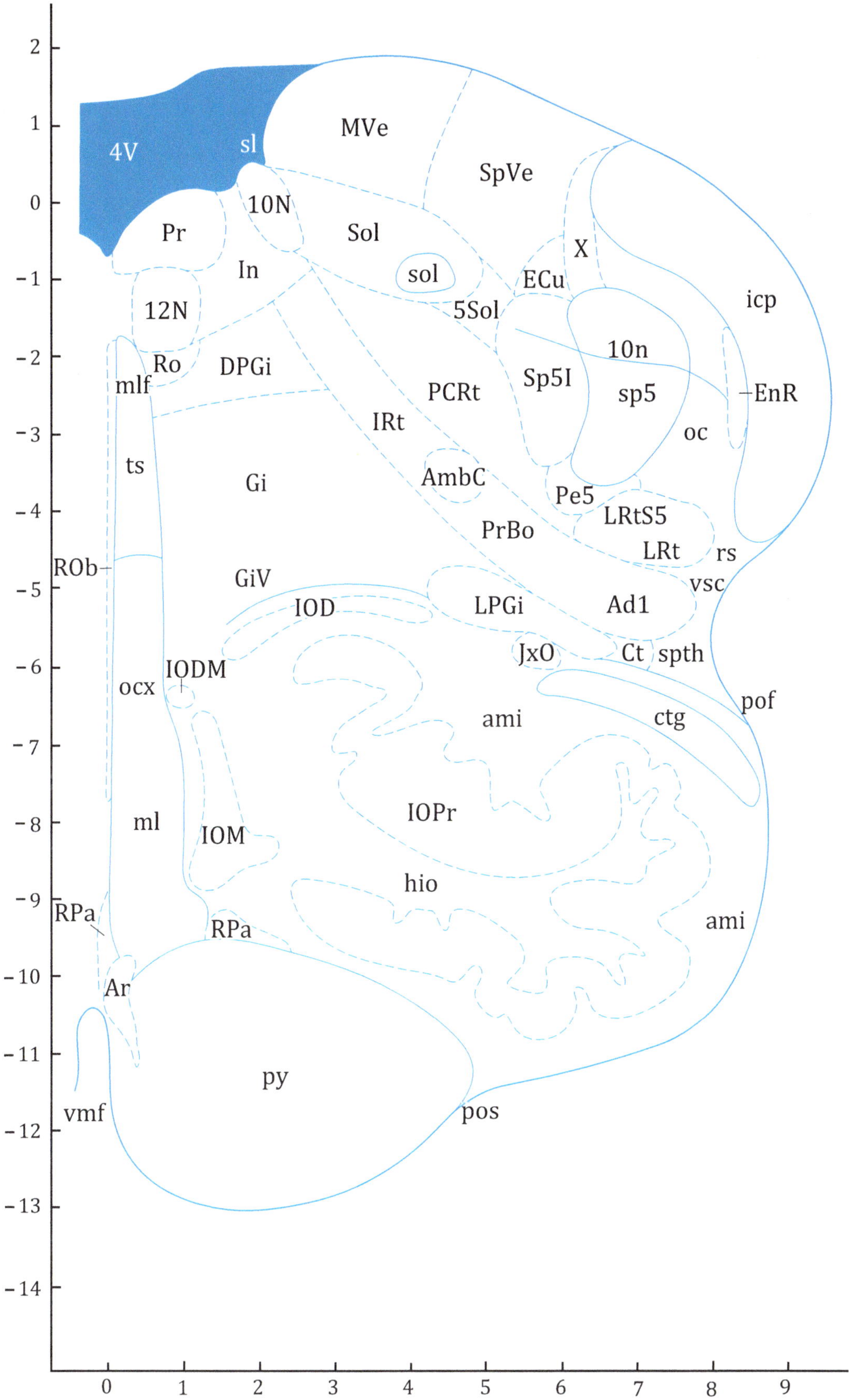

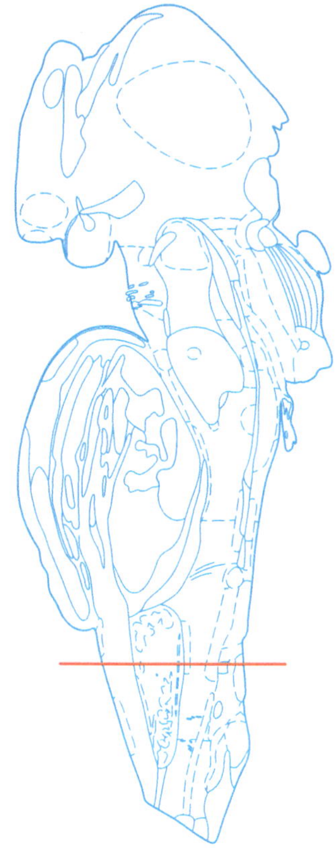

10n vagus nerve
10N dorsal motor nu of vagus
12N hypoglossal nu
4V 4th ventricle
5Sol trigeminal-solitary trans
Ad1 Ad1 adrenalin cells
AmbC ambiguus nu, compact part
ami amiculum of the olive
Ar arcuate nu of the caudal hindbrain
Ct conterminal nu
ctg central tegmental tract
DPGi dors paragigantocellular nu
ECu external cuneate nu

EnR endorestiform nucleus
Gi gigantocellular reticular nu
GiV gigantocellular retic, vent
hio hilus of the inferior olive
icp inf cerebellar peduncle
In intercalated nu medulla
IOD inf olive, dorsal nu
IODM inf olive, dorsomed cell group
IOM inf olive, med nu
IOPr inf olive, principal nu
IRt intermed reticular nu
JxO juxtaolivary nu
LPGi lat paragigantocellular nu

LRt lateral reticular nu
LRtS5 lat reticular nu, subtrigeminal
ml medial lemniscus
mlf med longitudinal fasciculus
MVe medial vestibular nu
oc olivocerebellar tract
ocx olivocerebellar bundle decussation
PCRt parvicell reticular nu
Pe5 peritrigeminal nucleus
pof post-olivary fissure
pos preolivary sulcus
Pr prepositus nu
PrBo pre-Botzinger complex

py pyramidal tr
Ro nu of Roller
ROb raphe obscurus nu
RPa raphe pallidus nu
rs rubrospinal tract
sl sulcus limitans
Sol nu of solitary tract
sol solitary tract
sp5 spinal trigem tract
Sp5I spinal trigem nu, interpolar
spth spinothalamic tract
SpVe spinal vestibular nu
ts tectospinal tr

Fig 15b
Obex +7.0mm
Transverse 15

vmf ventral median fissure
vsc ventral spinocerebellar tr
X nu X

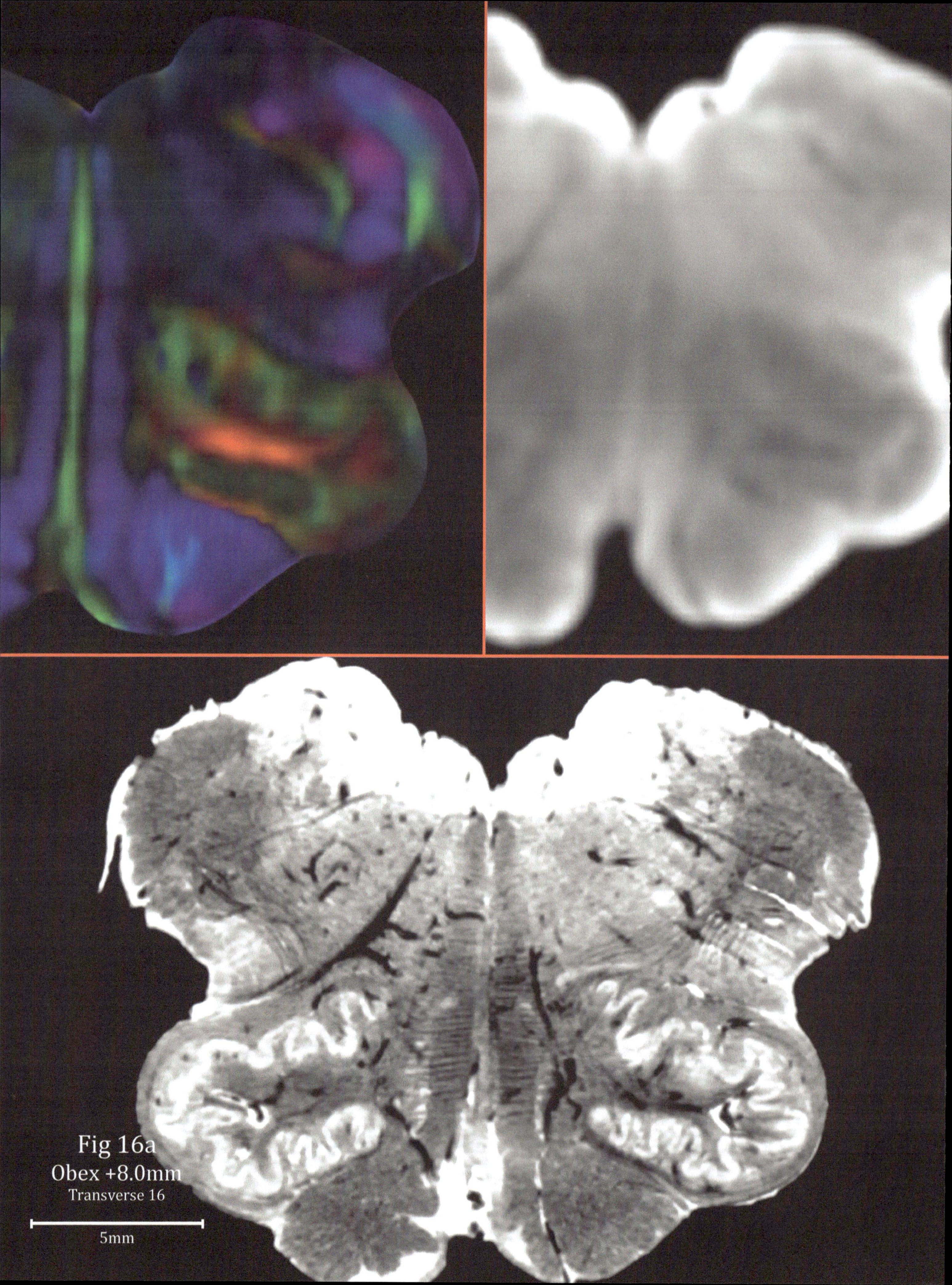

Fig 16a
Obex +8.0mm
Transverse 16

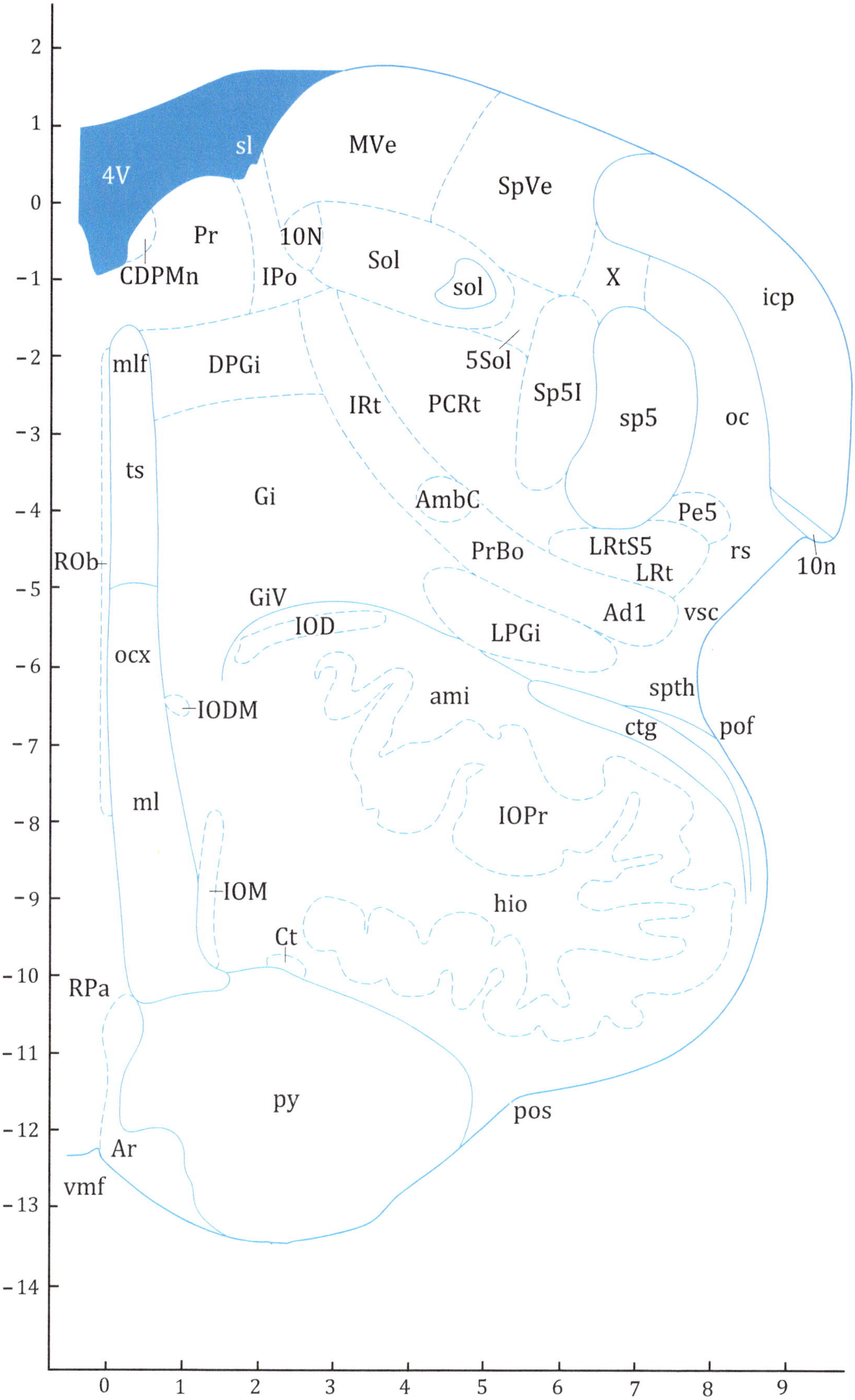

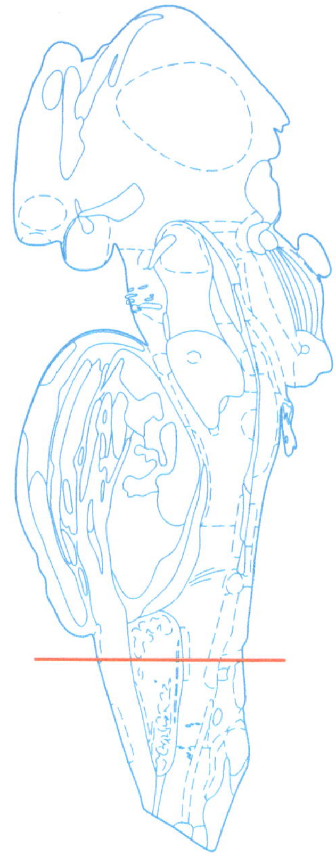

10N dorsal motor nu of vagus
10n vagus nerve
4V 4th ventricle
5Sol trigeminal-solitary trans
Ad1 Ad1 adrenalin cells
AmbC ambiguus nu, compact part
ami amiculum of the olive
Ar arcuate nu of the caudal hindbrain
CDPMn caudal dorsal paramedian nu
Ct conterminal nu
ctg central tegmental tract
DPGi dors paragigantocellular nu
Gi gigantocellular reticular nu
GiV gigantocellular retic, vent
hio hilus of the inferior olive
icp inf cerebellar peduncle
IOD inf olive, dorsal nu
IODM inf olive, dorsomed cell group
IOM inf olive, med nu
IOPr inf olive, principal nu
IPo interpositus
IRt intermed reticular nu
LPGi lat paragigantocellular nu
LRt lateral reticular nu
LRtS5 lat reticular nu, subtrigeminal
ml medial lemniscus
mlf med longitudinal fasciculus
MVe medial vestibular nu
oc olivocerebellar tract
ocx olivocerebellar bundle decussation
PCRt parvicell reticular nu
Pe5 peritrigeminal nucleus
pof post-olivary fissure
pos preolivary sulcus
Pr prepositus nu
PrBo pre-Botzinger complex
py pyramidal tr
ROb raphe obscurus nu
RPa raphe pallidus nu
rs rubrospinal tract
sl sulcus limitans
Sol nu of solitary tract
sol solitary tract
sp5 spinal trigem tract
Sp5I spinal trigem nu, interpolar
spth spinothalamic tract
SpVe spinal vestibular nu
ts tectospinal tr
vmf ventral median fissure
vsc ventral spinocerebellar tr
X nu X

Fig 16b
Obex +8.0mm
Transverse 16

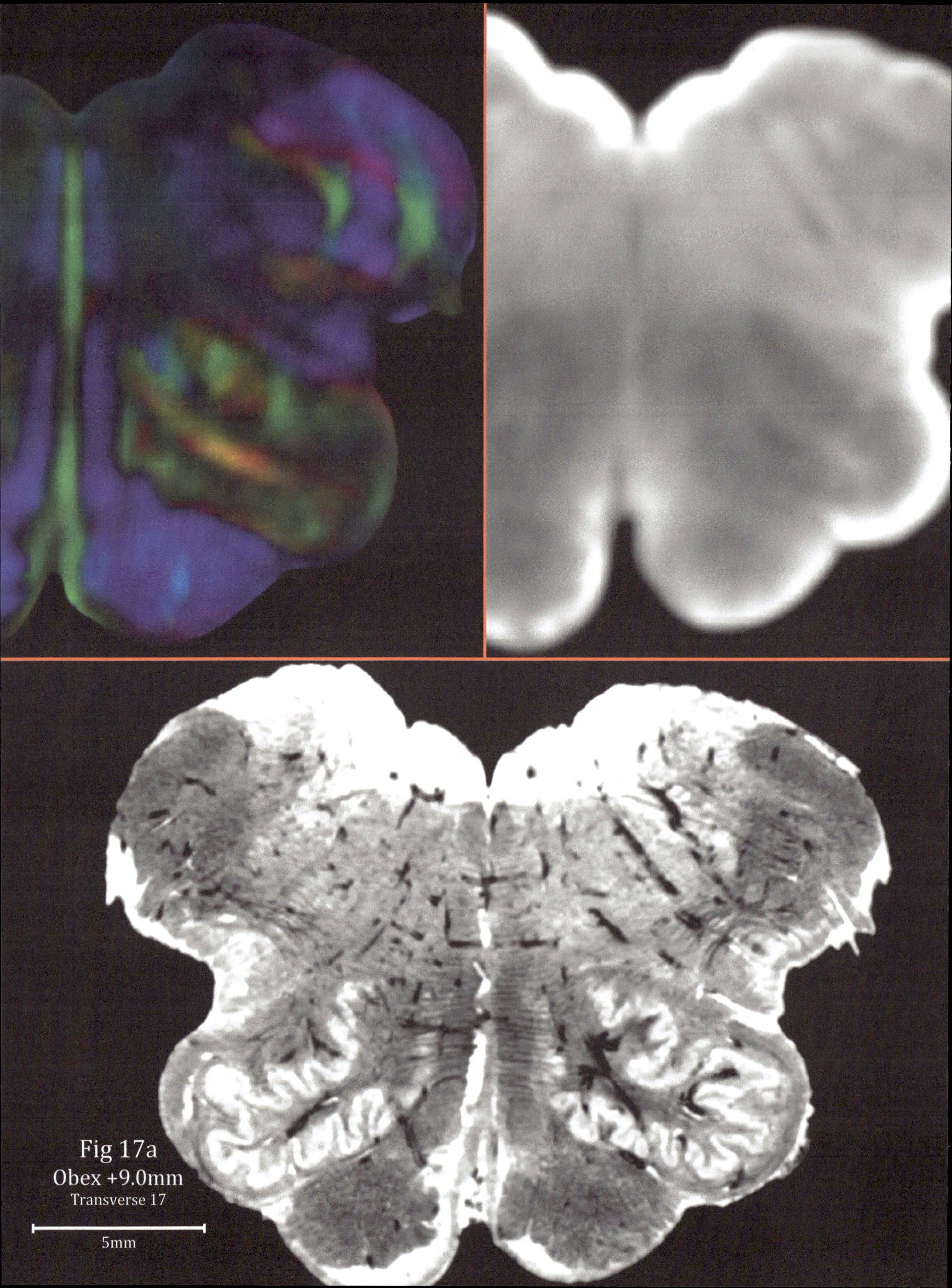

Fig 17a
Obex +9.0mm
Transverse 17

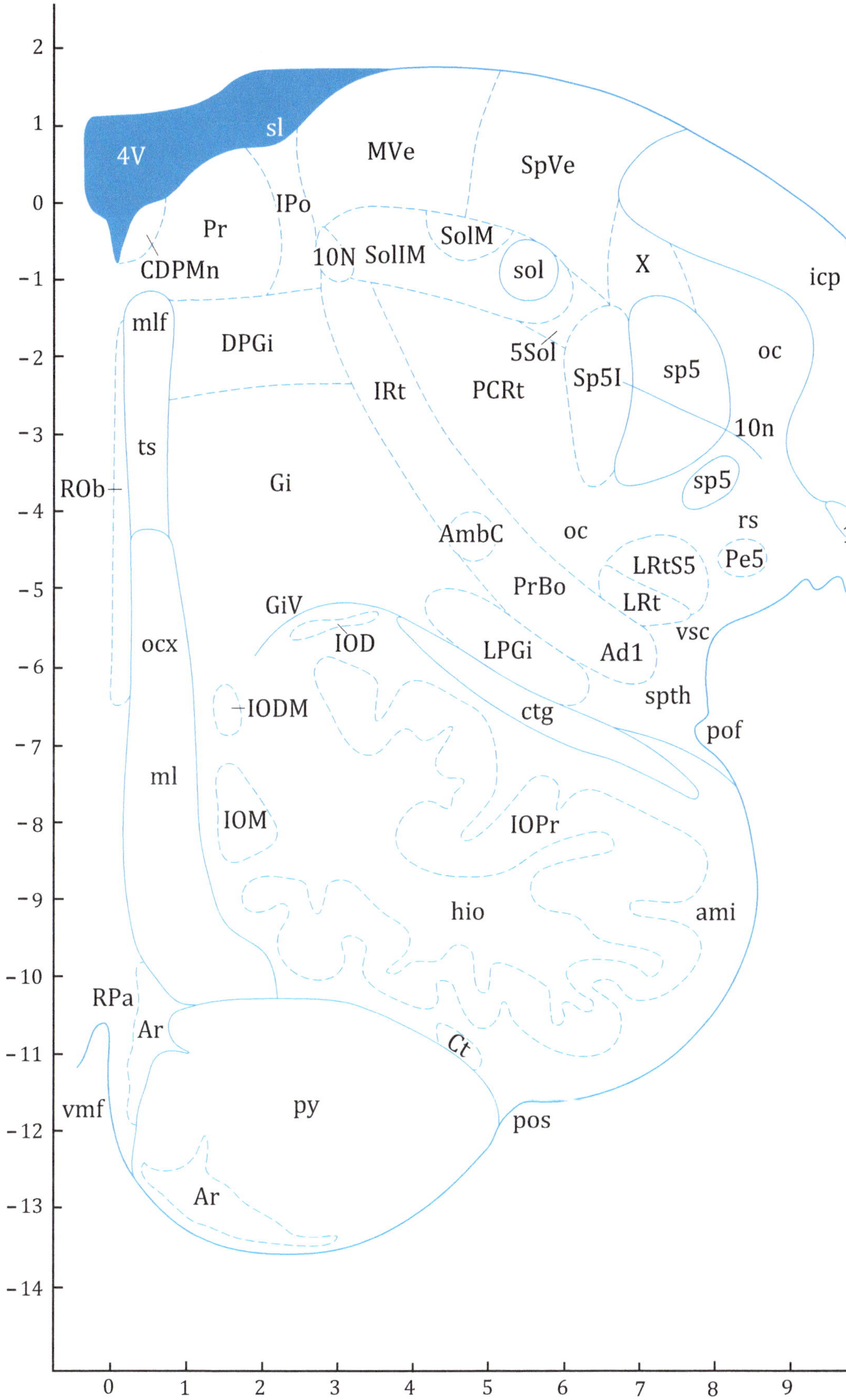

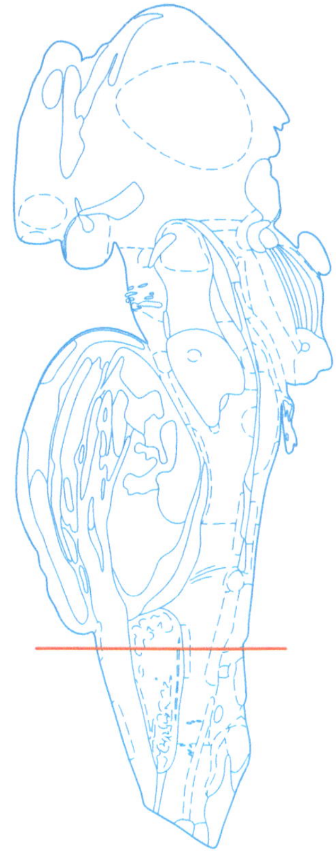

10N dorsal motor nu of vagus
10n vagus nerve
4V 4th ventricle
5Sol trigeminal-solitary trans
Ad1 Ad1 adrenalin cells
AmbC ambiguus nu, compact part
ami amiculum of the olive
Ar arcuate nu of the caudal hindbrain
CDPMn caudal dorsal paramedian nu
Ct conterminal nu
ctg central tegmental tract
DPGi dors paragigantocellular nu
Gi gigantocellular reticular nu
GiV gigantocellular retic, vent
hio hilus of the inferior olive
icp inf cerebellar peduncle
IOD inf olive, dorsal nu
IODM inf olive, dorsomed cell group
IOM inf olive, med nu
IOPr inf olive, principal nu
IPo interpositus
IRt intermed reticular nu
LPGi lat paragigantocellular nu
LRt lateral reticular nu
LRtS5 lat reticular nu, subtrigeminal
ml medial lemniscus
mlf med longitudinal fasciculus
MVe medial vestibular nu
oc olivocerebellar tract
ocx olivocerebellar bundle decussation
PCRt parvicell reticular nu
Pe5 peritrigeminal nucleus
pof post-olivary fissure
pos preolivary sulcus
Pr prepositus nu
PrBo pre-Botzinger complex
py pyramidal tr
ROb raphe obscurus nu
RPa raphe pallidus nu
rs rubrospinal tract
sl sulcus limitans
sol solitary tract
SolIM nu sol tract, intermediate
SolM nu sol tract, medial
sp5 spinal trigem tract
Sp5I spinal trigem nu, interpolar
spth spinothalamic tract
SpVe spinal vestibular nu
ts tectospinal tr
vmf ventral median fissure
vsc ventral spinocerebellar tr
X nu X

Fig 17b
Obex +9.0mm
Transverse 17

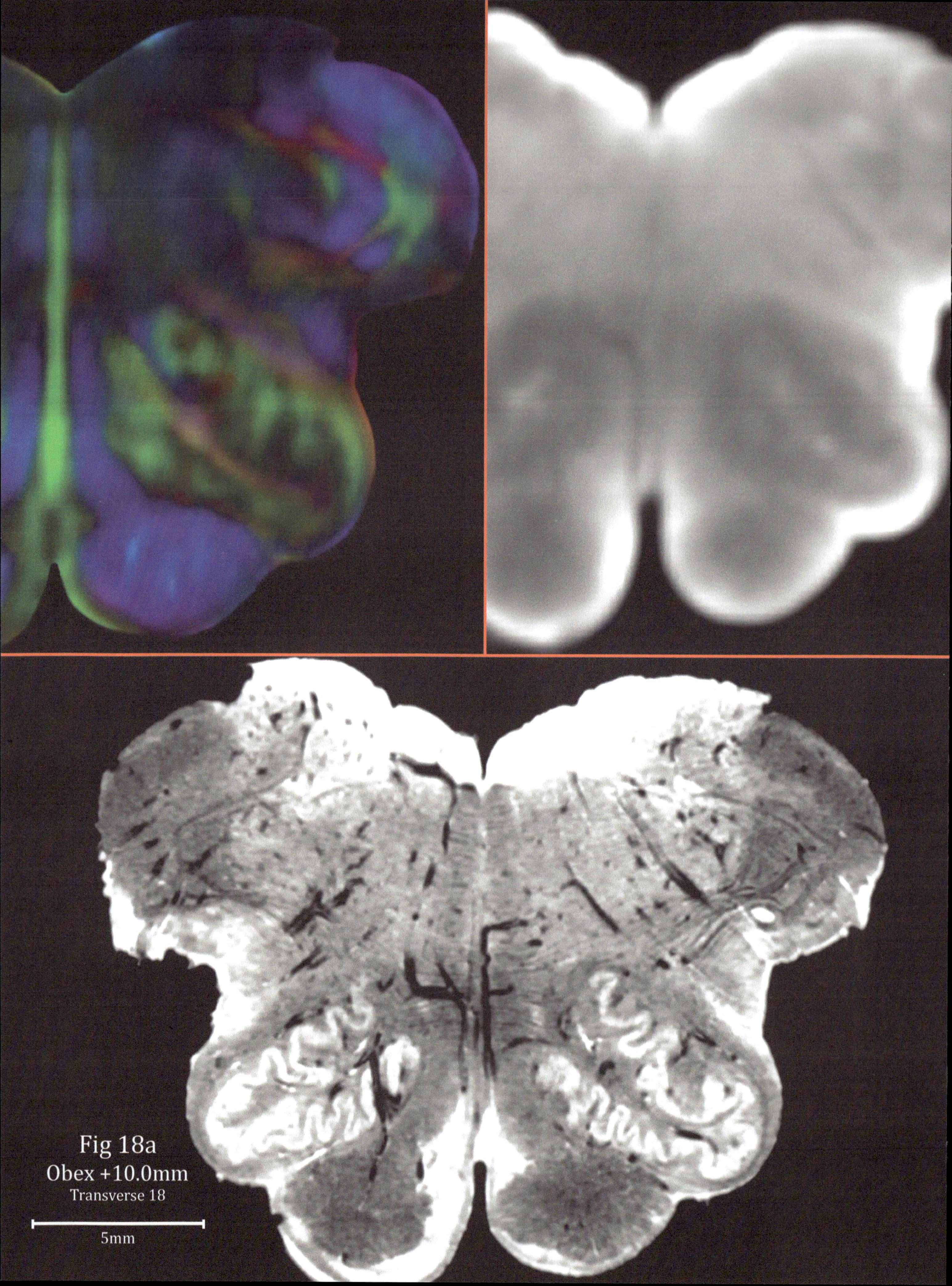

Fig 18a
Obex +10.0mm
Transverse 18

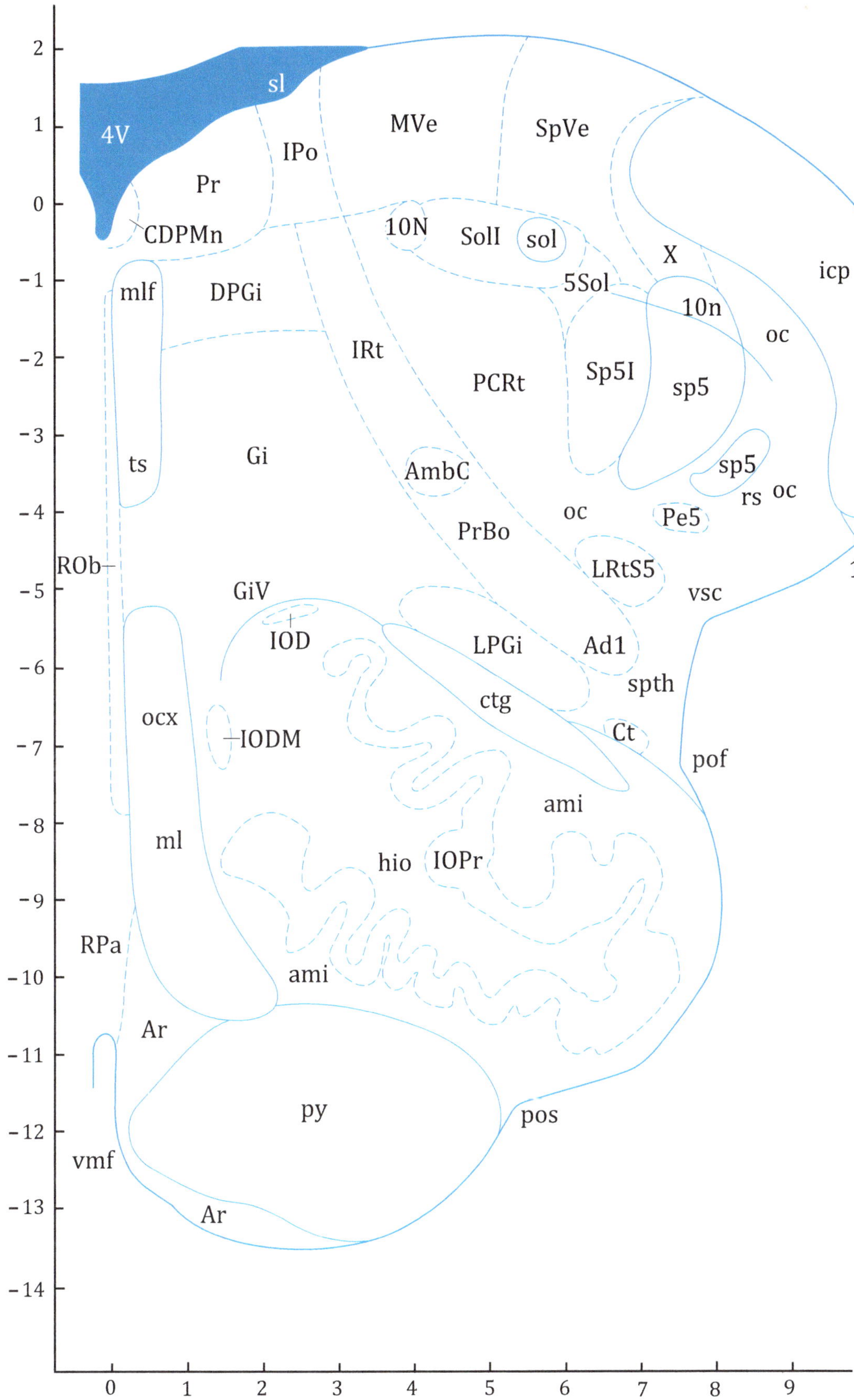

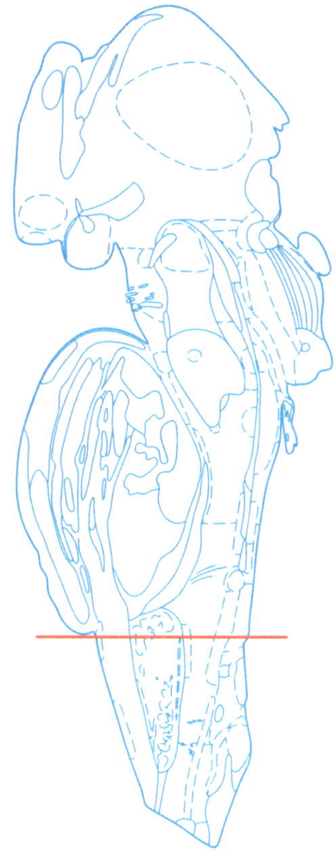

10N dorsal motor nu of vagus
10n vagus nerve
4V 4th ventricle
5Sol trigeminal-solitary trans
Ad1 Ad1 adrenalin cells
AmbC ambiguus nu, compact part
ami amiculum of the olive
Ar arcuate nu of the caudal hindbrain
CDPMn caudal dorsal paramedian nu
Ct conterminal nu
ctg central tegmental tract
DPGi dors paragigantocellular nu
Gi gigantocellular reticular nu
GiV gigantocellular retic, vent
hio hilus of the inferior olive
icp inf cerebellar peduncle
IOD inf olive, dorsal nu
IODM inf olive, dorsomed cell group
IOPr inf olive, principal nu
IPo interpositus
IRt intermed reticular nu
LPGi lat paragigantocellular nu
LRtS5 lat reticular nu, subtrigeminal
ml medial lemniscus
mlf med longitudinal fasciculus
MVe medial vestibular nu
oc olivocerebellar tract
ocx olivocerebellar bundle decussation
PCRt parvicell reticular nu
Pe5 peritrigeminal nucleus
pof post-olivary fissure
pos preolivary sulcus
Pr prepositus nu
PrBo pre-Botzinger complex
py pyramidal tr
ROb raphe obscurus nu
RPa raphe pallidus nu
rs rubrospinal tract
sl sulcus limitans
sol solitary tract
SolI nu sol tract, interstitial
sp5 spinal trigem tract
Sp5I spinal trigem nu, interpolar
spth spinothalamic tract
SpVe spinal vestibular nu
ts tectospinal tr
vmf ventral median fissure
vsc ventral spinocerebellar tr
X nu X

Fig 18b
Obex +10.0mm
Transverse 18

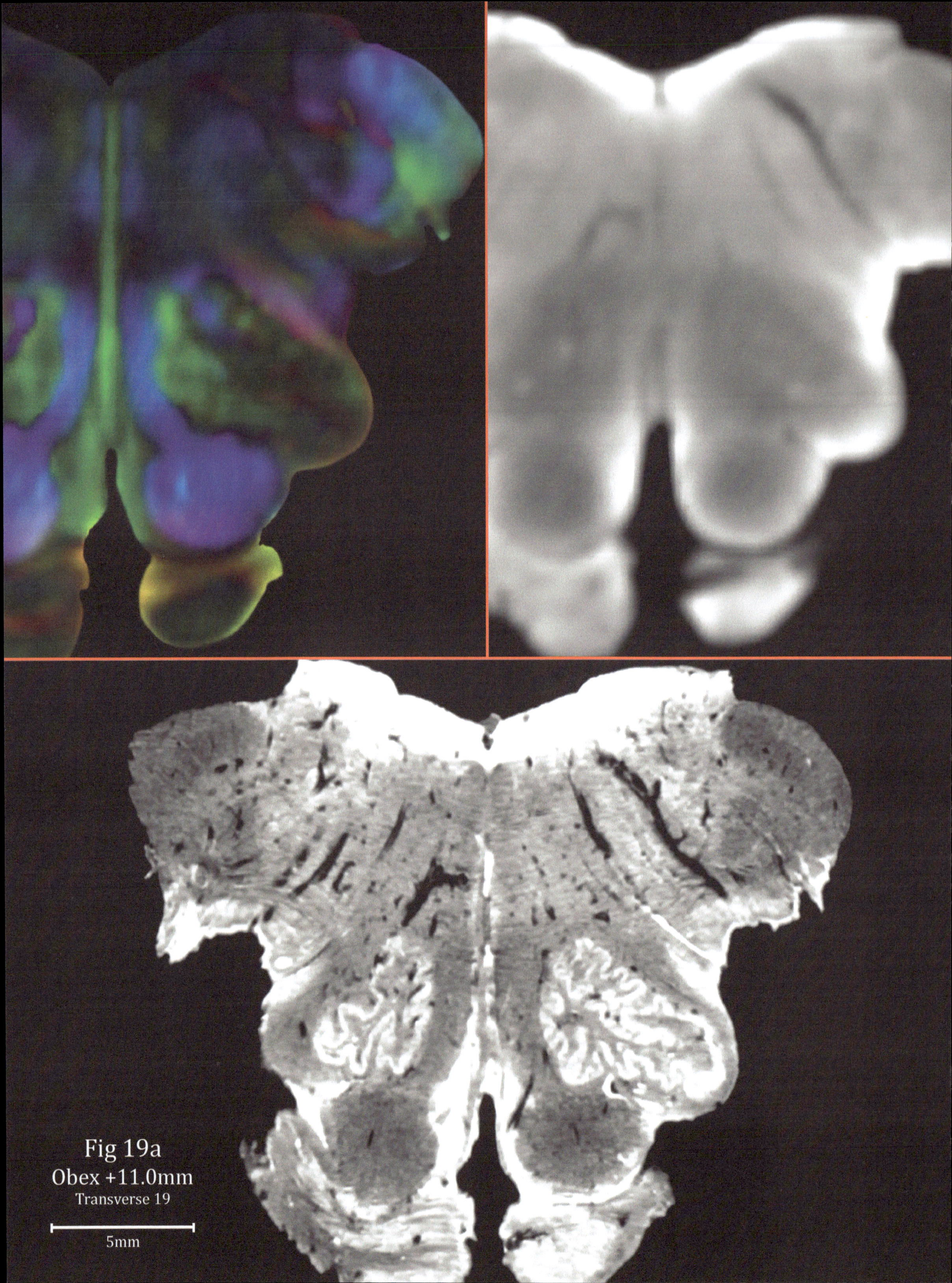

Fig 19a
Obex +11.0mm
Transverse 19

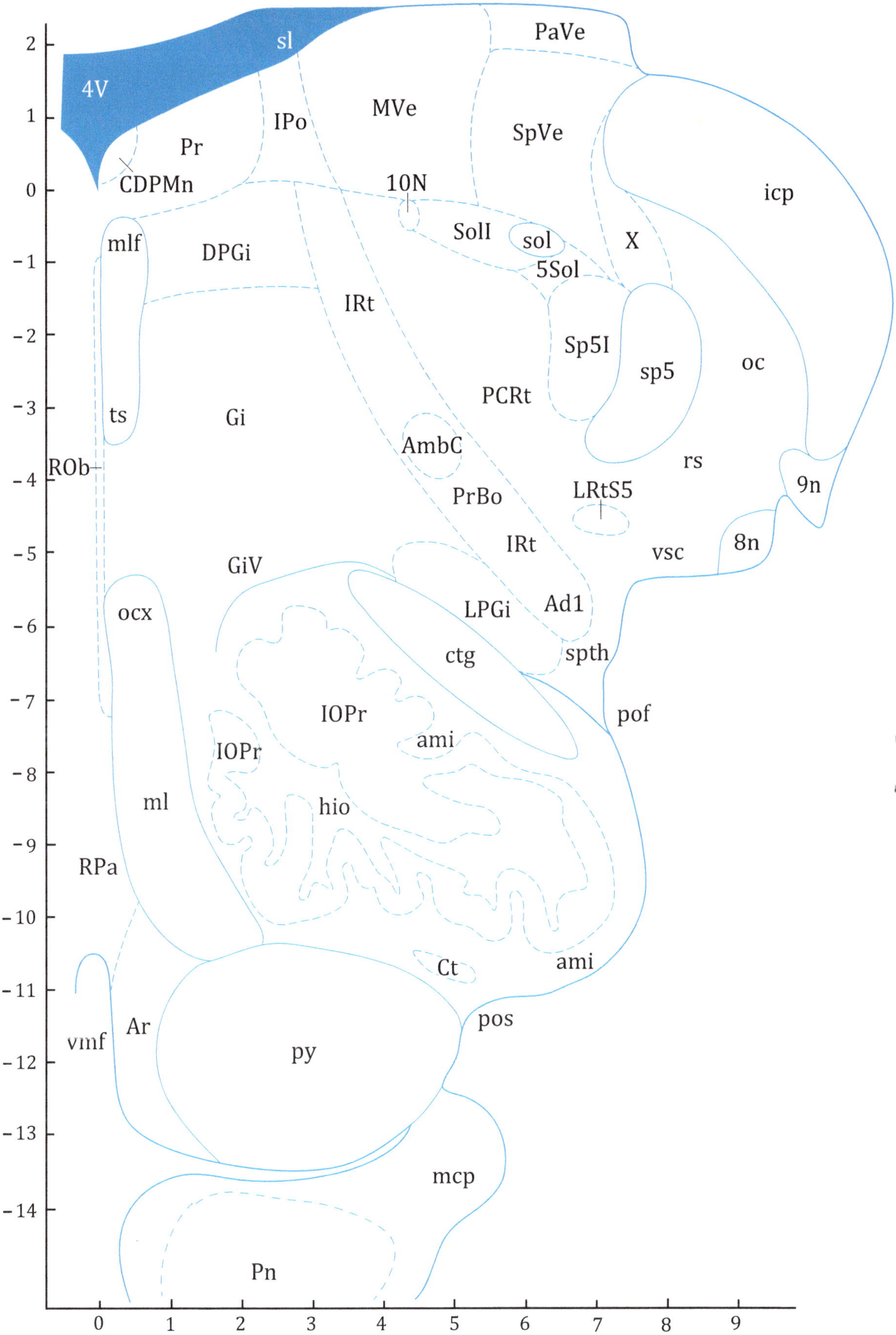

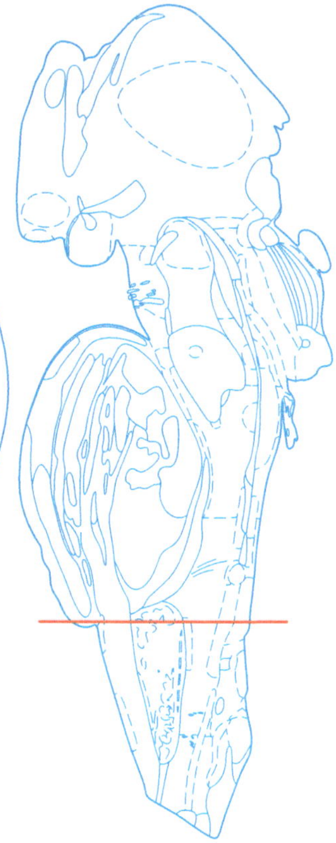

0N dorsal motor nu of vagus
V 4th ventricle
Sol trigeminal-solitary trans
n vestibulocochlear nerve
n glossopharyngeal nerve
d1 Ad1 adrenalin cells
mbC ambiguus nu, compact part
mi amiculum of the olive
r arcuate nu of the caudal hindbrain
DPMn caudal dorsal paramedian nu
t conterminal nu
g central tegmental tract
PGi dors paragigantocellular nu

Gi gigantocellular reticular nu
GiV gigantocellular retic, vent
hio hilus of the inferior olive
icp inf cerebellar peduncle
IOPr inf olive, principal nu
IPo interpositus
IRt intermed reticular nu
LPGi lat paragigantocellular nu
LRtS5 lat reticular nu, subtrigeminal
mcp middle cerebellar peduncle
ml medial lemniscus
mlf med longitudinal fasciculus
MVe medial vestibular nu

oc olivocerebellar tract
ocx olivocerebellar bundle decussation
PaVe paravestibular nucleus
PCRt parvicell reticular nu
Pn pontine nuclei
pof post-olivary fissure
pos preolivary sulcus
Pr prepositus nu
PrBo pre-Botzinger complex
py pyramidal tr
ROb raphe obscurus nu
RPa raphe pallidus nu
rs rubrospinal tract

sl sulcus limitans
sol solitary tract
SolI nu sol tract, interstitial
sp5 spinal trigem tract
Sp5I spinal trigem nu, interpolar
spth spinothalamic tract
SpVe spinal vestibular nu
ts tectospinal tr
vmf ventral median fissure
vsc ventral spinocerebellar tr
X nu X

Fig 19b
Obex +11.0mm
Transverse 19

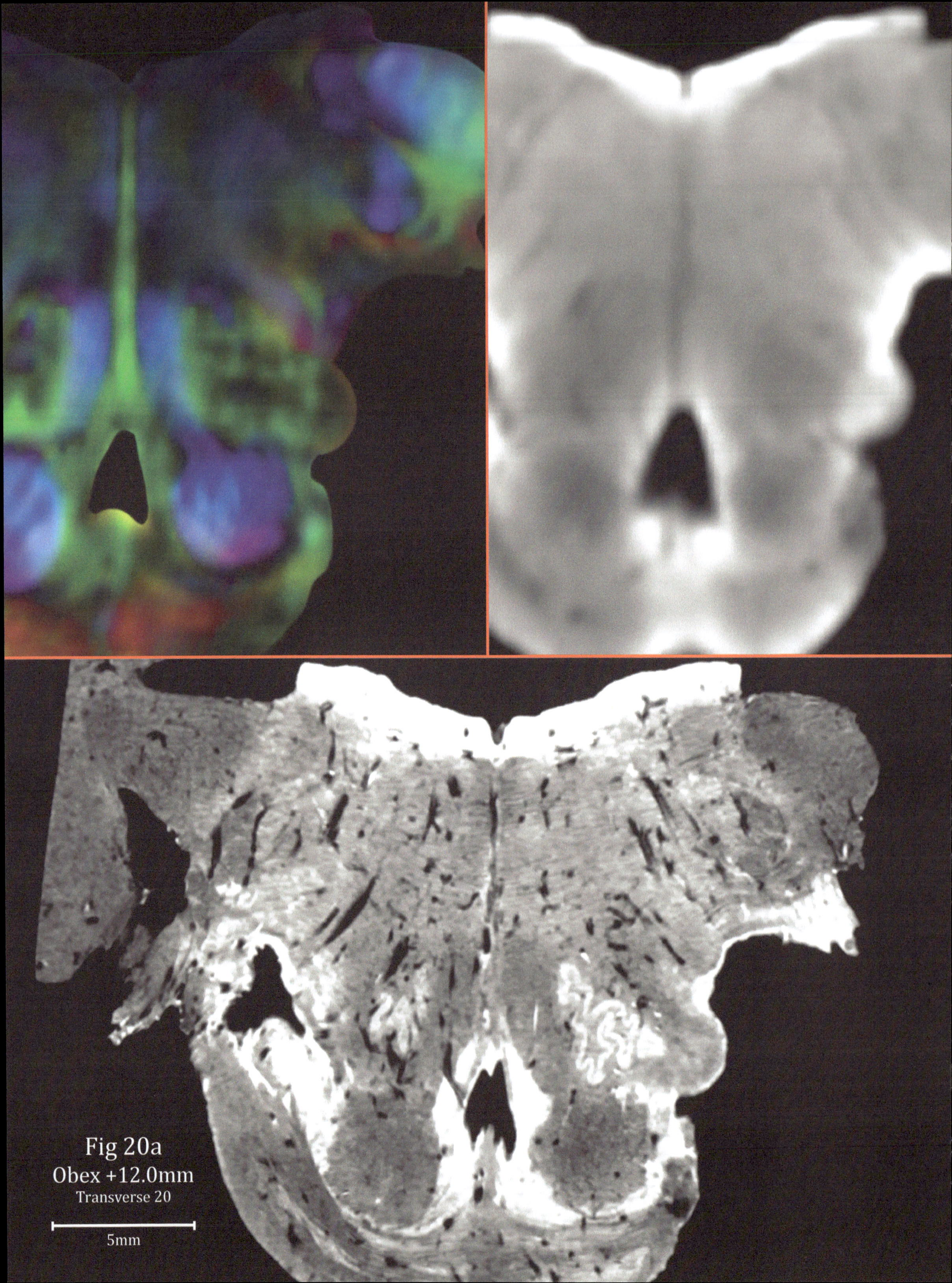
Fig 20a
Obex +12.0mm
Transverse 20
5mm

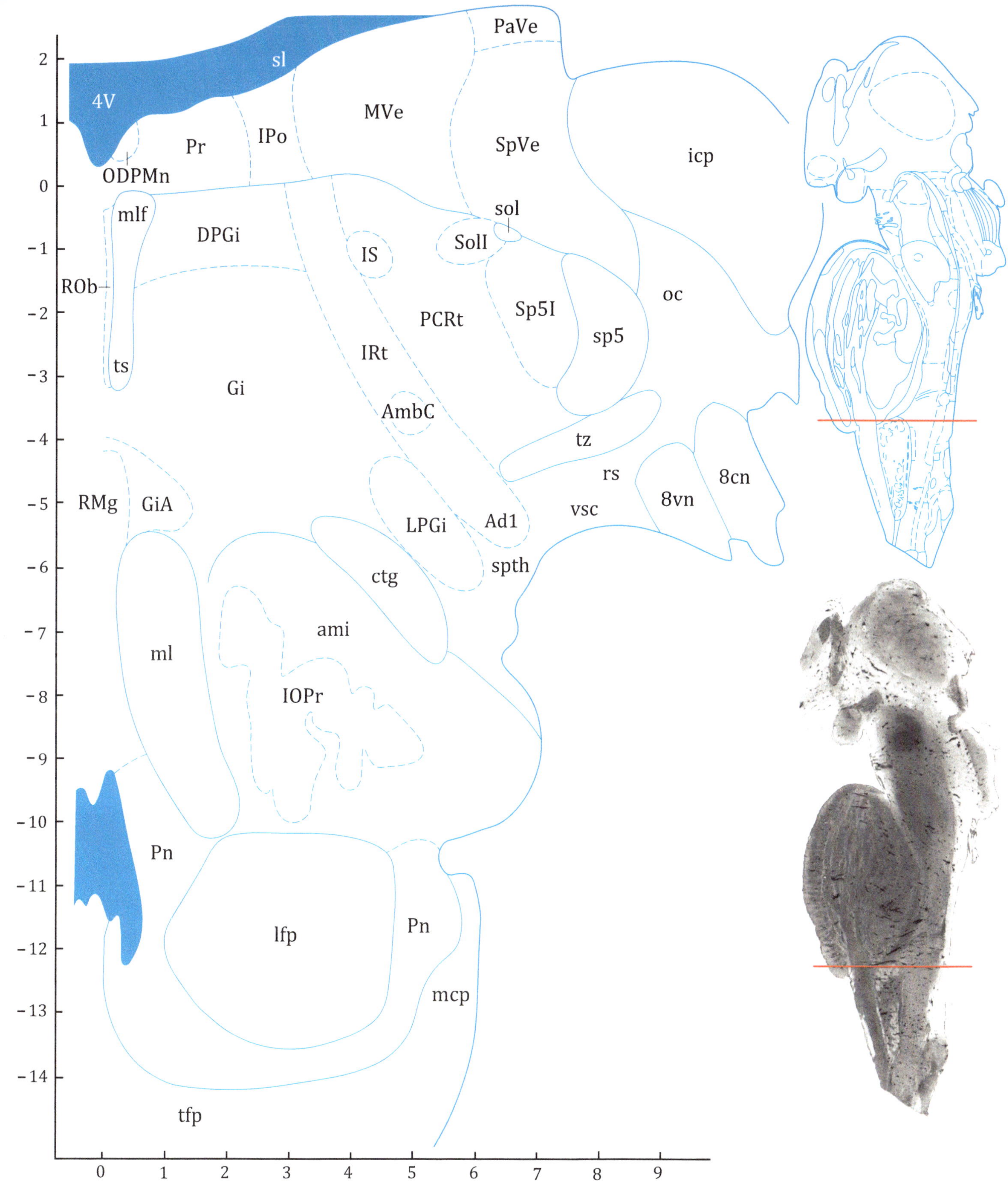

4V 4th ventricle
8cn cochlear root 8th nerve
8vn vestibular root of 8th nerve
Ad1 Ad1 adrenalin cells
AmbC ambiguus nu, compact part
ami amiculum of the olive
ctg central tegmental tract
DPGi dors paragigantocellular nu
Gi gigantocellular reticular nu
GiA gigantocellular retic, alpha
icp inf cerebellar peduncle
IOPr inf olive, principal nu
IRt intermed reticular nu
IS inferior salivatory nu
lfp longitudinal fasciculus pons
LPGi lat paragigantocellular nu
mcp middle cerebellar peduncle
ml medial lemniscus
mlf med longitudinal fasciculus
MVe medial vestibular nu
oc olivocerebellar tract
ODPMn oral dorsal paramedian nu
PaVe paravestibular nucleus
PCRt parvicell reticular nu
Pr prepositus nu
RMg raphe magnus nu
ROb raphe obscurus nu
rs rubrospinal tract
sl sulcus limitans
sol solitary tract
SolI nu sol tract, interstitial
sp5 spinal trigem tract
Sp5I spinal trigem nu, interpolar
spth spinothalamic tract
SpVe spinal vestibular nu
tfp transverse fibers pons
tz trapezoid body
vsc ventral spinocerebellar tr

Fig 20b
Obex +12.0mm
Transverse 20

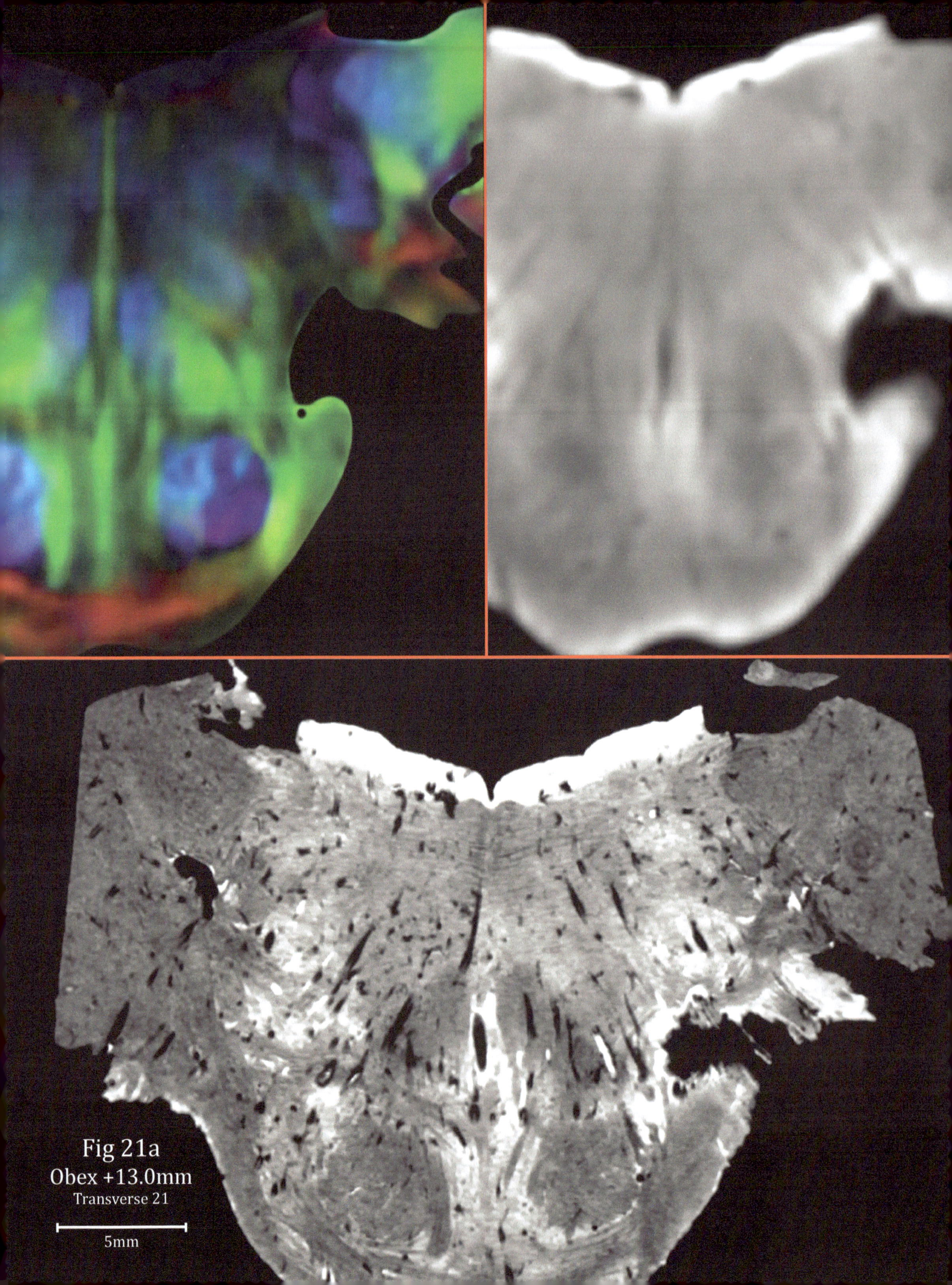
Fig 21a
Obex +13.0mm
Transverse 21
5mm

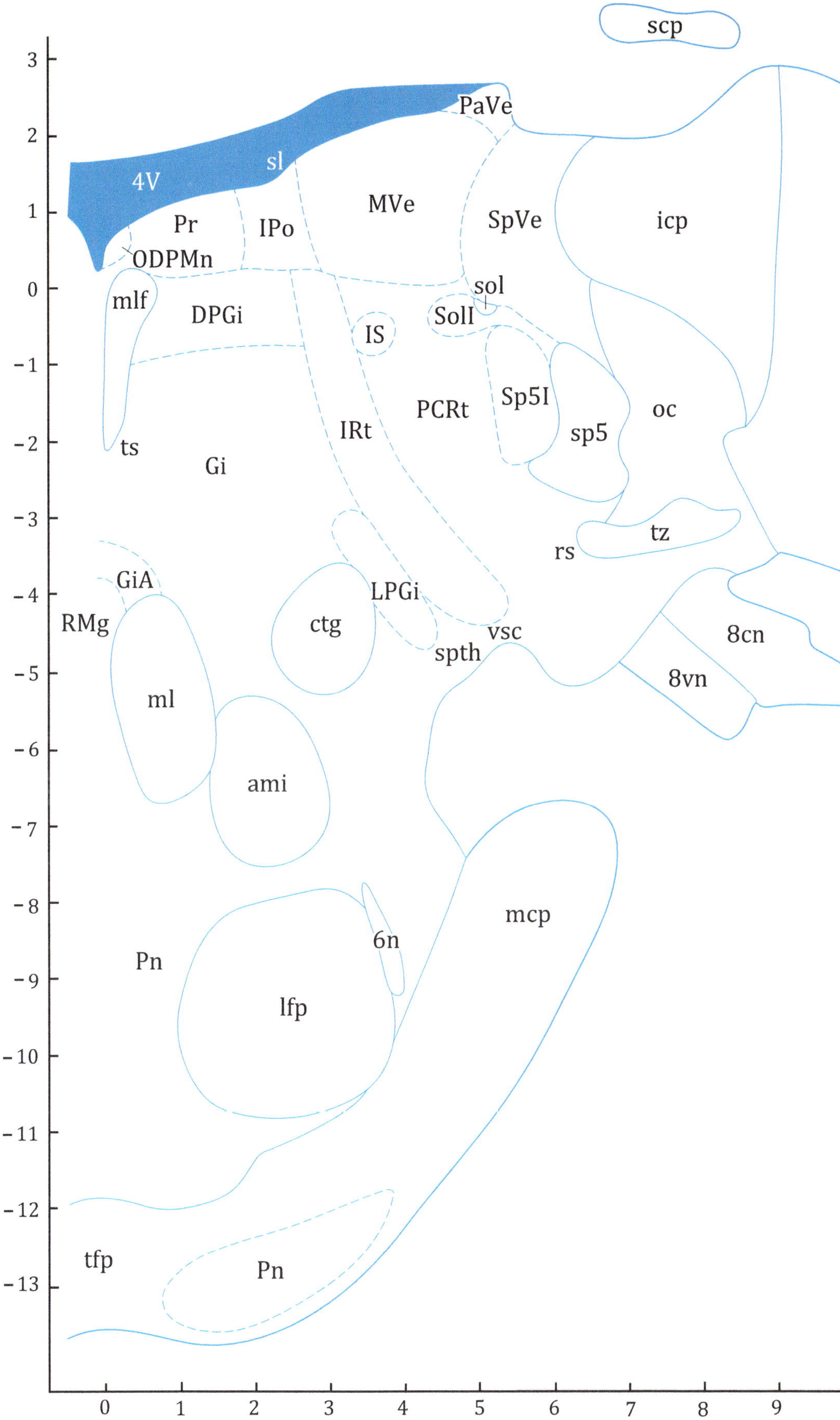

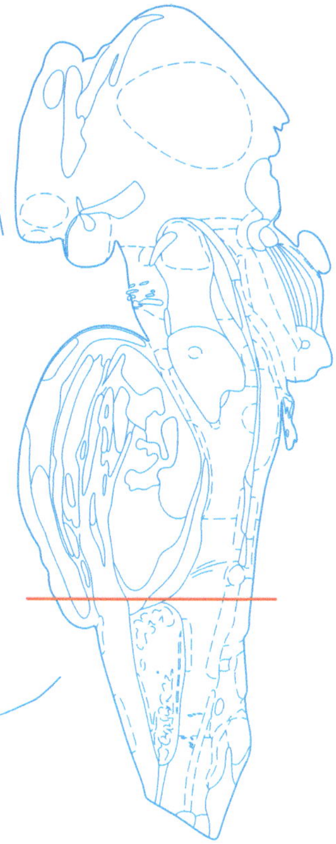

4V 4th ventricle
6n root of abducens nerve
8cn cochlear root 8th nerve
8vn vestibular root of 8th nerve
ami amiculum of the olive
ctg central tegmental tract
DPGi dors paragigantocellular nu
Gi gigantocellular reticular nu
GiA gigantocellular retic, alpha
icp inf cerebellar peduncle
IPo interpositus
IRt intermed reticular nu

lfp longitudinal fasciculus pons
LPGi lat paragigantocellular nu
mcp middle cerebellar peduncle
ml medial lemniscus
mlf med longitudinal fasciculus
MVe medial vestibular nu
oc olivocerebellar tract
ODPMn oral dorsal paramedian nu
PaVe paravestibular nucleus
PCRt parvicell reticular nu
Pn pontine nuclei
Pr prepositus nu

rs rubrospinal tract
scp superior Cb peduncle
sl sulcus limitans
sol solitary tract
SolI nu sol tract, interstitial
sp5 spinal trigem tract
Sp5I spinal trigem nu, interpolar
spth spinothalamic tract
SpVe spinal vestibular nu
tfp transverse fibers pons
ts tectospinal tr
tz trapezoid body

Fig 21b
Obex +13.0mm
Transverse 21

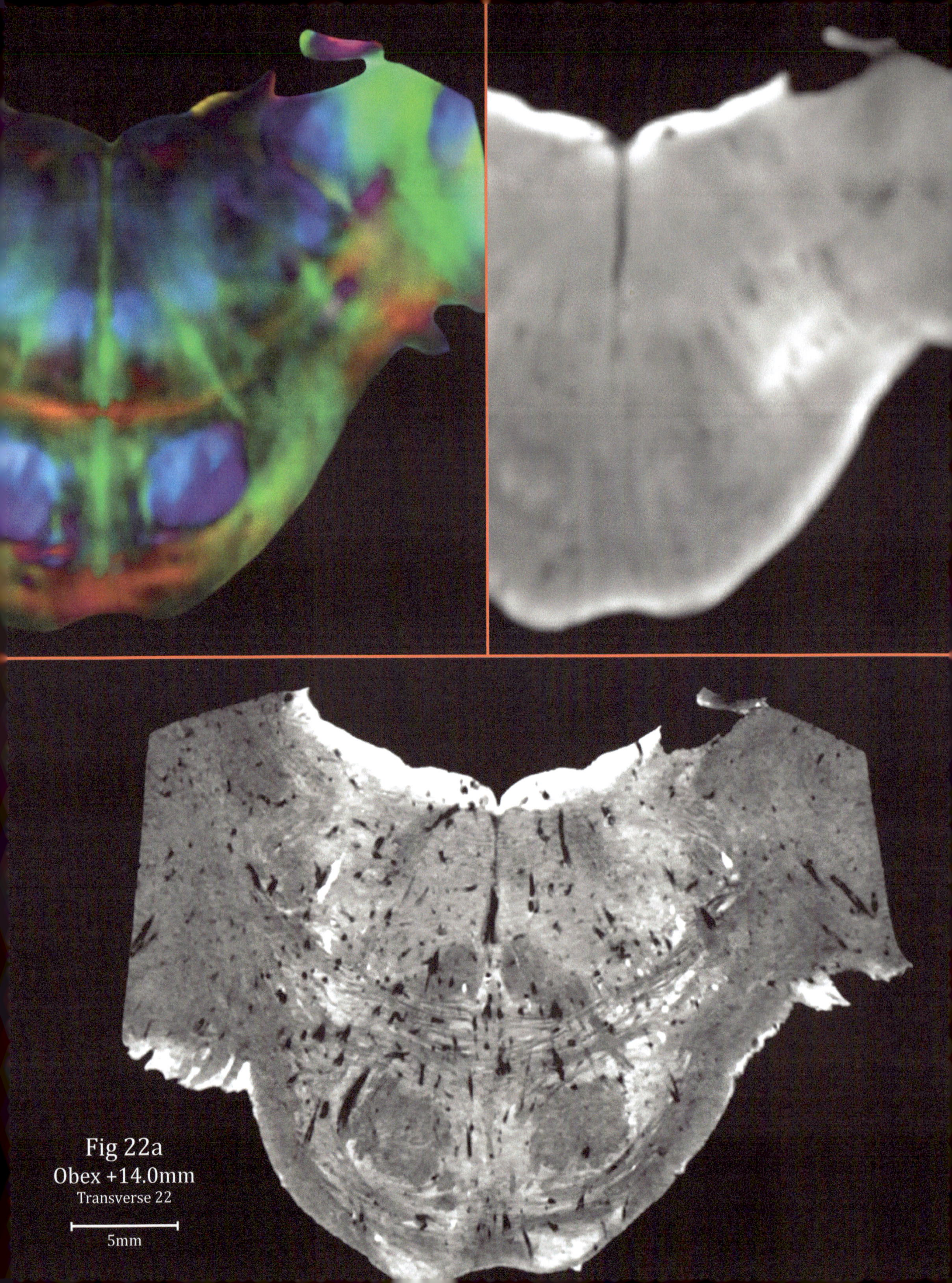

Fig 22a
Obex +14.0mm
Transverse 22

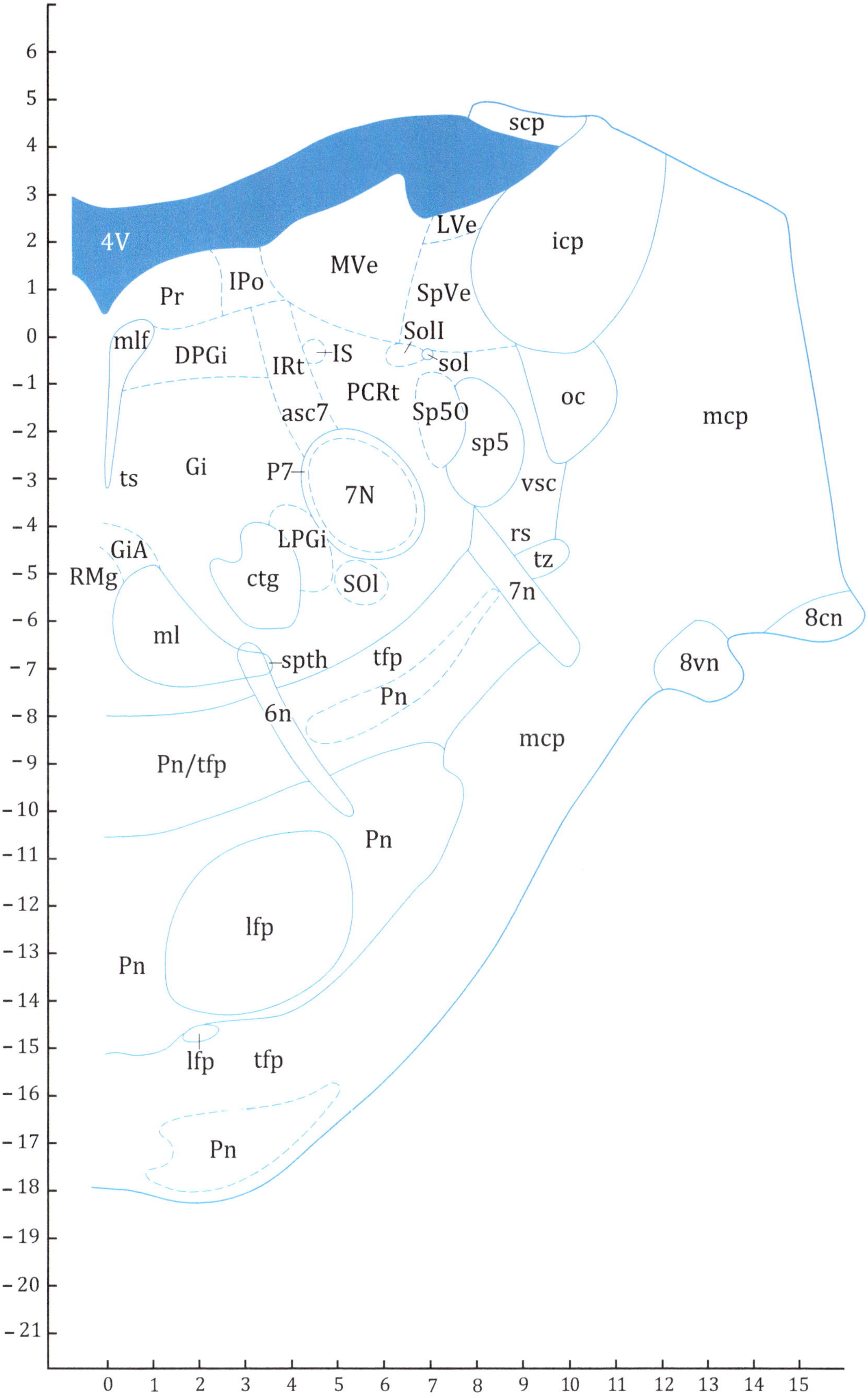

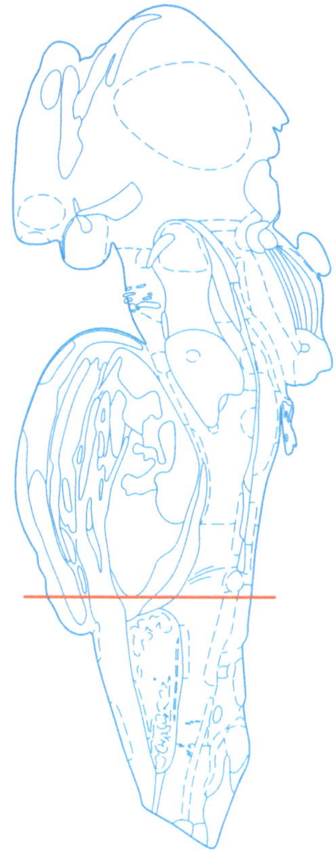

4V 4th ventricle
6n root of abducens nerve
7N facial nu
7n facial nerve
8cn cochlear root 8th nerve
8vn vestibular root of 8th nerve
asc7 ascending fibers facial n
ctg central tegmental tract
DPGi dors paragigantocellular nu
Gi gigantocellular reticular nu
GiA gigantocellular retic, alpha
icp inf cerebellar peduncle
IPo interpositus
IRt intermed reticular nu
IS inferior salivatory nu
lfp longitudinal fasciculus pons
LPGi lat paragigantocellular nu
LVe lateral vestibular nu
mcp middle cerebellar peduncle
ml medial lemniscus
mlf med longitudinal fasciculus
MVe medial vestibular nu
oc olivocerebellar tract
P7 perifacial zone
PCRt parvicell reticular nu
Pn pontine nuclei
Pn/tfp pontine nu/transv fibres pons
Pr prepositus nu
RMg raphe magnus nu
rs rubrospinal tract
scp superior Cb peduncle
sol solitary tract
SOl superior olive, lateral
SolI nu sol tract, interstitial
sp5 spinal trigem tract
Sp5O spinal trigeml nu, oral
spth spinothalamic tract
SpVe spinal vestibular nu
tfp transverse fibers pons
ts tectospinal tr
tz trapezoid body
vsc ventral spinocerebellar tr

Fig 22b
Obex +14.0mm
Transverse 22

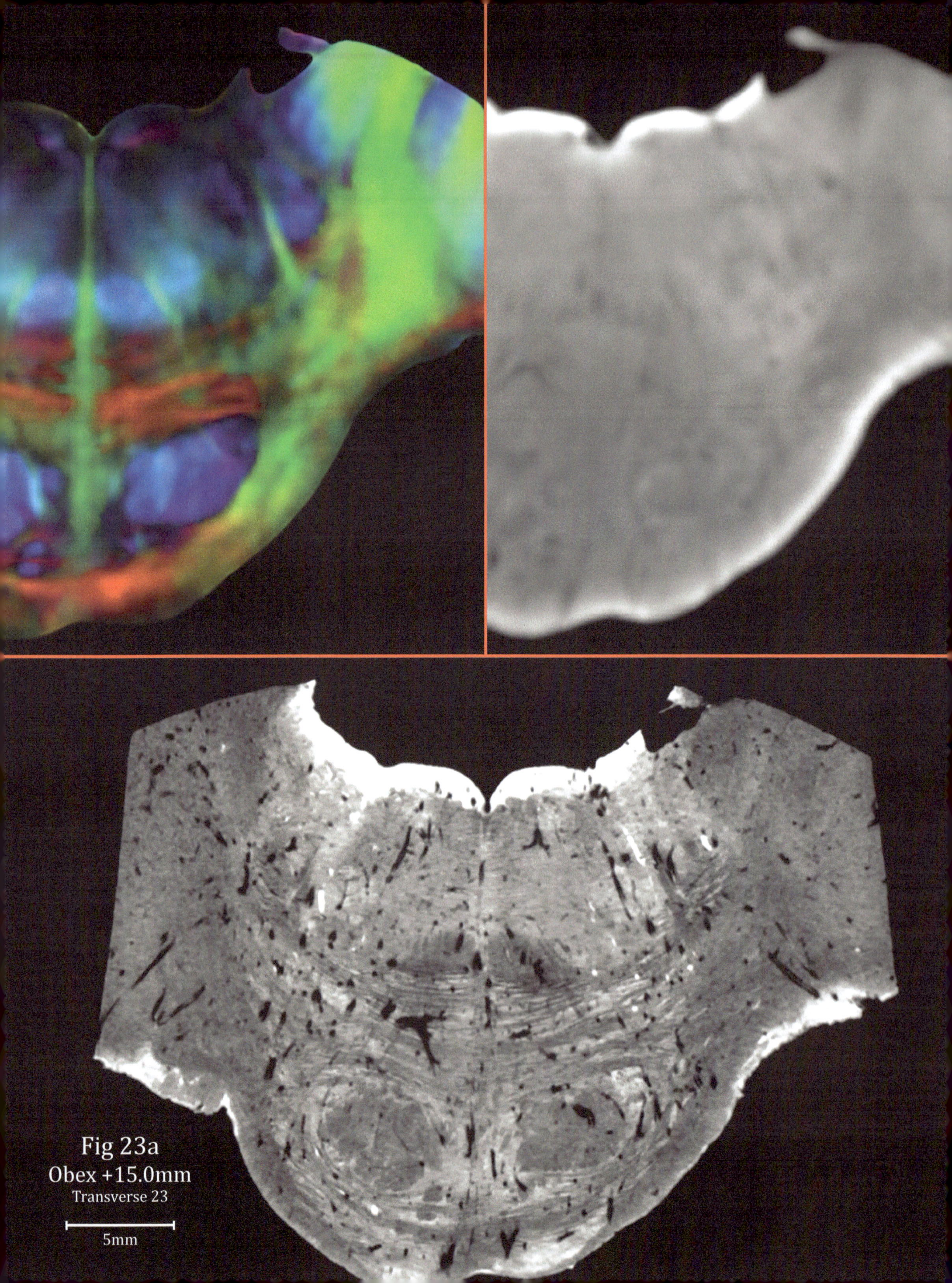

Fig 23a
Obex +15.0mm
Transverse 23

5mm

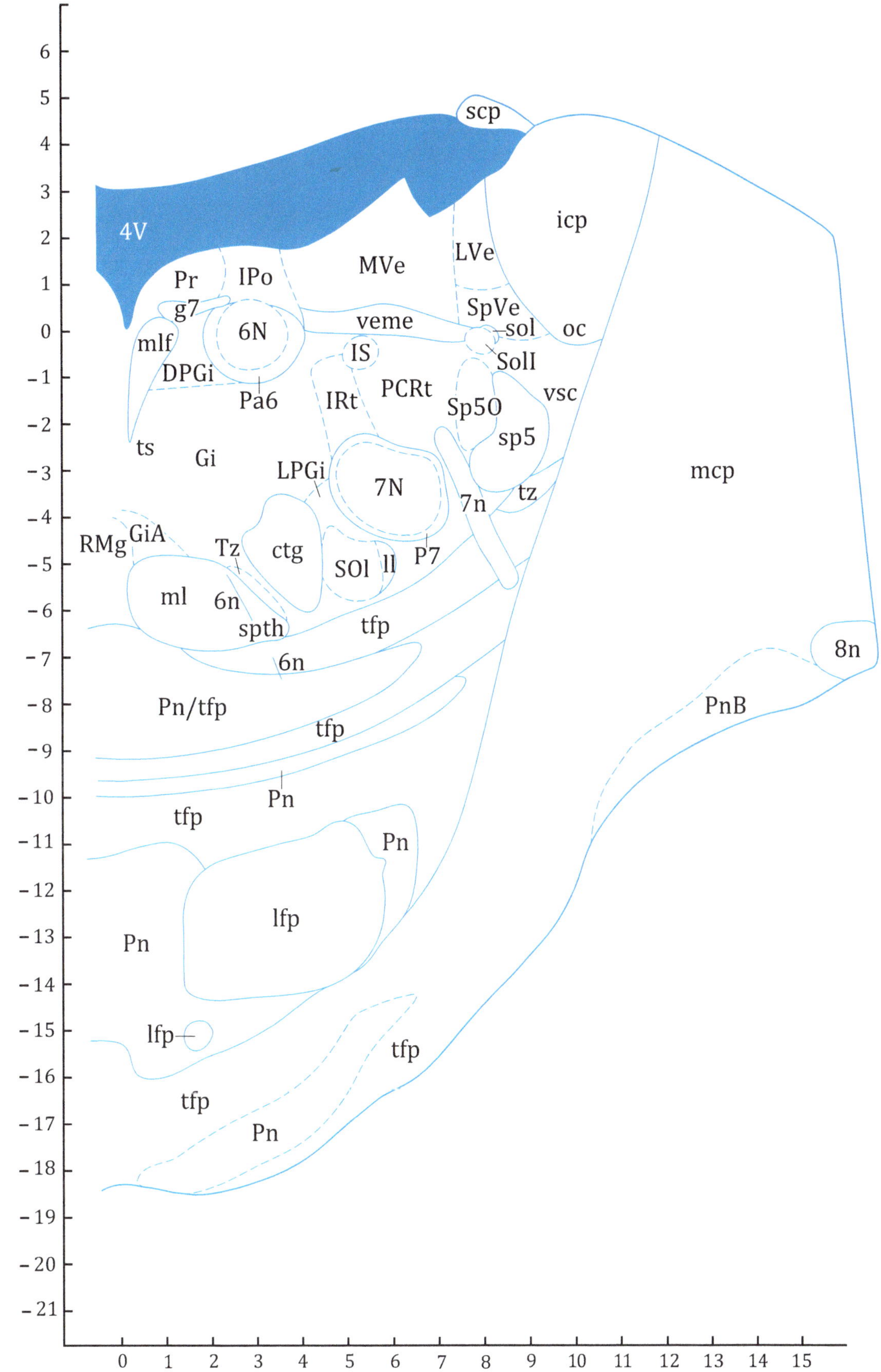

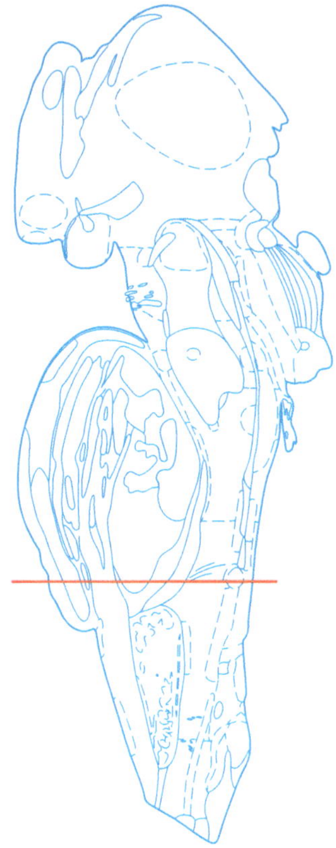

4V 4th ventricle
6n root of abducens nerve
6N abducens nu
7n facial nerve
7N facial nu
8n vestibulocochlear nerve
ctg central tegmental tract
DPGi dors paragigantocellular nu
g7 genu facial nerve
Gi gigantocellular reticular nu
GiA gigantocellular retic, alpha
icp inf cerebellar peduncle
IPo interpositus

IRt intermed reticular nu
IS inferior salivatory nu
lfp longitudinal fasciculus pons
ll lateral lemniscus
LPGi lat paragigantocellular nu
LVe lateral vestibular nu
mcp middle cerebellar peduncle
ml medial lemniscus
mlf med longitudinal fasciculus
MVe medial vestibular nu
oc olivocerebellar tract
P7 perifacial zone
Pa6 paraabducens nu

PCRt parvicell reticular nu
Pn pontine nuclei
Pn/tfp pontine nu/transv fibres pons
PnB pontobulbar nucleus
Pr prepositus nu
RMg raphe magnus nu
scp superior Cb peduncle
sol solitary tract
SOl superior olive, lateral
SolI nu sol tract, interstitial
sp5 spinal trigem tract
Sp5O spinal trigeml nu, oral
spth spinothalamic tract

SpVe spinal vestibular nu
tfp transverse fibers pons
ts tectospinal tr
tz trapezoid body
Tz nu trapezoid body
veme vestibulomesencephalic tr
vsc ventral spinocerebellar tr

Fig 23b
Obex +15.0mm
Transverse 23

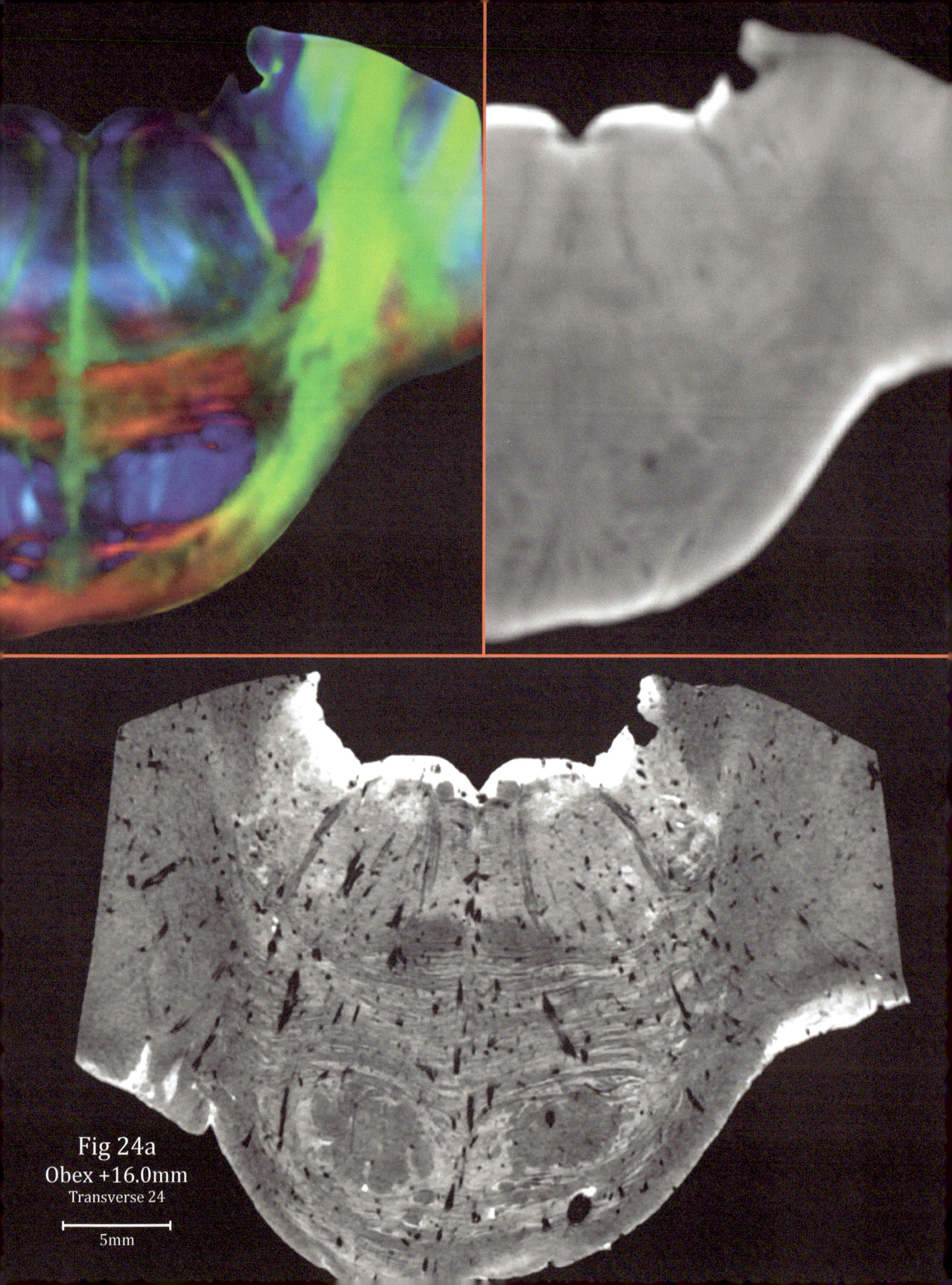

Fig 24a
Obex +16.0mm
Transverse 24

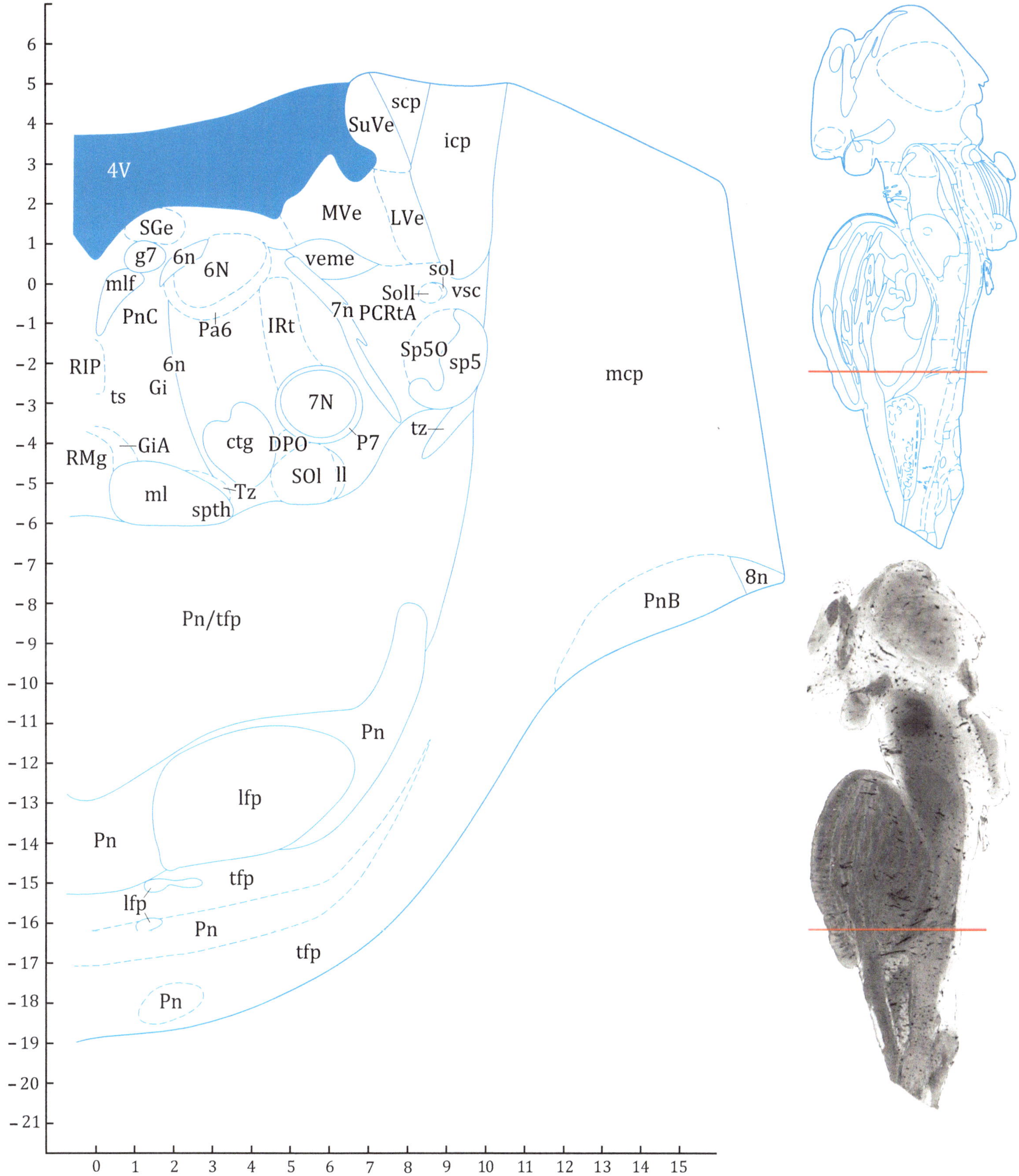

4V 4th ventricle
6n root of abducens nerve
6N abducens nu
7n facial nerve
7N facial nu
8n vestibulocochlear nerve
ctg central tegmental tract
DPO dorsal periolivary region
g7 genu facial nerve
Gi gigantocellular reticular nu
GiA gigantocellular retic, alpha
icp inf cerebellar peduncle
IRt intermed reticular nu
lfp longitudinal fasciculus pons
ll lateral lemniscus
LVe lateral vestibular nu
mcp middle cerebellar peduncle
ml medial lemniscus
mlf med longitudinal fasciculus
MVe medial vestibular nu
P7 perifacial zone
Pa6 paraabducens nu
PCRtA parvicell reticular nu, alpha
Pn pontine nuclei
Pn/tfp pontine nu/transv fibres pons
PnB pontobulbar nucleus
PnC pontine reticular nu, caudal
RIP raphe interpositus nu
RMg raphe magnus nu
scp superior Cb peduncle
SGe supragenual nu
sol solitary tract
SOl superior olive, lateral
SolI nu sol tract, interstitial
sp5 spinal trigem tract
Sp5O spinal trigeml nu, oral
spth spinothalamic tract
SuVe superior vestibular nu
tfp transverse fibers pons
ts tectospinal tr
Tz nu trapezoid body
tz trapezoid body
veme vestibulomesencephalic tr
vsc ventral spinocerebellar tr

Fig 24b
Obex +16.0mm
Transverse 24

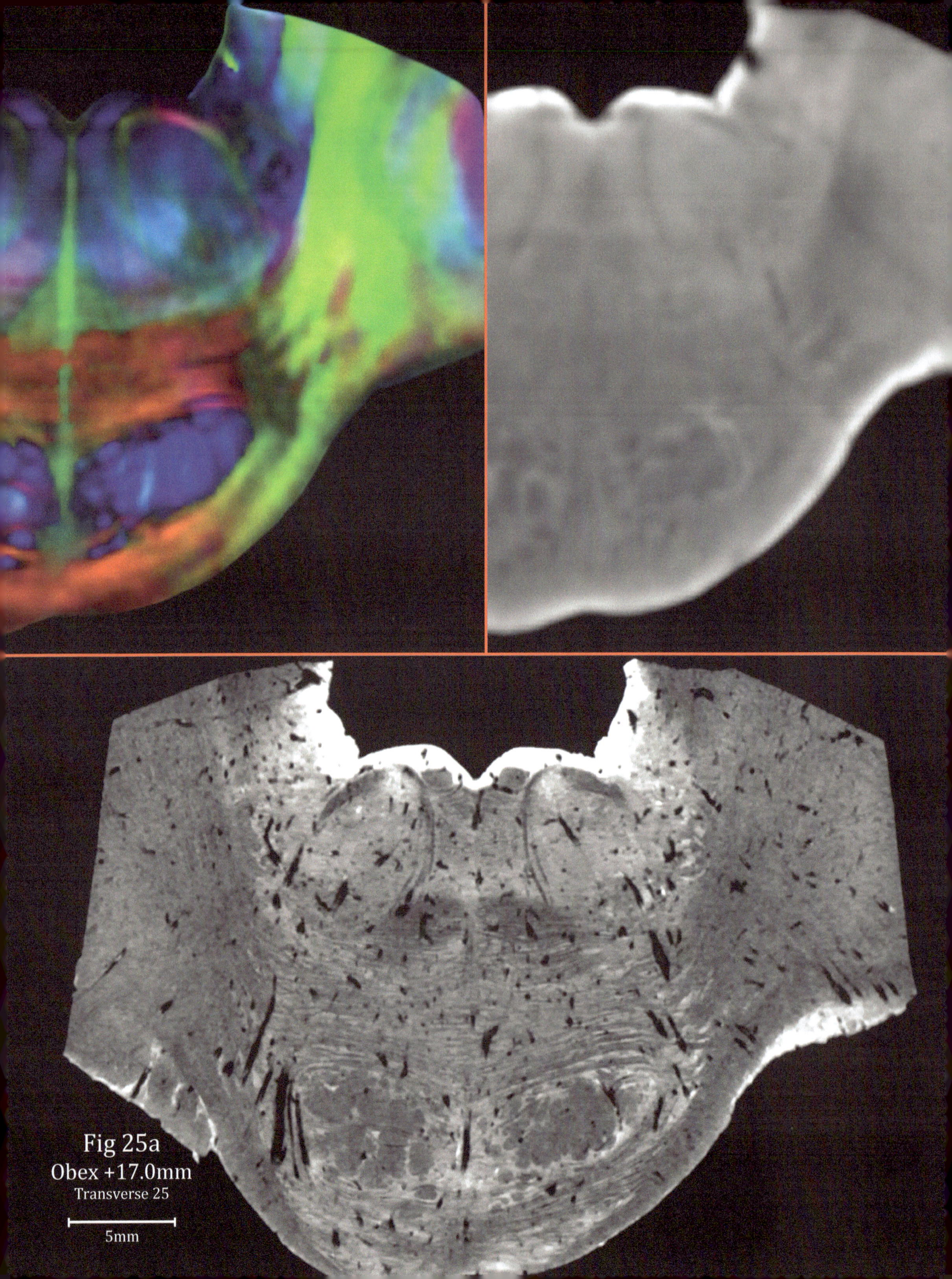

Fig 25a
Obex +17.0mm
Transverse 25
5mm

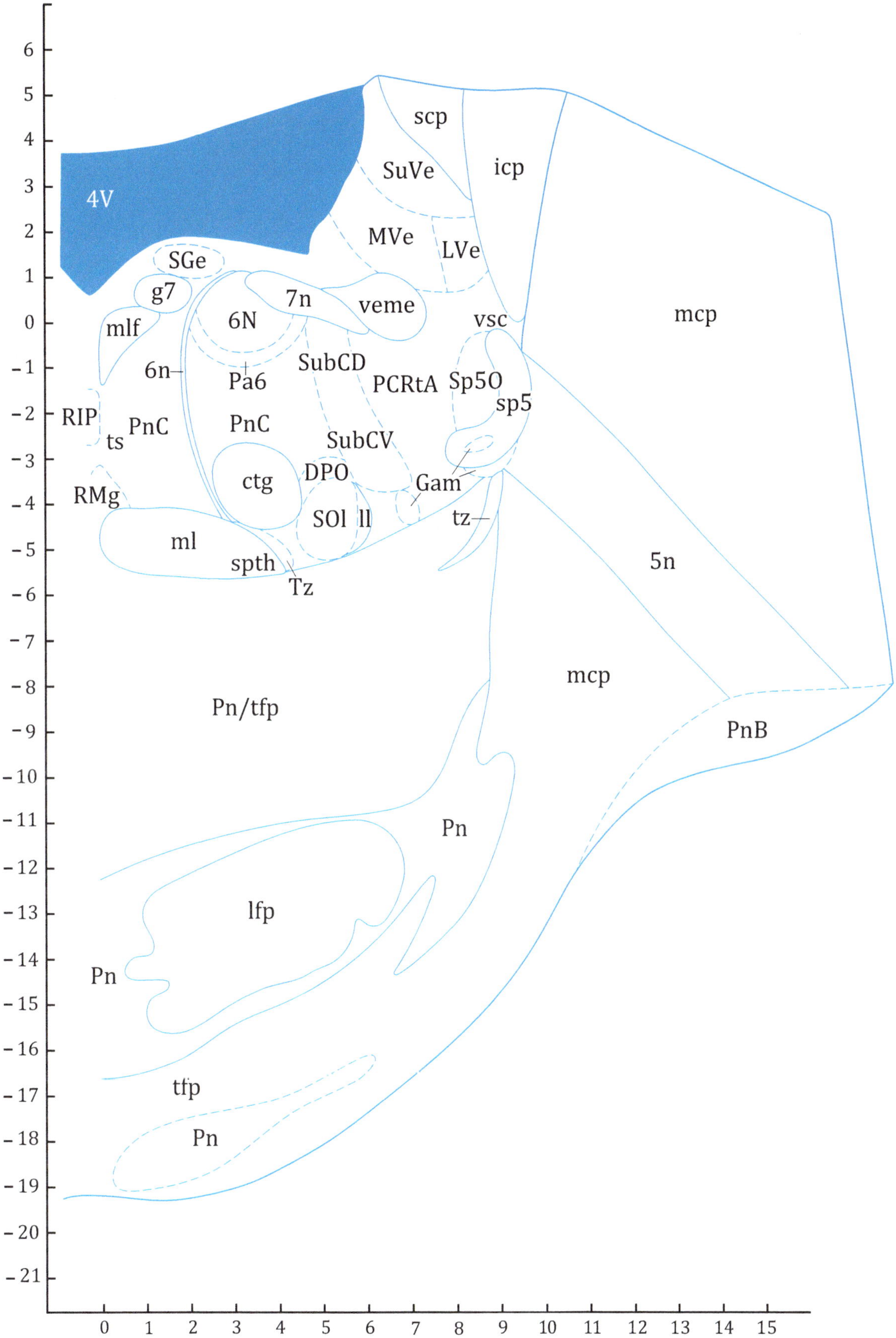

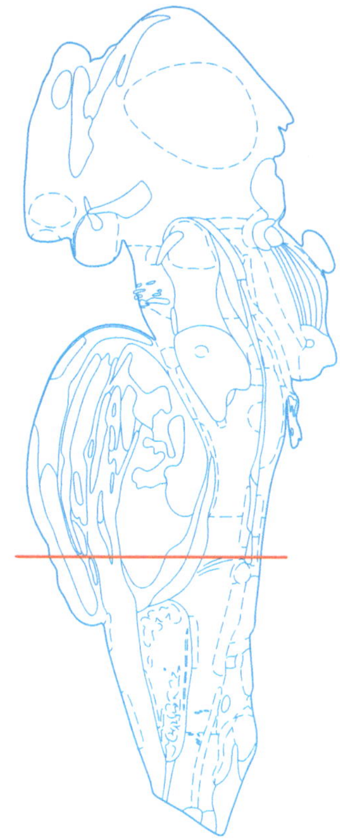

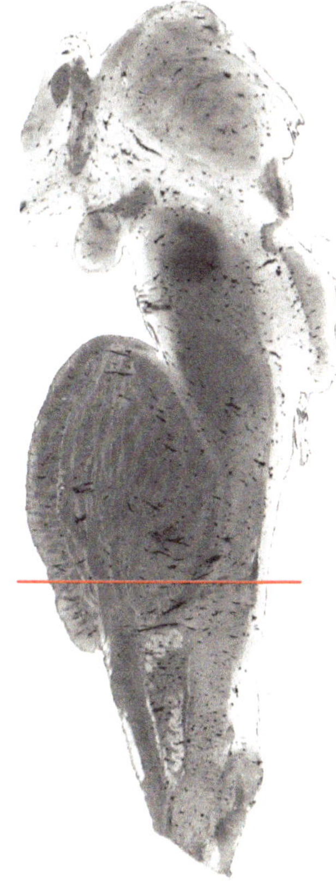

4V 4th ventricle
5n trigeminal nerve
6n root of abducens nerve
6N abducens nu
7n facial nerve
ctg central tegmental tract
DPO dorsal periolivary region
g7 genu facial nerve
Gam gamma pontine nucleus
icp inf cerebellar peduncle
lfp longitudinal fasciculus pons
ll lateral lemniscus
LVe lateral vestibular nu
mcp middle cerebellar peduncle
ml medial lemniscus
mlf med longitudinal fasciculus
MVe medial vestibular nu
Pa6 paraabducens nu
PCRtA parvicell reticular nu, alpha
Pn pontine nuclei
Pn/tfp pontine nu/transv fibres pons
PnB pontobulbar nucleus
PnC pontine reticular nu, caudal
RIP raphe interpositus nu
RMg raphe magnus nu
scp superior Cb peduncle
SGe supragenual nu
SOl superior olive, lateral
sp5 spinal trigem tract
Sp5O spinal trigeml nu, oral
spth spinothalamic tract
SubCD subcoeruleus nu, dorsal
SubCV subcoeruleus nu, ventral
SuVe superior vestibular nu
tfp transverse fibers pons
ts tectospinal tr
tz trapezoid body
Tz nu trapezoid body
veme vestibulomesencephalic tr
vsc ventral spinocerebellar tr

Fig 25b
Obex +17.0mm
Transverse 25

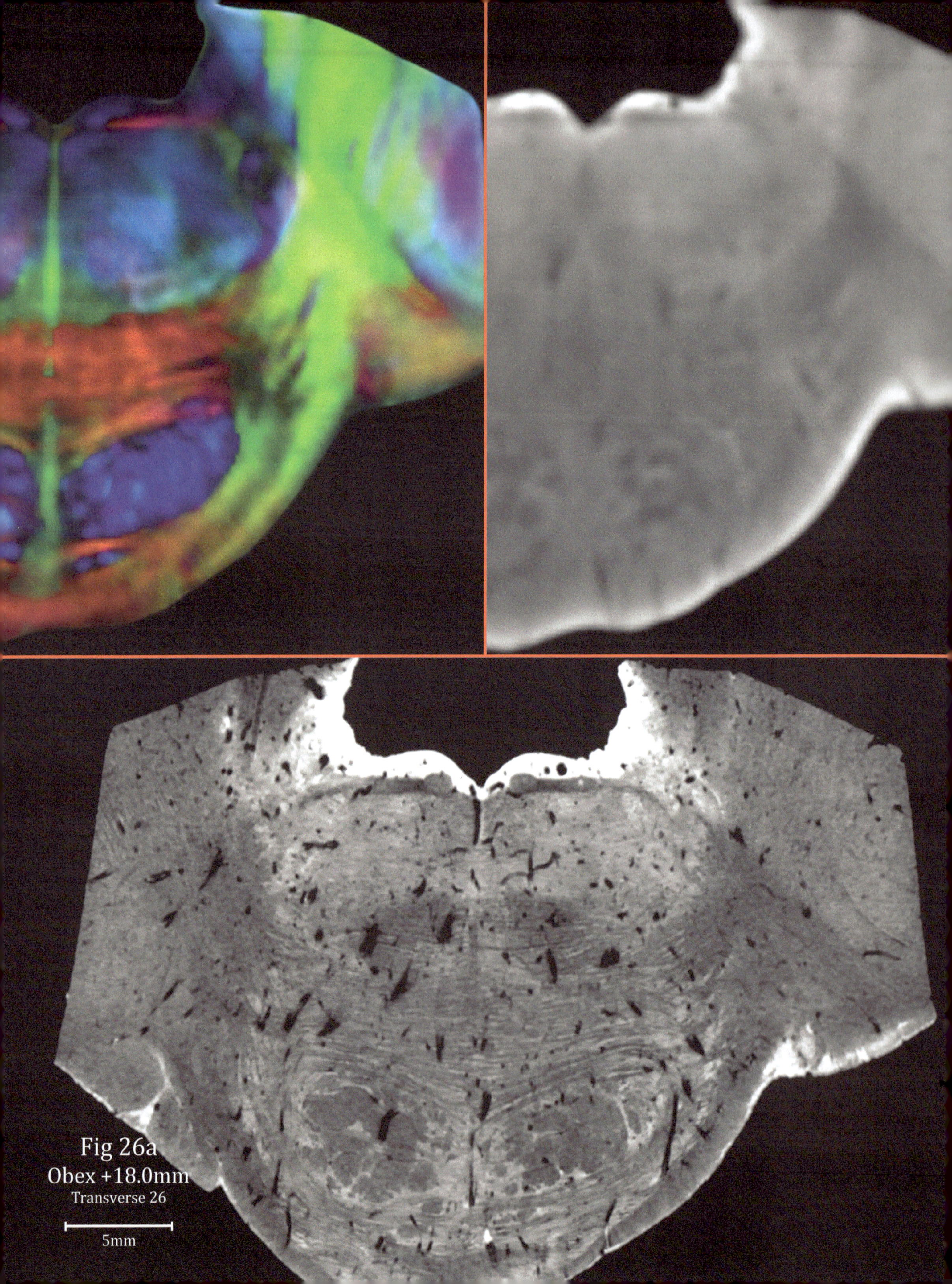

Fig 26a
Obex +18.0mm
Transverse 26

5mm

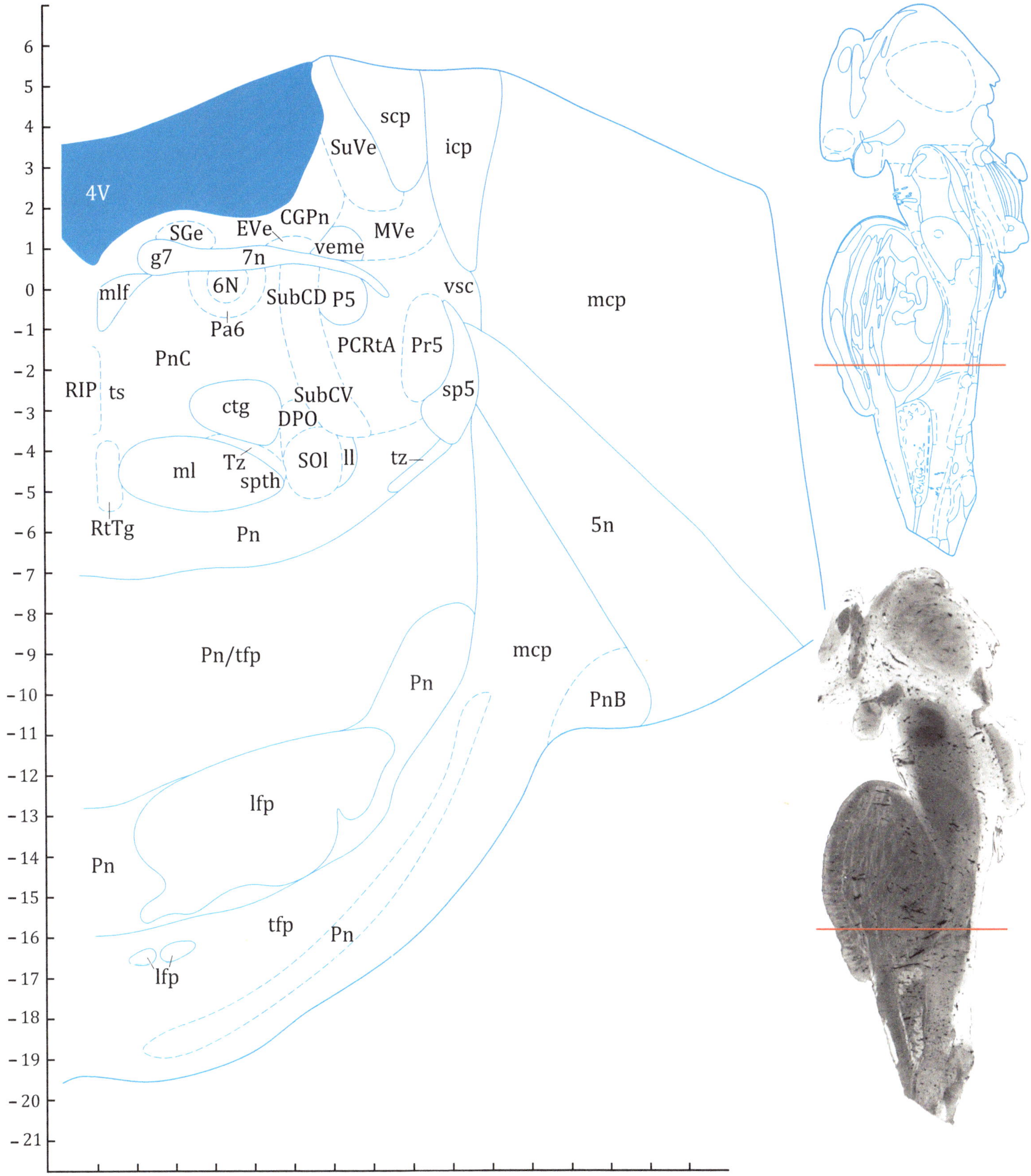

4V 4th ventricle
5n trigeminal nerve
6N abducens nu
7n facial nerve
CGPn central gray pons
ctg central tegmental tract
DPO dorsal periolivary region
EVe nu origin efferents ve n
g7 genu facial nerve
icp inf cerebellar peduncle
lfp longitudinal fasciculus pons
ll lateral lemniscus
mcp middle cerebellar peduncle
ml medial lemniscus
mlf med longitudinal fasciculus
MVe medial vestibular nu
P5 peritrigeminal zone
Pa6 paraabducens nu
PCRtA parvicell reticular nu, alpha
Pn pontine nuclei
Pn/tfp pontine nu/transv fibres pons
PnB pontobulbar nucleus
PnC pontine reticular nu, caudal
Pr5 principal sensory trigem nu
RIP raphe interpositus nu
RtTg reticulotegmental nu pons
scp superior Cb peduncle
SGe supragenual nu
SOl superior olive, lateral
sp5 spinal trigem tract
spth spinothalamic tract
SubCD subcoeruleus nu, dorsal
SubCV subcoeruleus nu, ventral
SuVe superior vestibular nu
tfp transverse fibers pons
ts tectospinal tr
tz trapezoid body
Tz nu trapezoid body
veme vestibulomesencephalic tr
vsc ventral spinocerebellar tr

Fig 26b
Obex +18.0mm
Transverse 26

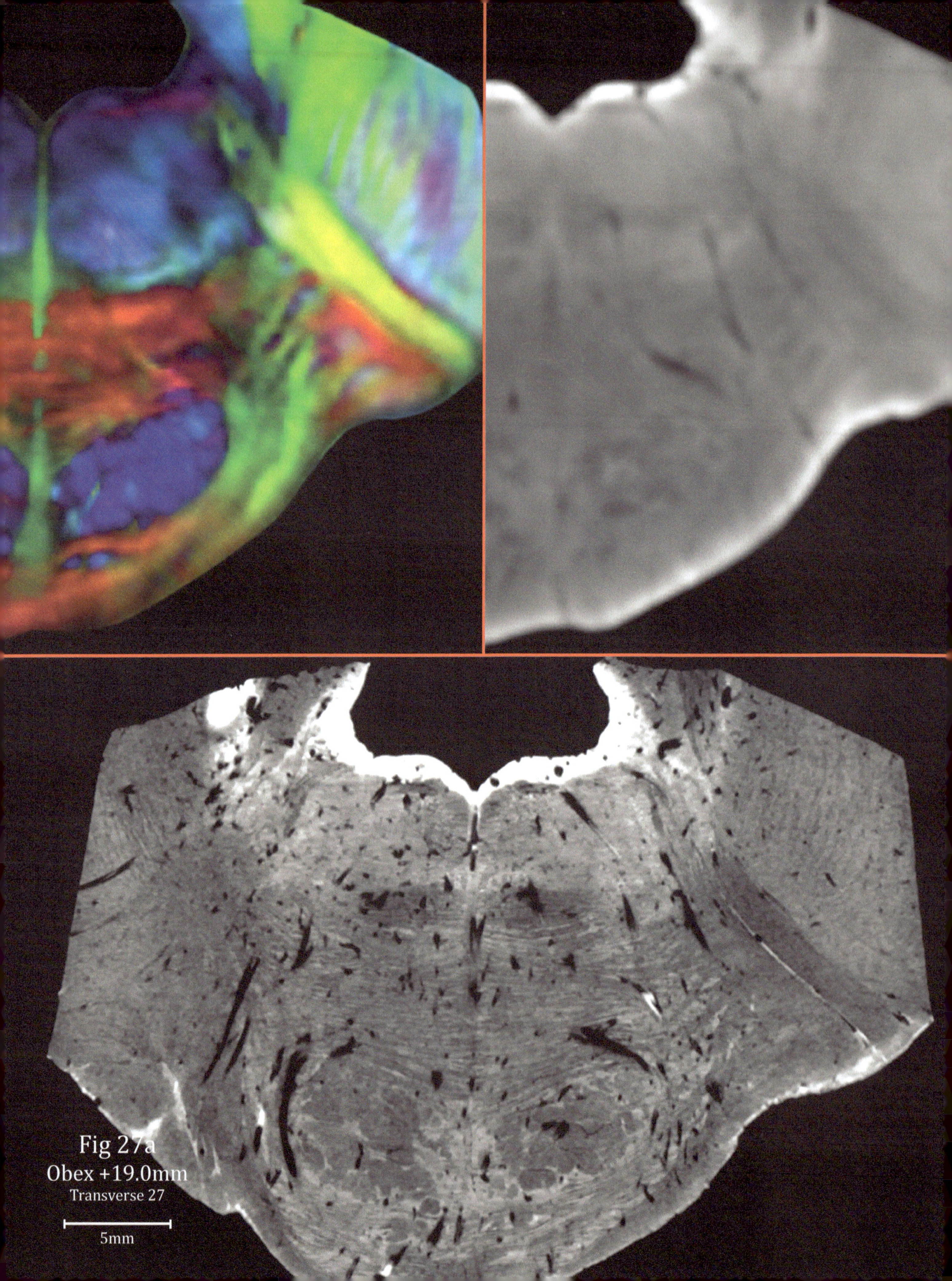

Fig 27a
Obex +19.0mm
Transverse 27

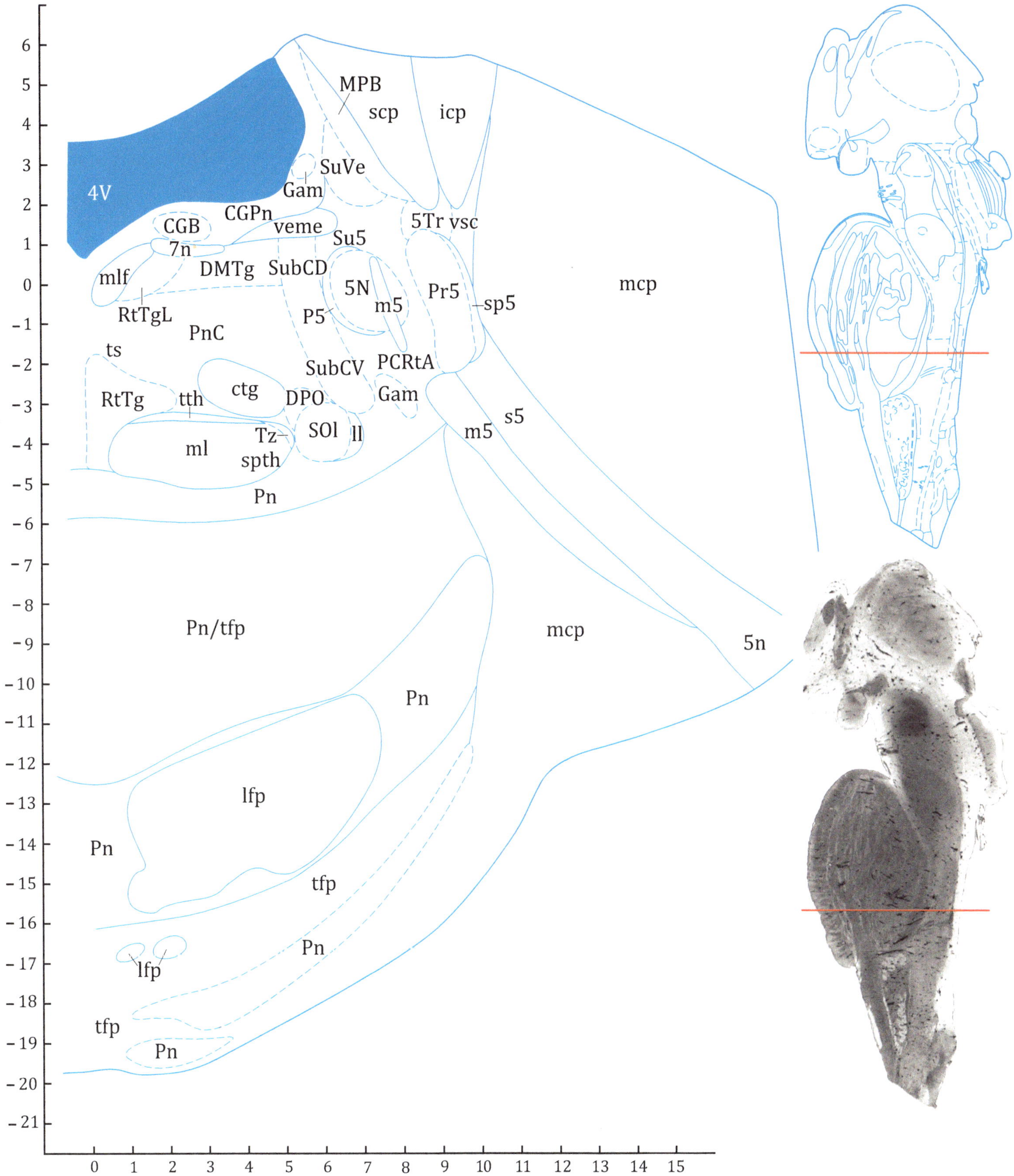

4V 4th ventricle
5n trigeminal nerve
5N motor trigeminal nu
5Tr trigem transition zone
7n facial nerve
CGB central gray, beta part
CGPn central gray pons
ctg central tegmental tract
DMTg dorsomed tegmental area
DPO dorsal periolivary region
Gam gamma pontine nucleus
icp inf cerebellar peduncle
lfp longitudinal fasciculus pons
ll lateral lemniscus
m5 motor root trigeminal nerve
mcp middle cerebellar peduncle
ml medial lemniscus
mlf med longitudinal fasciculus
MPB medial parabrachial nu
P5 peritrigeminal zone
PCRtA parvicell reticular nu, alpha
Pn pontine nuclei
Pn/tfp pontine nu/transv fibres pons
PnC pontine reticular nu, caudal
Pr5 principal sensory trigem nu
RtTg reticulotegmental nu pons
RtTgL reticulotegmental pons, lat
s5 sensory root trigem n
scp superior Cb peduncle
SOl superior olive, lateral
sp5 spinal trigem tract
spth spinothalamic tract
Su5 supratrigeminal nu
SubCD subcoeruleus nu, dorsal
SubCV subcoeruleus nu, ventral
SuVe superior vestibular nu
tfp transverse fibers pons
ts tectospinal tr
tth trigeminothalamic tr
Tz nu trapezoid body
veme vestibulomesencephalic tr
vsc ventral spinocerebellar tr

Fig 27b
Obex +19.0mm
Transverse 27

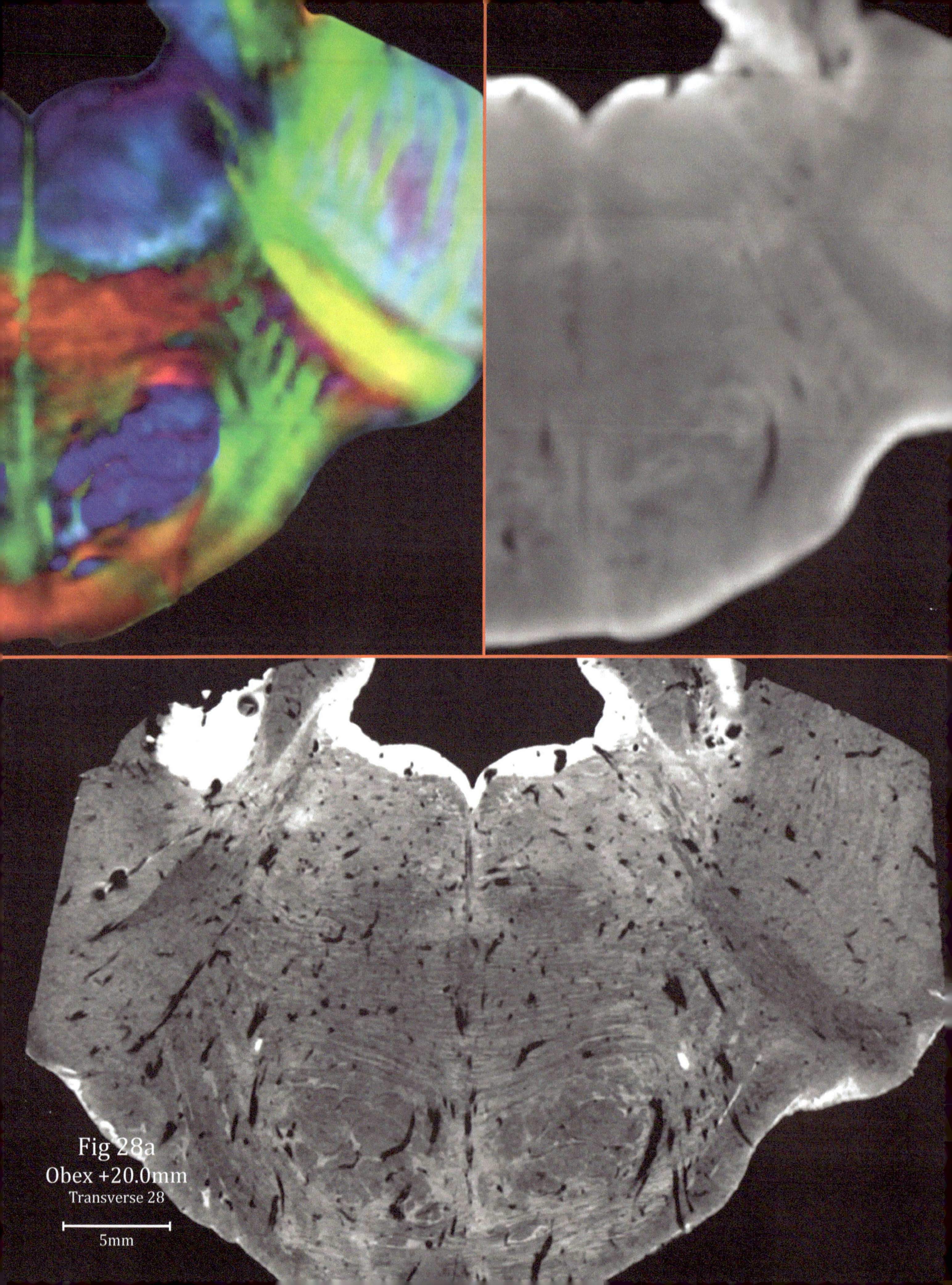
Fig 28a
Obex +20.0mm
Transverse 28
5mm

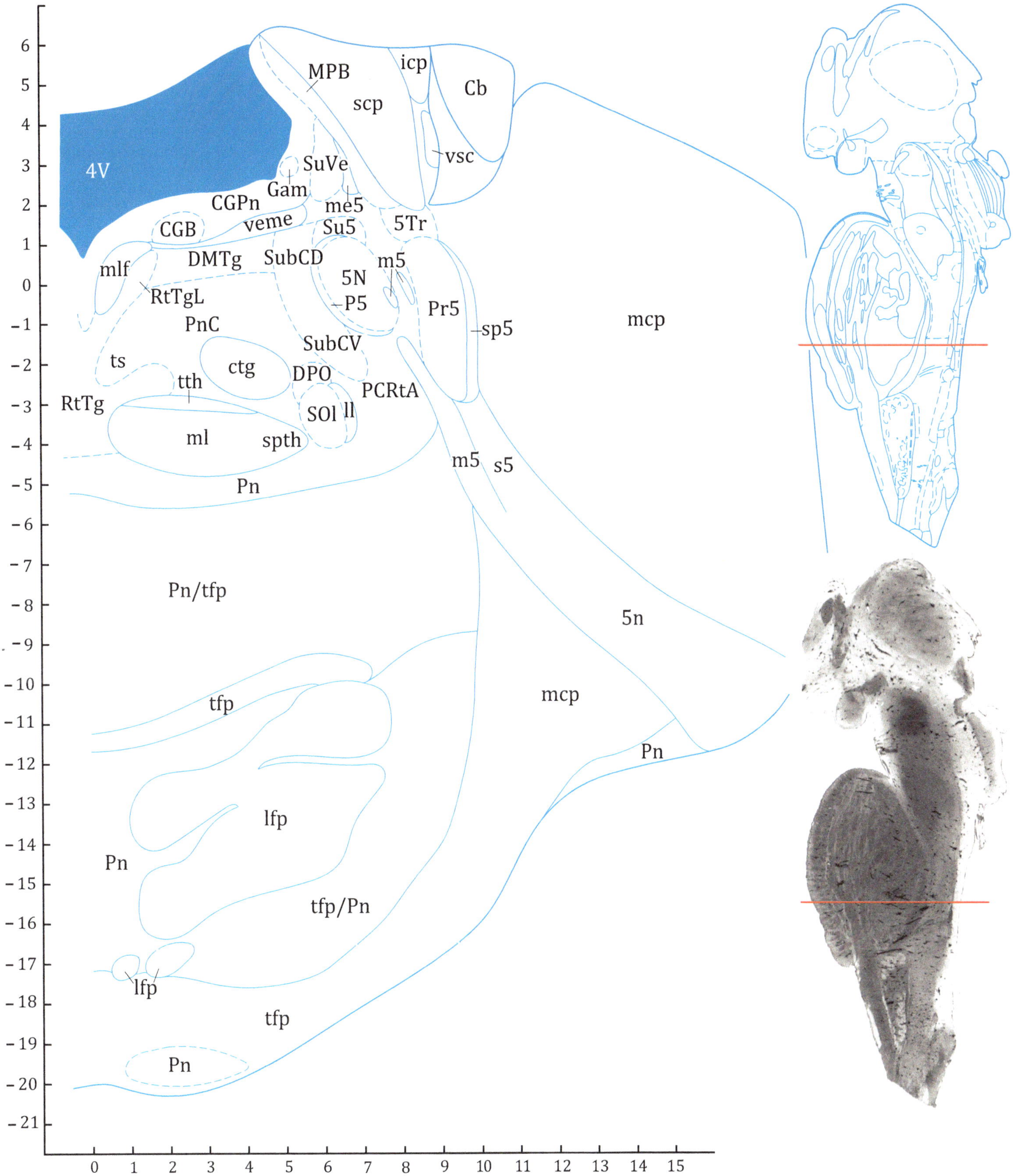

4V 4th ventricle
5N motor trigeminal nu
5n trigeminal nerve
5Tr trigem transition zone
Cb cerebellum
CGB central gray, beta part
CGPn central gray pons
ctg central tegmental tract
DMTg dorsomed tegmental area
DPO dorsal periolivary region
Gam gamma pontine nucleus
icp inf cerebellar peduncle
lfp longitudinal fasciculus pons

ll lateral lemniscus
m5 motor root trigeminal nerve
mcp middle cerebellar peduncle
me5 mesencephalic trigem tract
ml medial lemniscus
mlf med longitudinal fasciculus
MPB medial parabrachial nu
P5 peritrigeminal zone
PCRtA parvicell reticular nu, alpha
Pn pontine nuclei
Pn/tfp pontine nu/transv fibres pons
PnC pontine reticular nu, caudal
Pr5 principal sensory trigem nu

RtTg reticulotegmental nu pons
RtTgL reticulotegmental pons, lat
s5 sensory root trigem n
scp superior Cb peduncle
SOl superior olive, lateral
sp5 spinal trigem tract
spth spinothalamic tract
Su5 supratrigeminal nu
SubCD subcoeruleus nu, dorsal
SubCV subcoeruleus nu, ventral
SuVe superior vestibular nu
tfp transverse fibers pons
tfp/Pn transv fibers pons/pontine nu

ts tectospinal tr
tth trigeminothalamic tr
veme vestibulomesencephalic tr
vsc ventral spinocerebellar tr

Fig 28b
Obex +20.0mm
Transverse 28

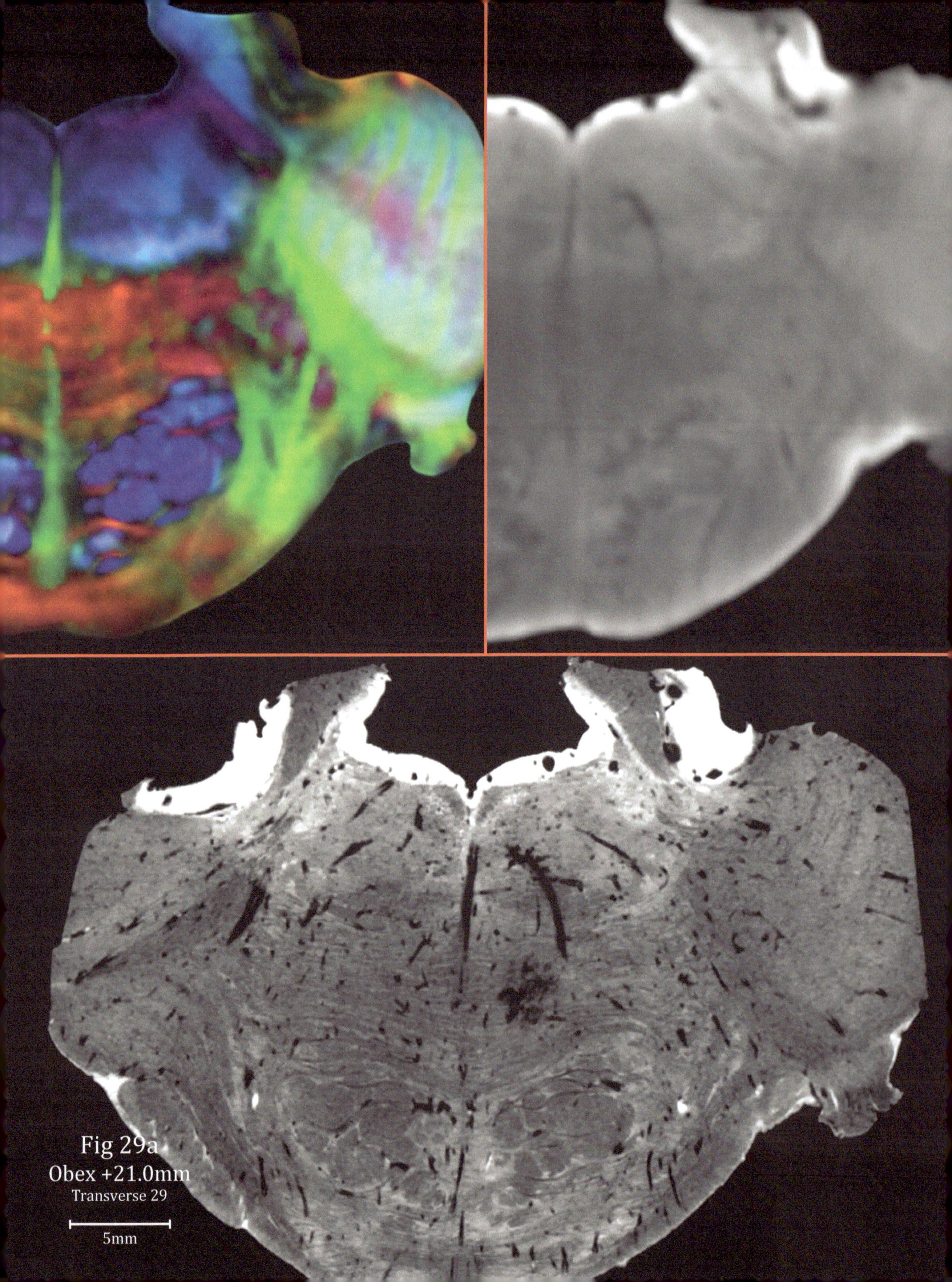

Fig 29a
Obex +21.0mm
Transverse 29

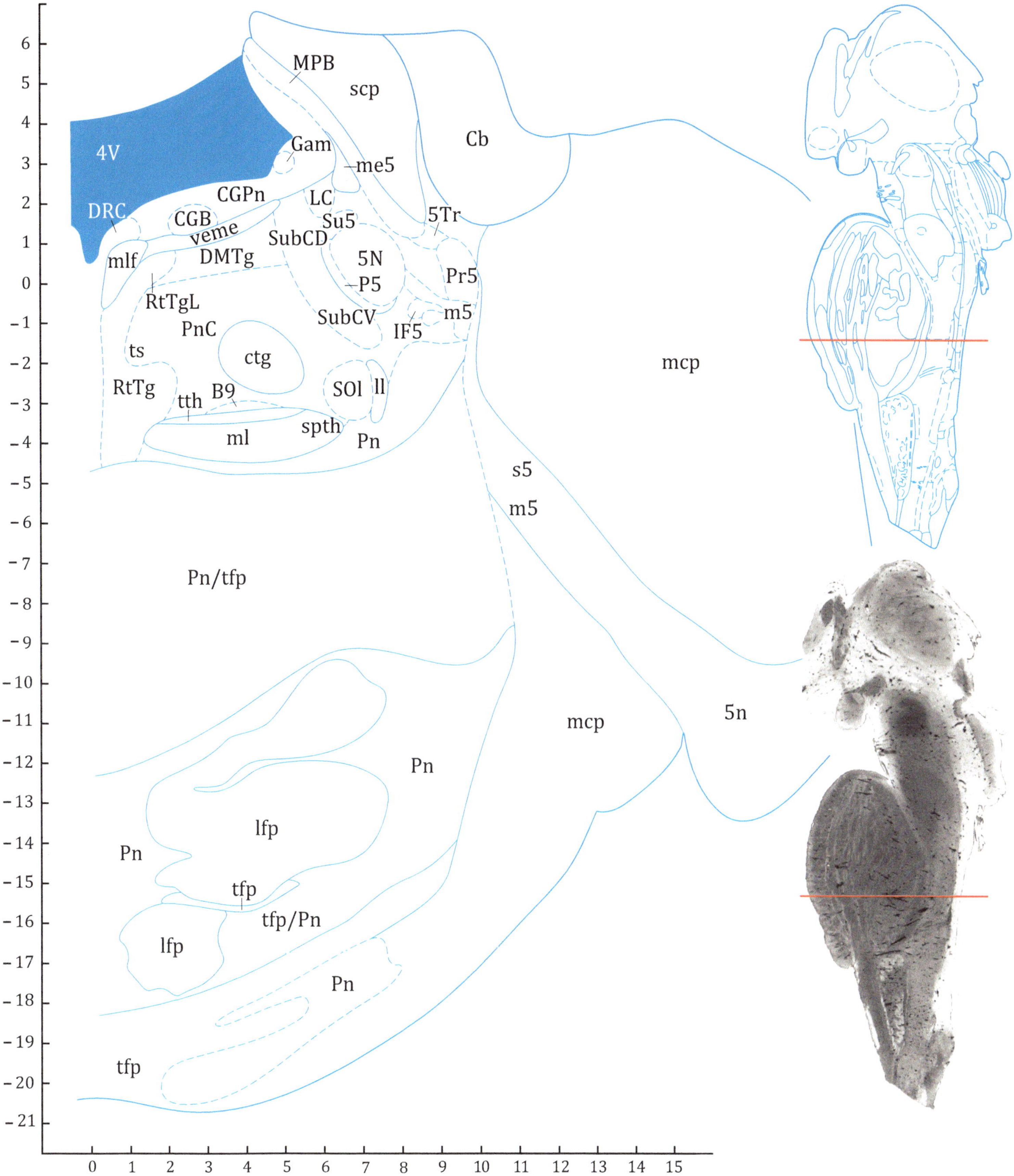

4V 4th ventricle
5N motor trigeminal nu
5n trigeminal nerve
5Tr trigem transition zone
B9 B9 serotonin cells
Cb cerebellum
CGB central gray, beta part
CGPn central gray pons
ctg central tegmental tract
DMTg dorsomed tegmental area
DRC dorsal raphe nu, caudal
Gam gamma pontine nucleus
IF5 interfascicular trigeminal nu
LC locus coeruleus
lfp longitudinal fasciculus pons
ll lateral lemniscus
m5 motor root trigeminal nerve
mcp middle cerebellar peduncle
me5 mesencephalic trigem tract
ml medial lemniscus
mlf med longitudinal fasciculus
MPB medial parabrachial nu
P5 peritrigeminal zone
Pn pontine nuclei
Pn/tfp pontine nu/transv fibres pons
PnC pontine reticular nu, caudal
Pr5 principal sensory trigem nu
RtTg reticulotegmental nu pons
RtTgL reticulotegmental pons, lat
s5 sensory root trigem n
scp superior Cb peduncle
SOl superior olive, lateral
spth spinothalamic tract
Su5 supratrigeminal nu
SubCD subcoeruleus nu, dorsal
SubCV subcoeruleus nu, ventral
tfp transverse fibers pons
tfp/Pn transv fibers pons/pontine nu
ts tectospinal tr
tth trigeminothalamic tr
veme vestibulomesencephalic tr

Fig 29b
Obex +21.0mm
Transverse 29

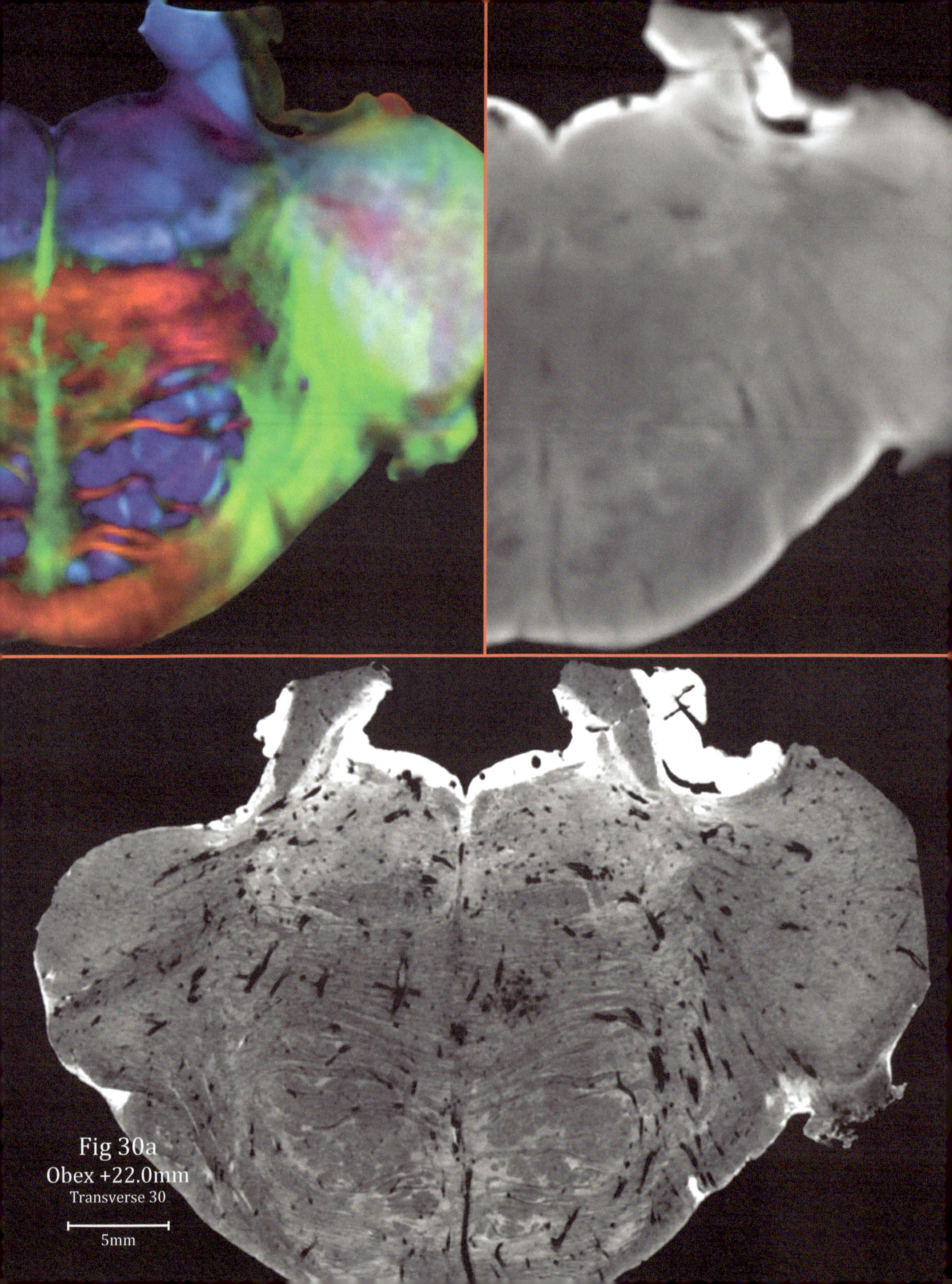

Fig 30a
Obex +22.0mm
Transverse 30

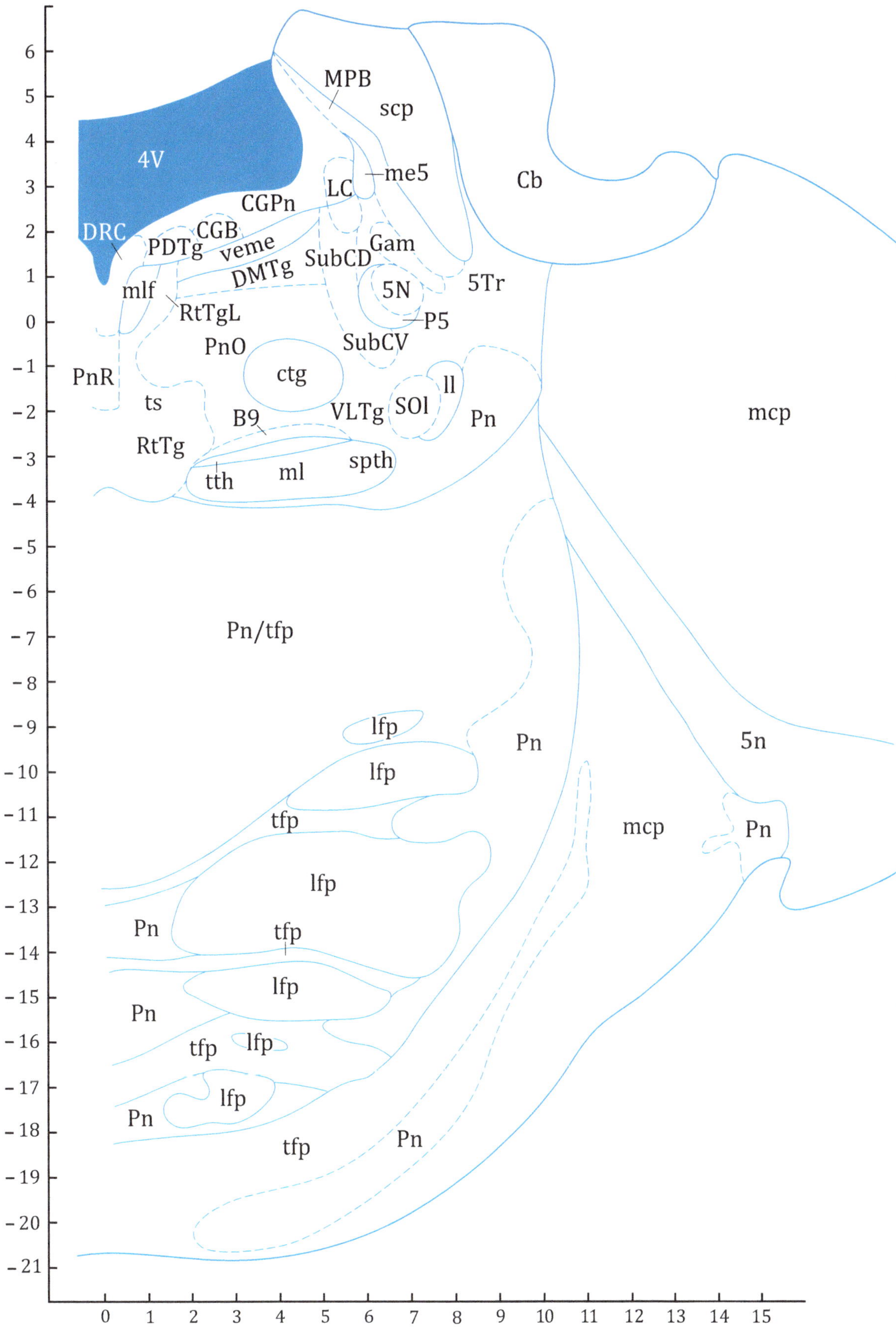

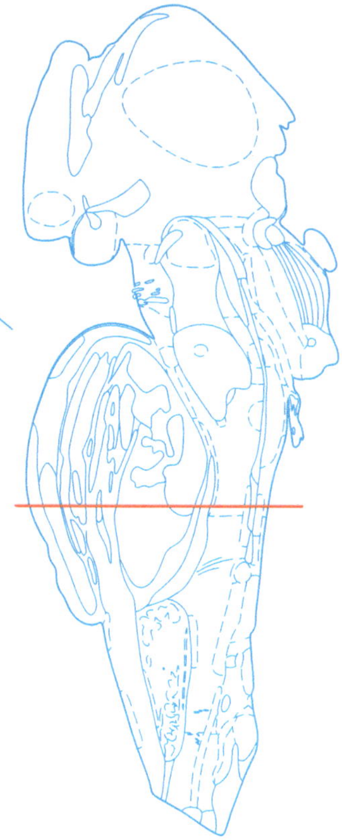

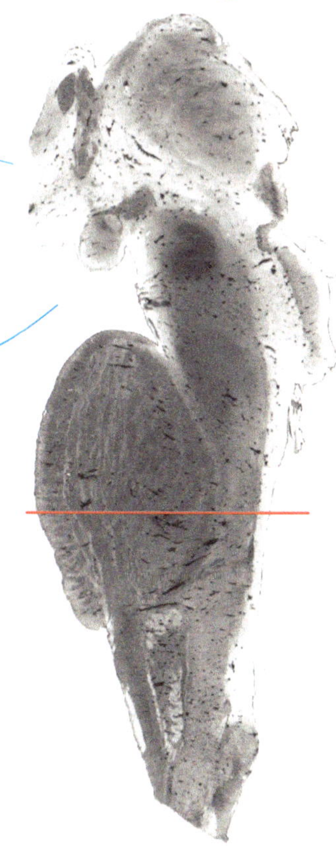

4V 4th ventricle
5N motor trigeminal nu
5n trigeminal nerve
5Tr trigem transition zone
B9 B9 serotonin cells
Cb cerebellum
CGB central gray, beta part
CGPn central gray pons
ctg central tegmental tract
DMTg dorsomed tegmental area
DRC dorsal raphe nu, caudal
Gam gamma pontine nucleus
LC locus coeruleus
lfp longitudinal fasciculus pons
ll lateral lemniscus
mcp middle cerebellar peduncle
me5 mesencephalic trigem tract
ml medial lemniscus
mlf med longitudinal fasciculus
MPB medial parabrachial nu
P5 peritrigeminal zone
PDTg posterodors tegmental nu
Pn pontine nuclei
Pn/tfp pontine nu/transv fibres pons
PnO pontine reticular nu, oral
PnR pontine raphe nu
RtTg reticulotegmental nu pons
RtTgL reticulotegmental pons, lat
scp superior Cb peduncle
SOl superior olive, lateral
spth spinothalamic tract
SubCD subcoeruleus nu, dorsal
SubCV subcoeruleus nu, ventral
tfp transverse fibers pons
ts tectospinal tr
tth trigeminothalamic tr
veme vestibulomesencephalic tr
VLTg ventrolateral tegmental area

Fig 30b
Obex +22.0mm
Transverse 30

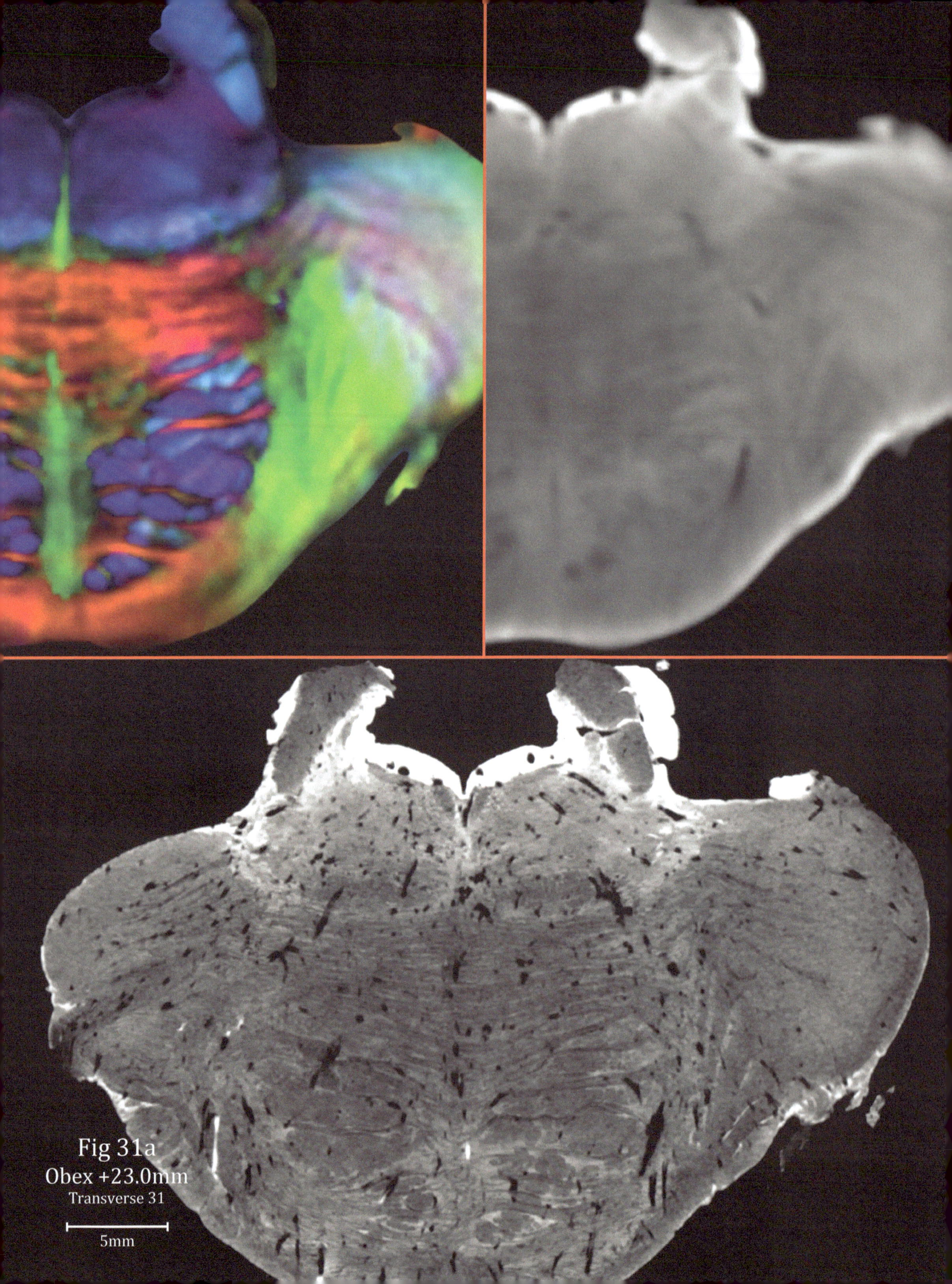
Fig 31a
Obex +23.0mm
Transverse 31
5mm

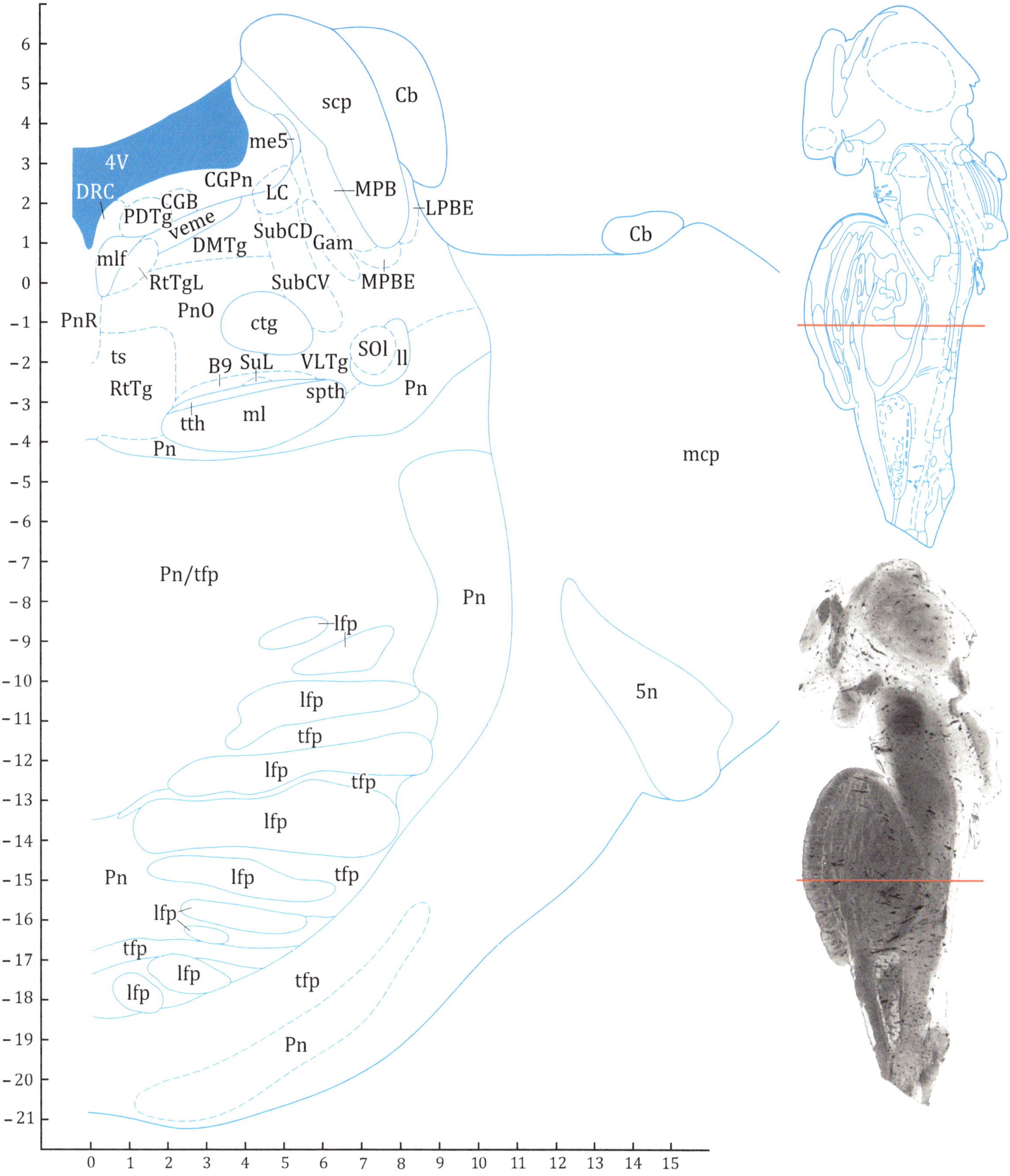

Fig 31b
Obex +23.0mm
Transverse 31

4V 4th ventricle
5n trigeminal nerve
B9 B9 serotonin cells
Cb cerebellum
CGB central gray, beta part
CGPn central gray pons
ctg central tegmental tract
DMTg dorsomed tegmental area
DRC dorsal raphe nu, caudal
Gam gamma pontine nucleus
LC locus coeruleus
lfp longitudinal fasciculus pons
ll lateral lemniscus
LPBE lat parabrachial nu, external
mcp middle cerebellar peduncle
me5 mesencephalic trigem tract
ml medial lemniscus
mlf med longitudinal fasciculus
MPB medial parabrachial nu
MPBE med parabrachial nu ext
PDTg posterodors tegmental nu
Pn pontine nuclei
Pn/tfp pontine nu/transv fibres pons
PnO pontine reticular nu, oral
PnR pontine raphe nu
RtTg reticulotegmental nu pons
RtTgL reticulotegmental pons, lat
scp superior Cb peduncle
SOl superior olive, lateral
spth spinothalamic tract
SubCD subcoeruleus nu, dorsal
SubCV subcoeruleus nu, ventral
SuL supralemniscal nucleus
tfp transverse fibers pons
ts tectospinal tr
tth trigeminothalamic tr
veme vestibulomesencephalic tr
VLTg ventrolateral tegmental area

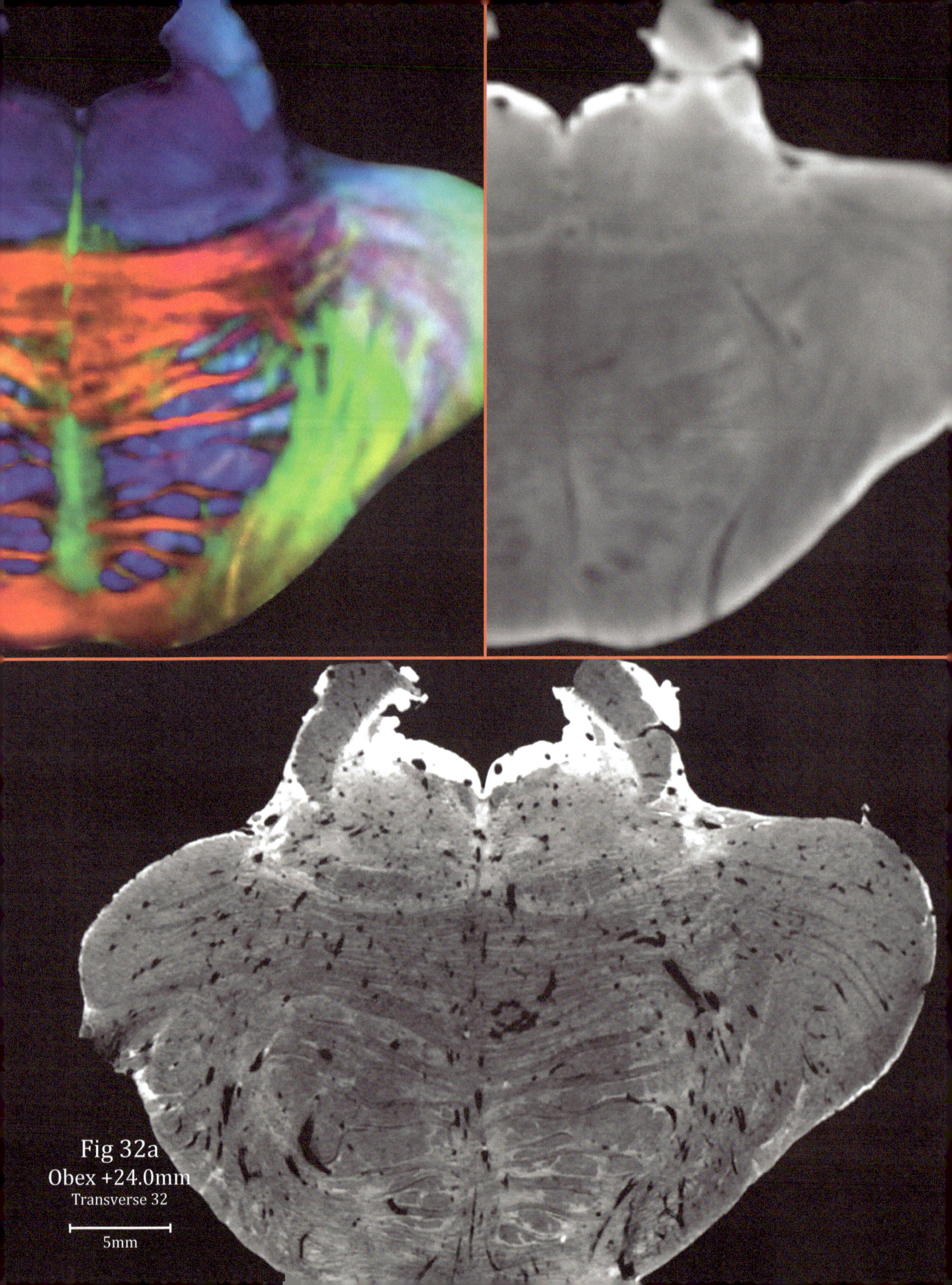

Fig 32a
Obex +24.0mm
Transverse 32

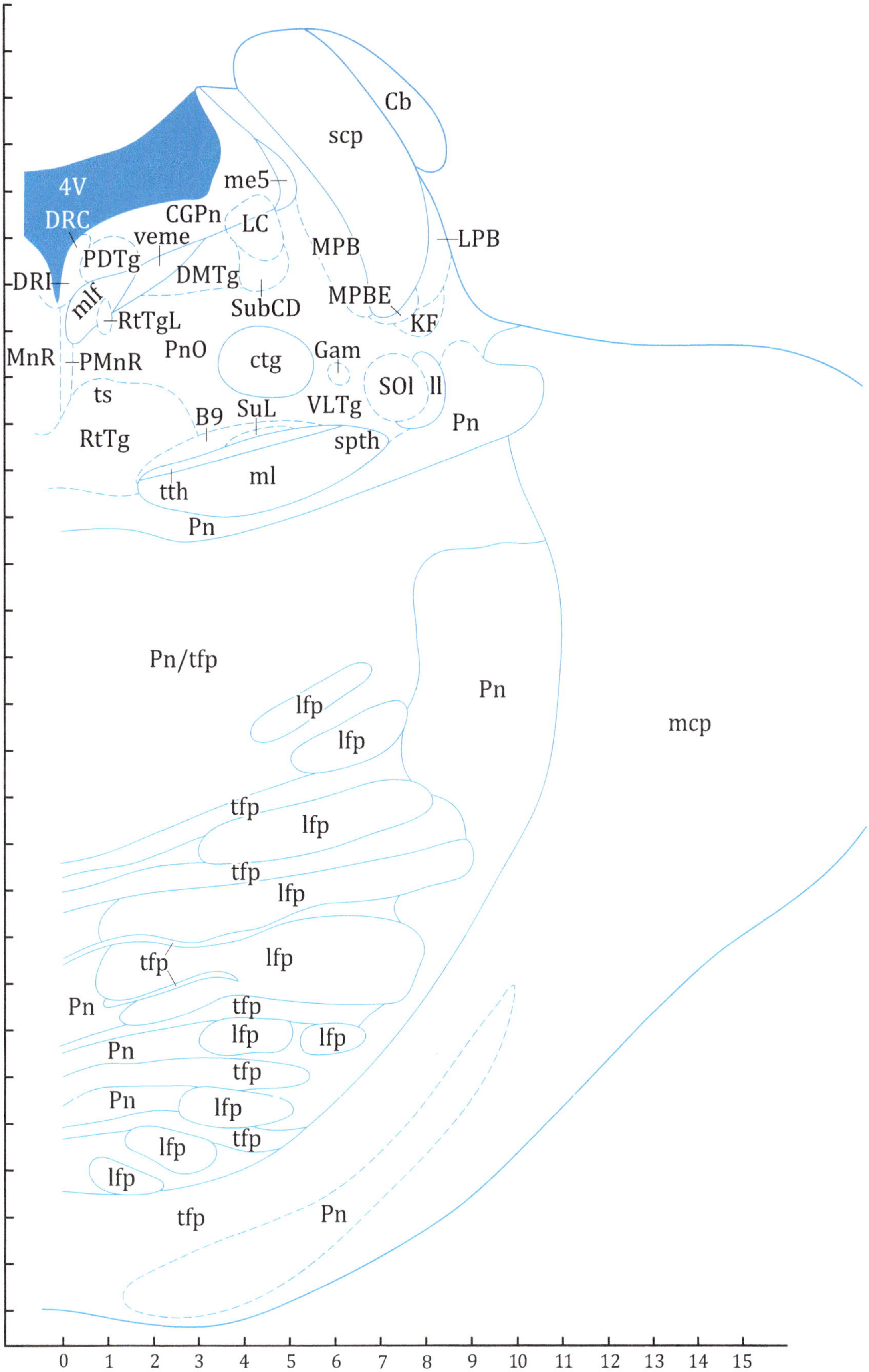

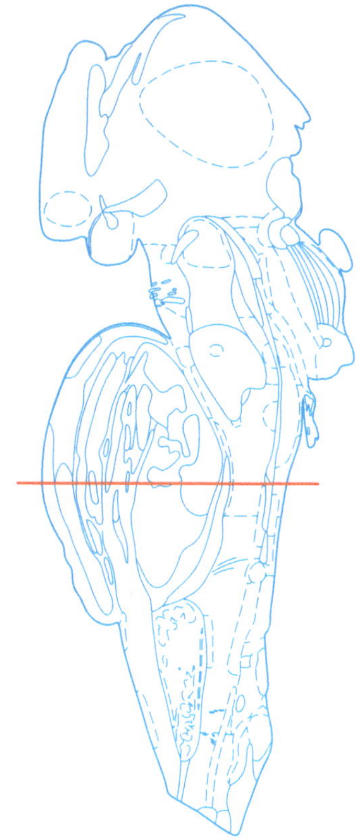

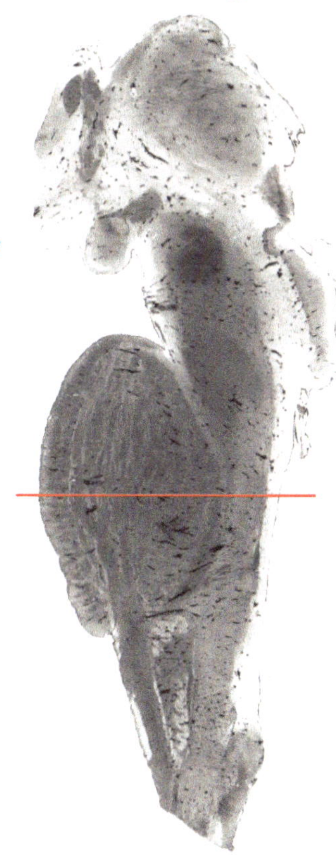

4V 4th ventricle
B9 B9 serotonin cells
Cb cerebellum
CGPn central gray pons
ctg central tegmental tract
DMTg dorsomed tegmental area
DRC dorsal raphe nu, caudal
DRI dorsal raphe, interfascicular
Gam gamma pontine nucleus
KF Kölliker-Fuse nu
LC locus coeruleus
lfp longitudinal fasciculus pons
ll lateral lemniscus
LPB lateral parabrachial nu
mcp middle cerebellar peduncle
me5 mesencephalic trigem tract
ml medial lemniscus
mlf med longitudinal fasciculus
MnR median raphe nu
MPB medial parabrachial nu
MPBE med parabrachial nu ext
PDTg posterodors tegmental nu
PMnR paramedian raphe nu
Pn pontine nuclei
Pn/tfp pontine nu/transv fibres pons
PnO pontine reticular nu, oral
RtTg reticulotegmental nu pons
RtTgL reticulotegmental pons, lat
scp superior Cb peduncle
SOl superior olive, lateral
spth spinothalamic tract
SubCD subcoeruleus nu, dorsal
SuL supralemniscal nucleus
tfp transverse fibers pons
ts tectospinal tr
tth trigeminothalamic tr
veme vestibulomesencephalic tr
VLTg ventrolateral tegmental area

Fig 32b
Obex +24.0mm
Transverse 32

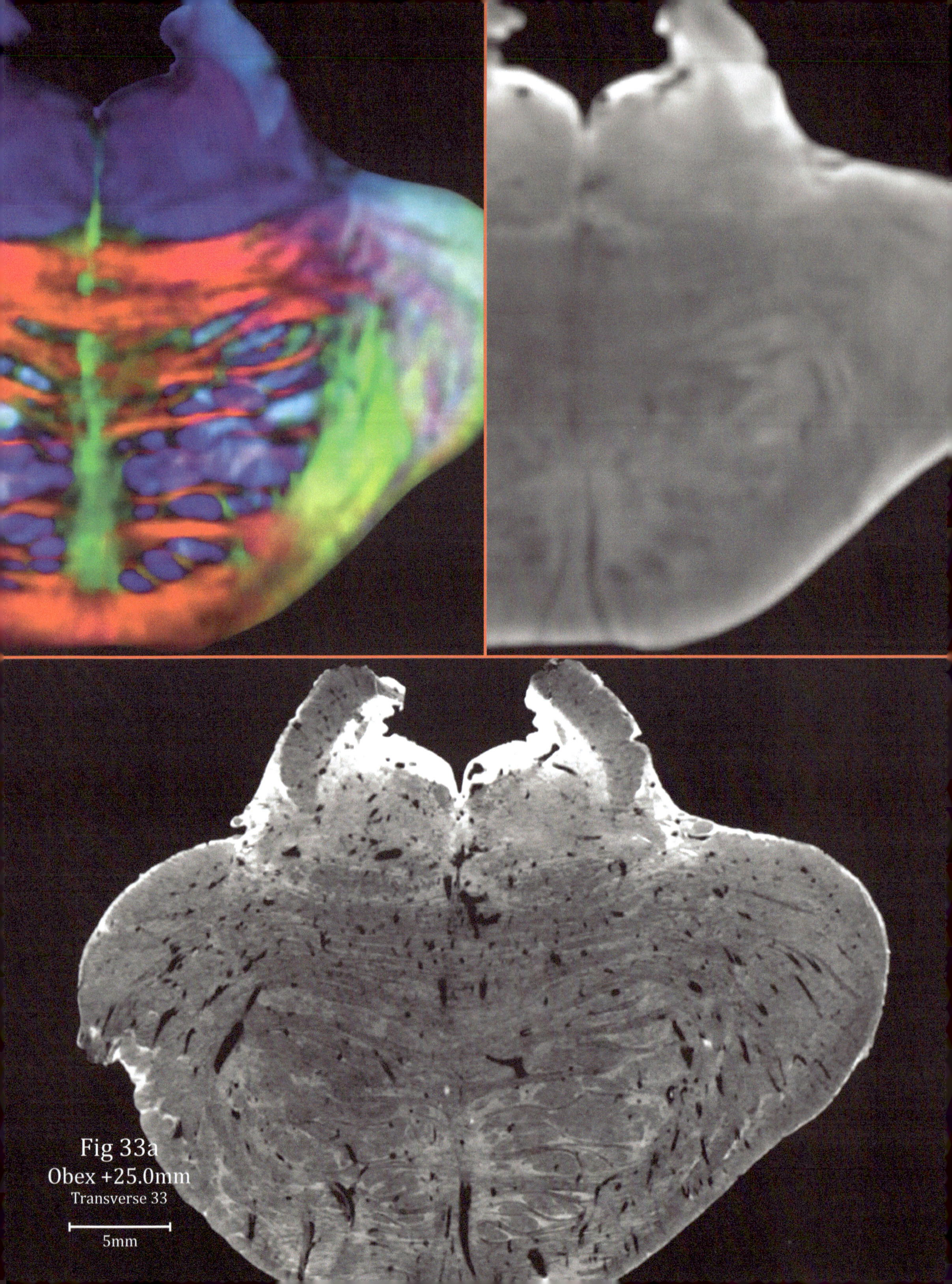
Fig 33a
Obex +25.0mm
Transverse 33
5mm

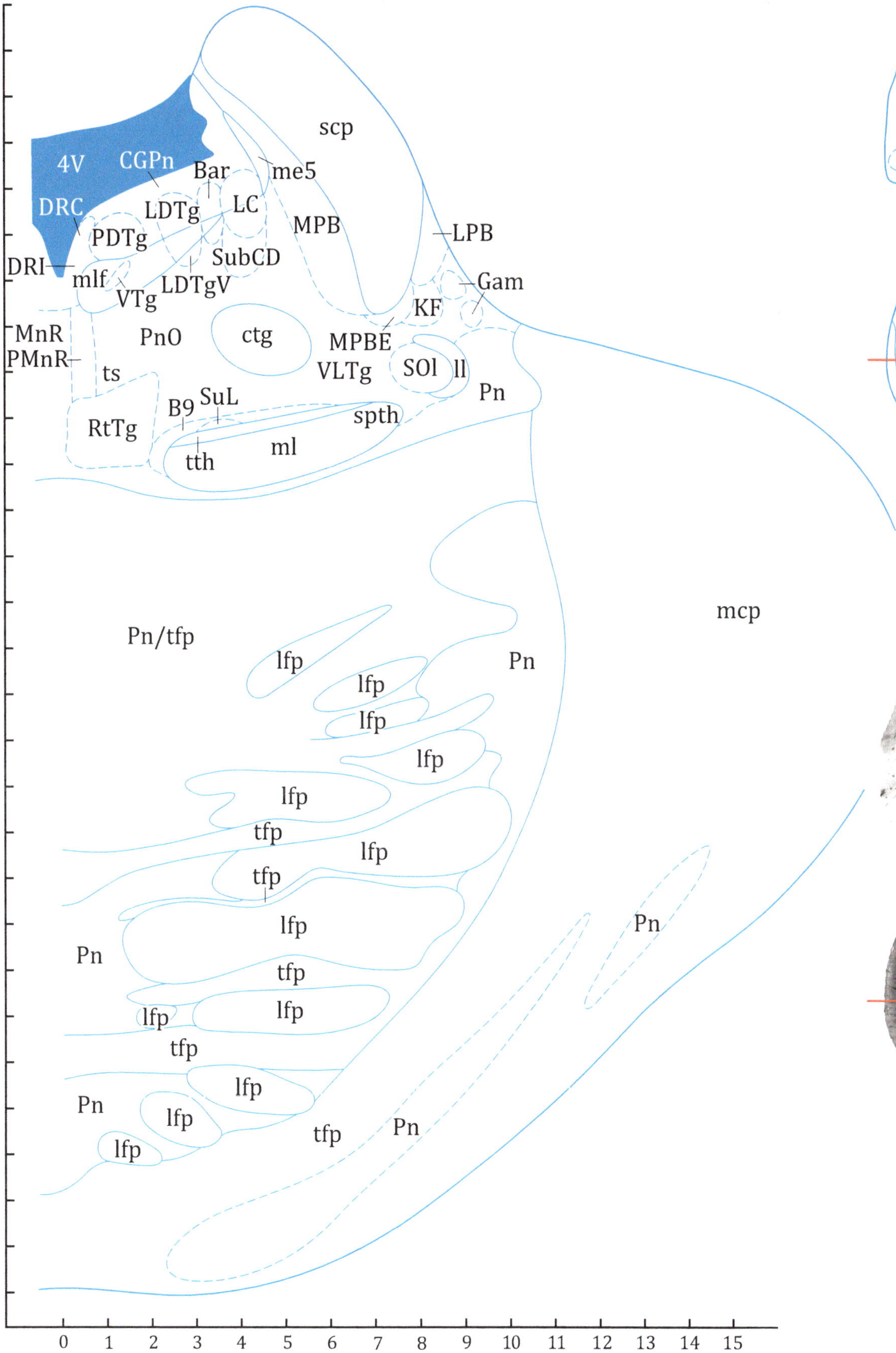

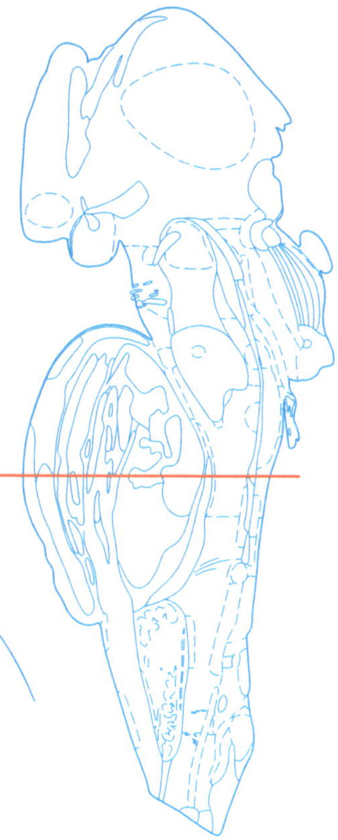

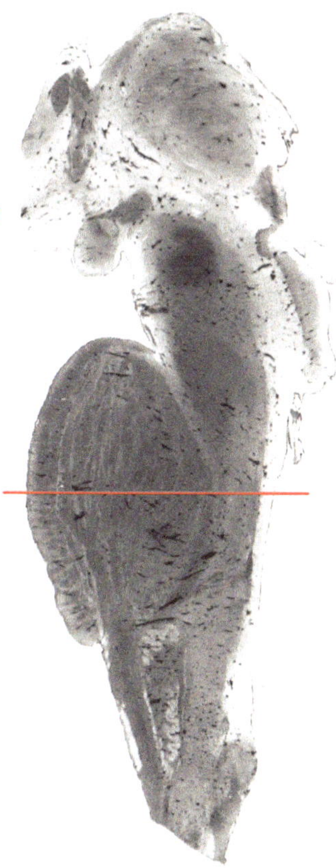

4V 4th ventricle
B9 B9 serotonin cells
Bar Barrington's nu
CGPn central gray pons
ctg central tegmental tract
DRC dorsal raphe nu, caudal
DRI dorsal raphe, interfascicular
Gam gamma pontine nucleus
KF Kölliker-Fuse nu
LC locus coeruleus
LDTg laterodor tegmental nu
LDTgV laterodor tegmental nu, vent
lfp longitudinal fasciculus pons
ll lateral lemniscus
LPB lateral parabrachial nu
mcp middle cerebellar peduncle
me5 mesencephalic trigem tract
ml medial lemniscus
mlf med longitudinal fasciculus
MnR median raphe nu
MPB medial parabrachial nu
MPBE med parabrachial nu ext
PDTg posterodors tegmental nu
PMnR paramedian raphe nu
Pn pontine nuclei
Pn/tfp pontine nu/transv fibres pons
PnO pontine reticular nu, oral
RtTg reticulotegmental nu pons
scp superior Cb peduncle
SOl superior olive, lateral
spth spinothalamic tract
SubCD subcoeruleus nu, dorsal
SuL supralemniscal nucleus
tfp transverse fibers pons
ts tectospinal tr
tth trigeminothalamic tr
VLTg ventrolateral tegmental area
VTg vent tegmental nu

Fig 33b
Obex +25.0mm
Transverse 33

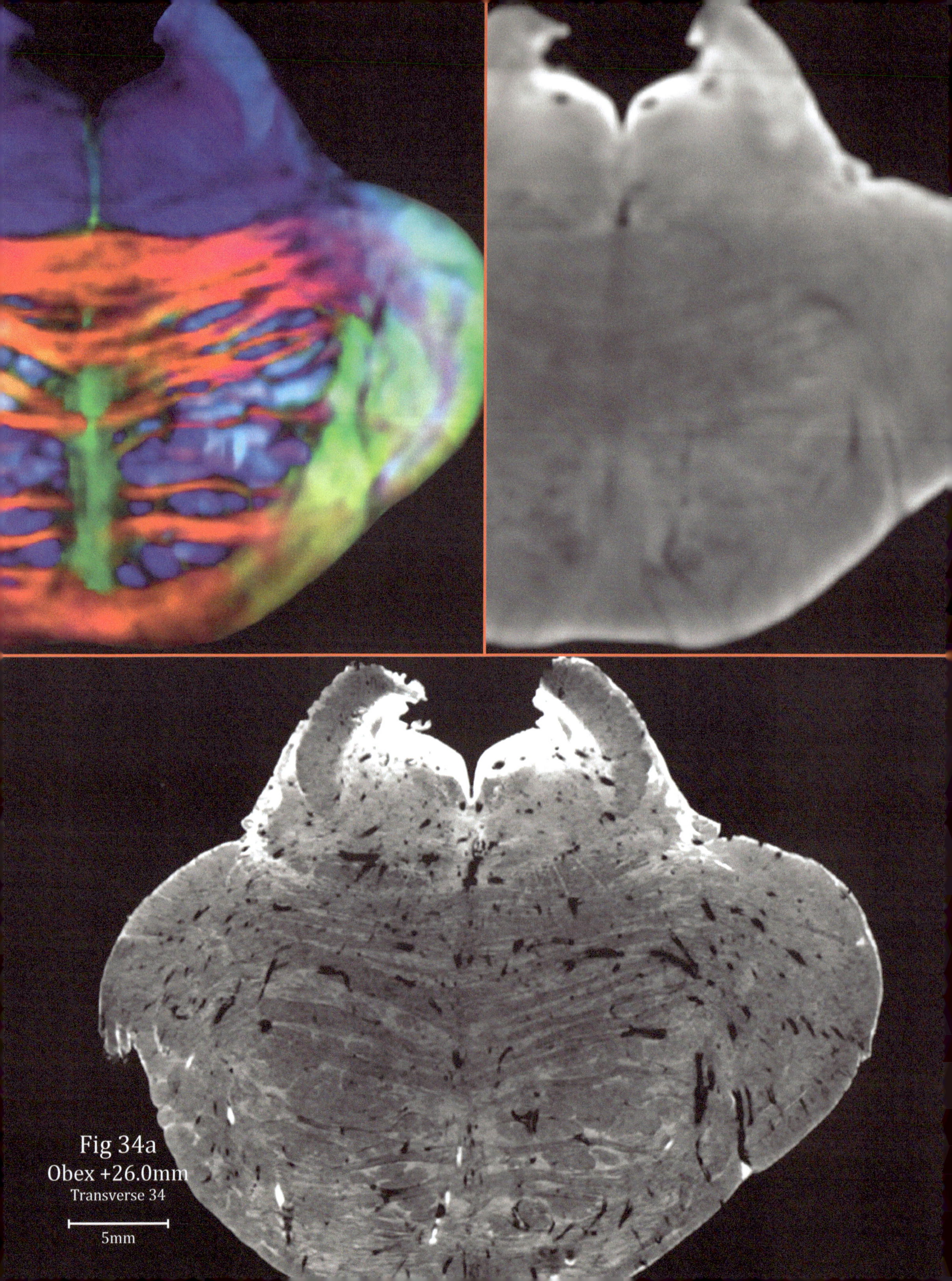

Fig 34a
Obex +26.0mm
Transverse 34

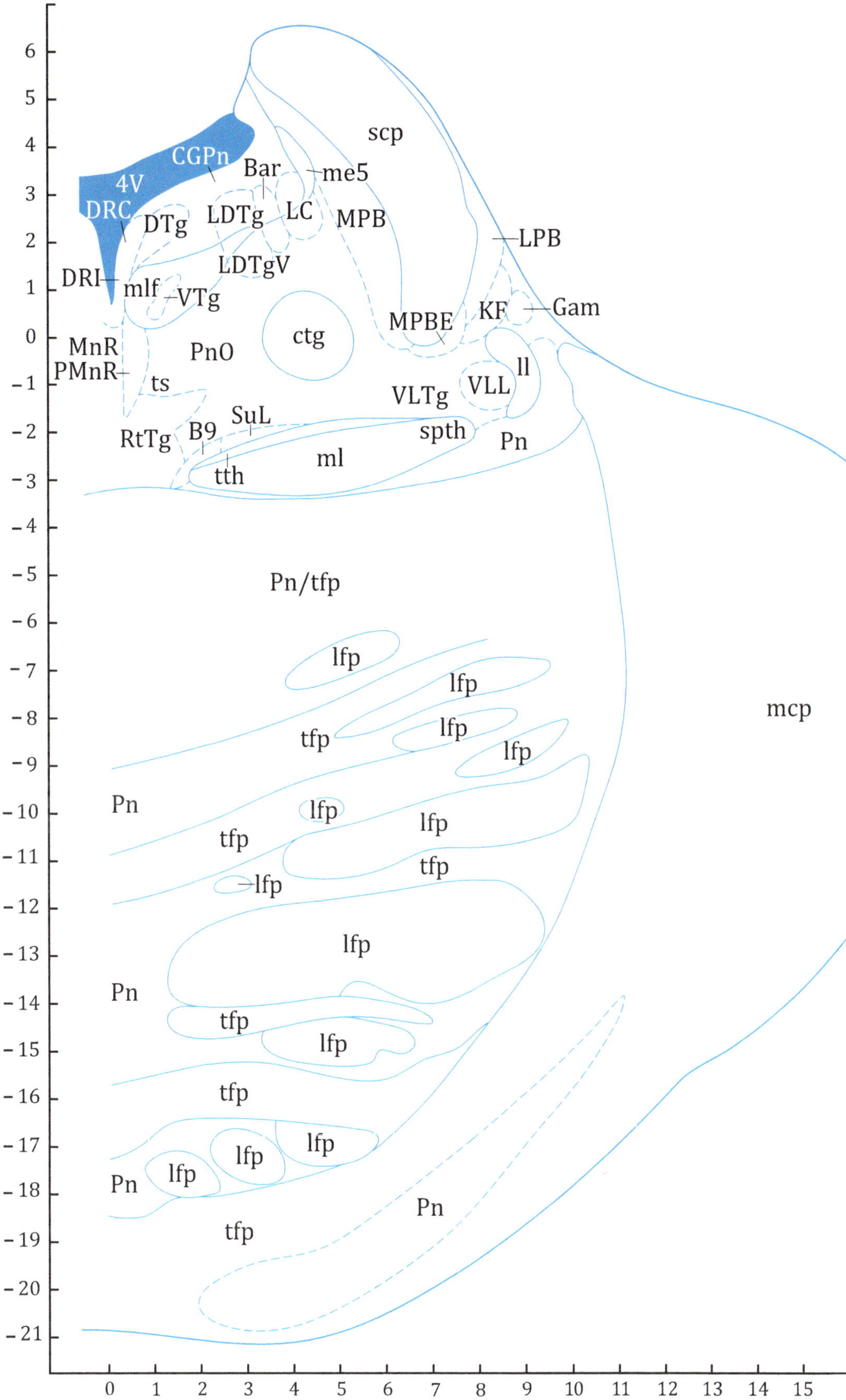

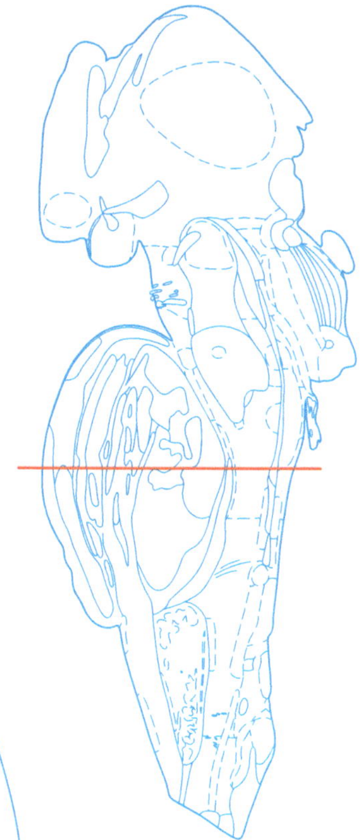

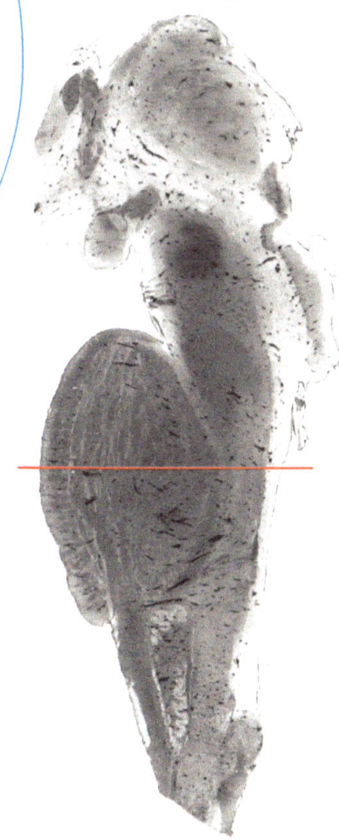

4V 4th ventricle
B9 B9 serotonin cells
Bar Barrington's nu
CGPn central gray pons
ctg central tegmental tract
DRC dorsal raphe nu, caudal
DRI dorsal raphe, interfascicular
DTg dorsal tegmental nu
Gam gamma pontine nucleus
KF Kölliker-Fuse nu
LC locus coeruleus
LDTg laterodor tegmental nu
LDTgV laterodor tegmental nu, vent
lfp longitudinal fasciculus pons
ll lateral lemniscus
LPB lateral parabrachial nu
mcp middle cerebellar peduncle
me5 mesencephalic trigem tract
ml medial lemniscus
mlf med longitudinal fasciculus
MnR median raphe nu
MPB medial parabrachial nu
MPBE med parabrachial nu ext
PMnR paramedian raphe nu
Pn pontine nuclei
Pn/tfp pontine nu/transv fibres pons
PnO pontine reticular nu, oral
RtTg reticulotegmental nu pons
scp superior Cb peduncle
spth spinothalamic tract
SuL supralemniscal nucleus
tfp transverse fibers pons
ts tectospinal tr
tth trigeminothalamic tr
VLL ventral nu lat lemniscus
VLTg ventrolateral tegmental area
VTg vent tegmental nu

Fig 34b
Obex +26.0mm
Transverse 34

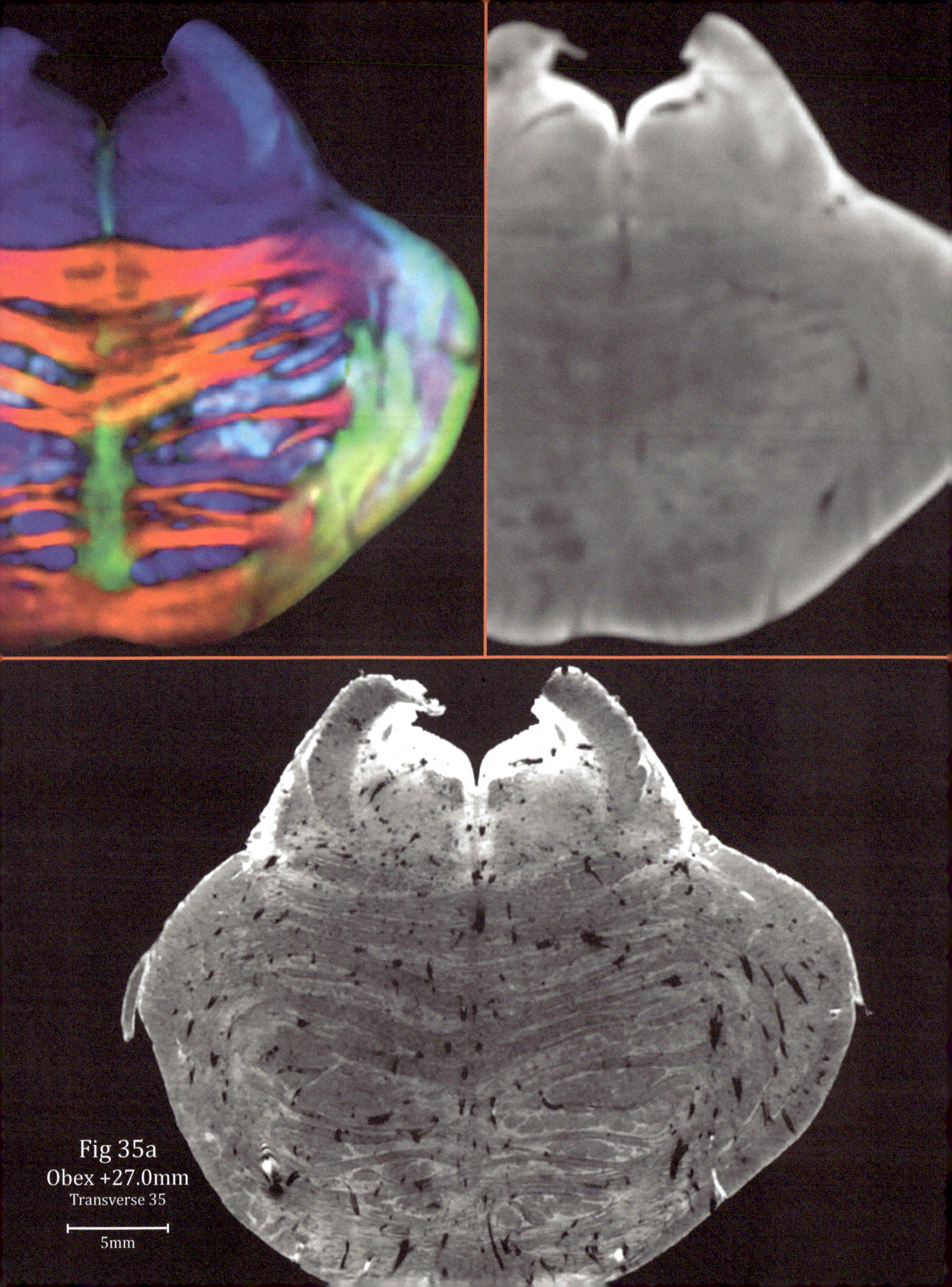

Fig 35a
Obex +27.0mm
Transverse 35
5mm

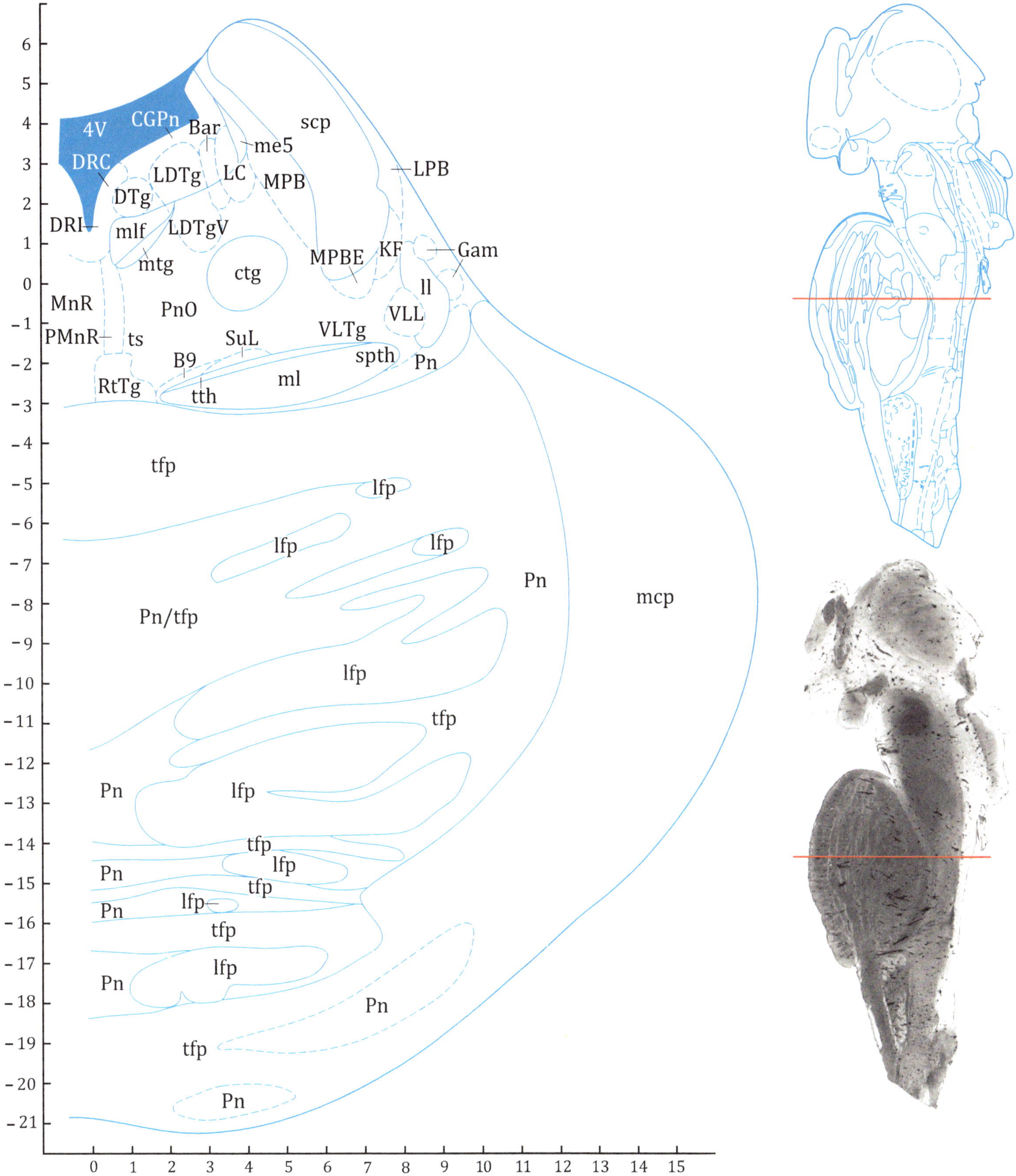

4V 4th ventricle
B9 B9 serotonin cells
Bar Barrington's nu
CGPn central gray pons
ctg central tegmental tract
DRC dorsal raphe nu, caudal
DRI dorsal raphe, interfascicular
DTg dorsal tegmental nu
Gam gamma pontine nucleus
KF Kölliker-Fuse nu
LC locus coeruleus
LDTg laterodor tegmental nu
LDTgV laterodor tegmental nu, vent

lfp longitudinal fasciculus pons
ll lateral lemniscus
LPB lateral parabrachial nu
mcp middle cerebellar peduncle
me5 mesencephalic trigem tract
ml medial lemniscus
mlf med longitudinal fasciculus
MnR median raphe nu
MPB medial parabrachial nu
MPBE med parabrachial nu ext
mtg mammillotegmental tract
PMnR paramedian raphe nu
Pn pontine nuclei

Pn/tfp pontine nu/transv fibres pons
PnO pontine reticular nu, oral
RtTg reticulotegmental nu pons
scp superior Cb peduncle
spth spinothalamic tract
SuL supralemniscal nucleus
tfp transverse fibers pons
ts tectospinal tr
tth trigeminothalamic tr
VLL ventral nu lat lemniscus
VLTg ventrolateral tegmental area

Fig 35b
Obex +27.0mm
Transverse 35

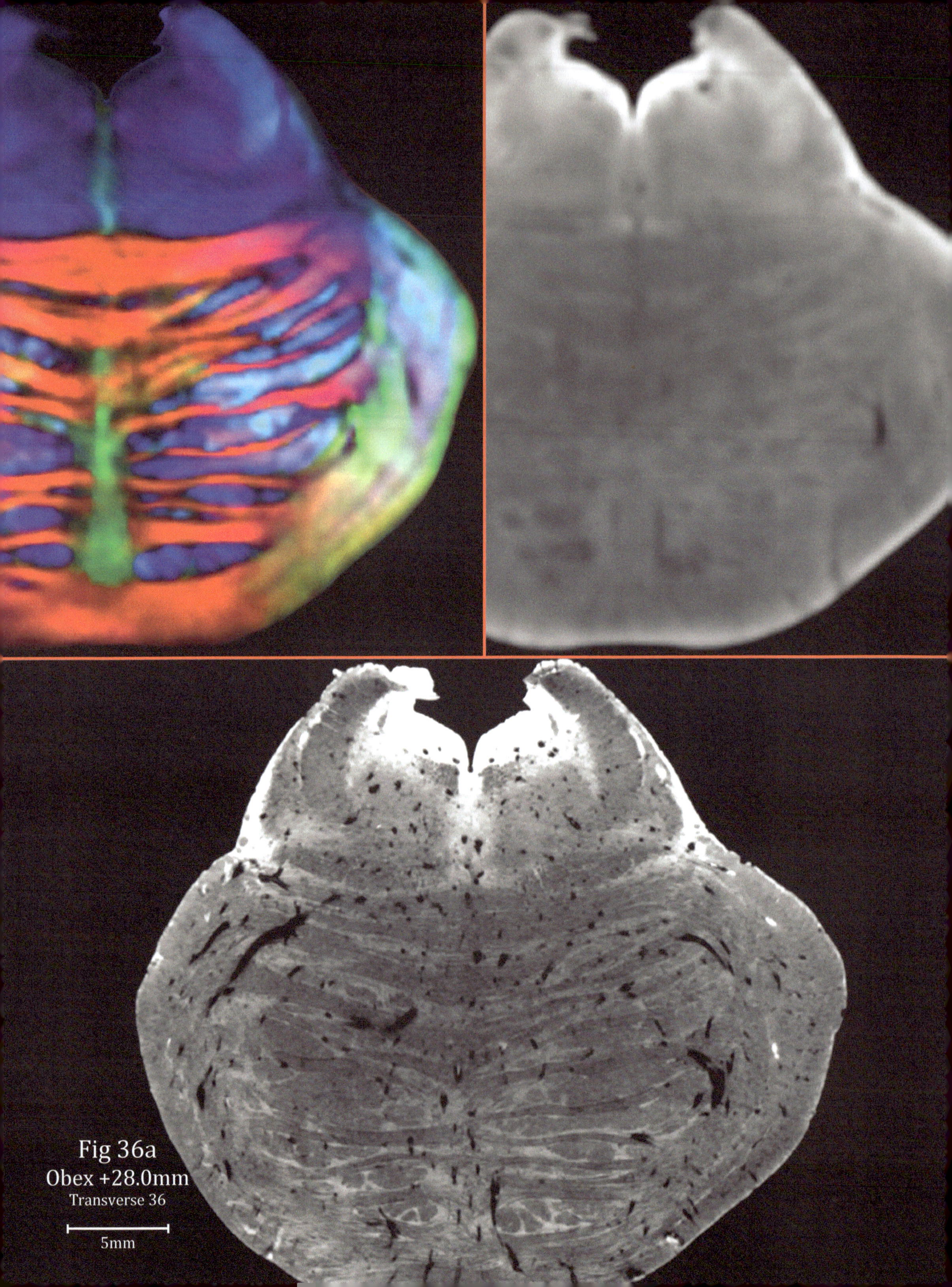

Fig 36a
Obex +28.0mm
Transverse 36

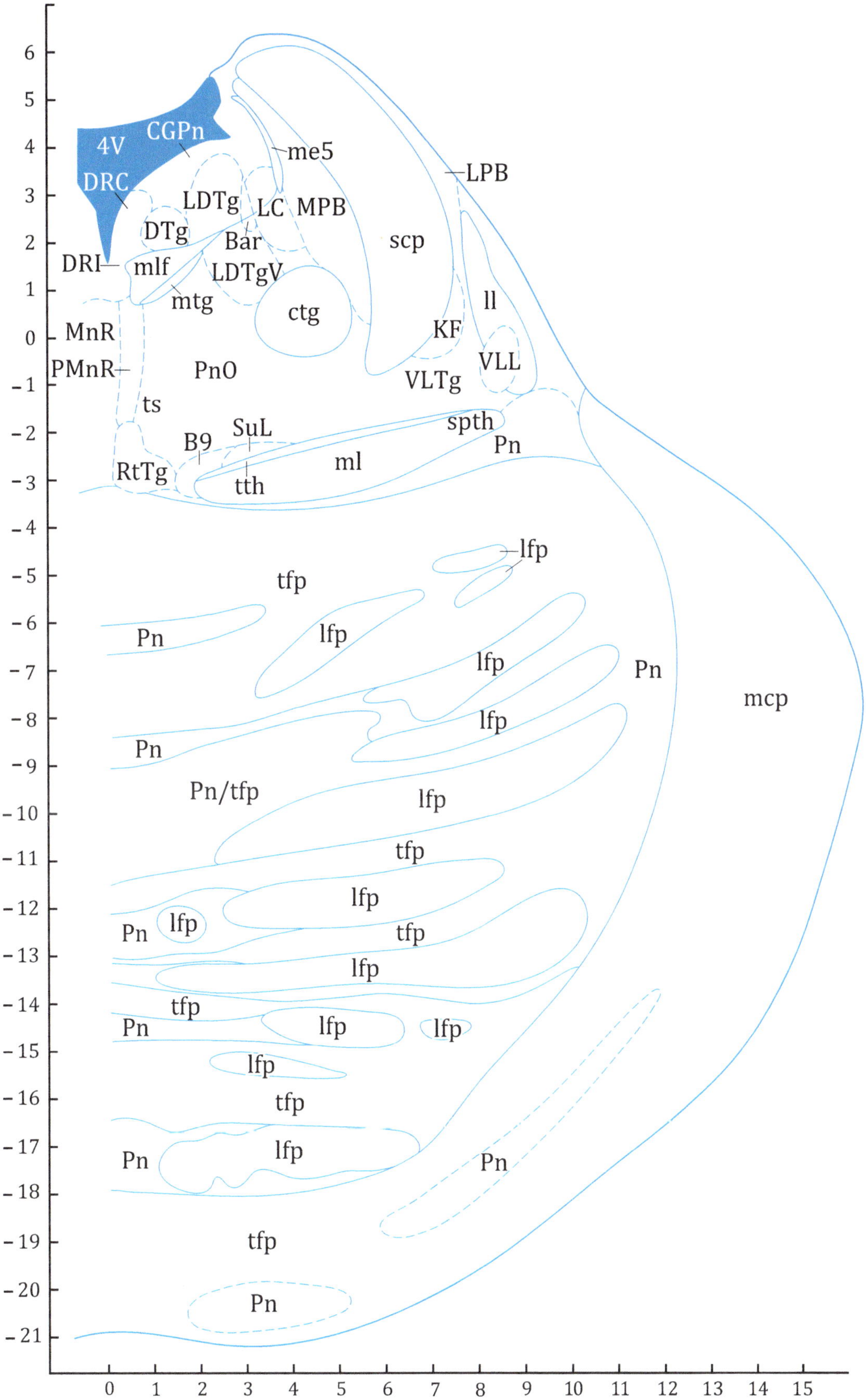

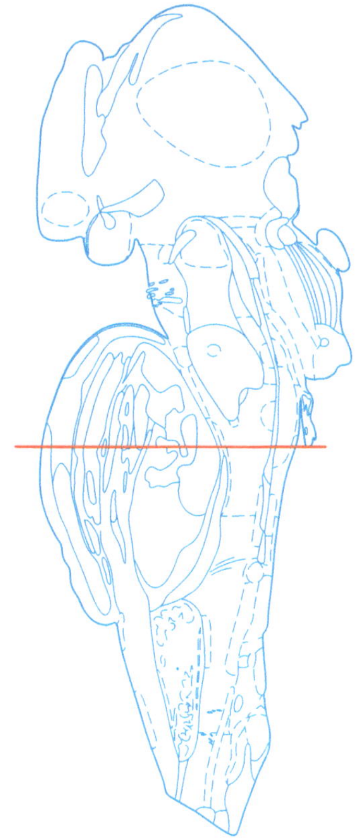

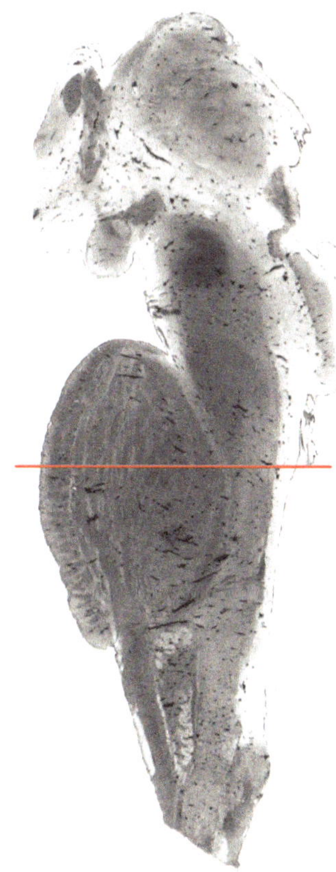

4V 4th ventricle
B9 B9 serotonin cells
Bar Barrington's nu
CGPn central gray pons
ctg central tegmental tract
DRC dorsal raphe nu, caudal
DRI dorsal raphe, interfascicular
DTg dorsal tegmental nu
KF Kölliker-Fuse nu
LC locus coeruleus
LDTg laterodor tegmental nu
LDTgV laterodor tegmental nu, vent
lfp longitudinal fasciculus pons

ll lateral lemniscus
LPB lateral parabrachial nu
mcp middle cerebellar peduncle
me5 mesencephalic trigem tract
ml medial lemniscus
mlf med longitudinal fasciculus
MnR median raphe nu
MPB medial parabrachial nu
mtg mammillotegmental tract
PMnR paramedian raphe nu
Pn pontine nuclei
Pn/tfp pontine nu/transv fibres pons
PnO pontine reticular nu, oral

RtTg reticulotegmental nu pons
scp superior Cb peduncle
spth spinothalamic tract
SuL supralemniscal nucleus
tfp transverse fibers pons
ts tectospinal tr
tth trigeminothalamic tr
VLL ventral nu lat lemniscus
VLTg ventrolateral tegmental area

Fig 36b
Obex +28.0mm
Transverse 36

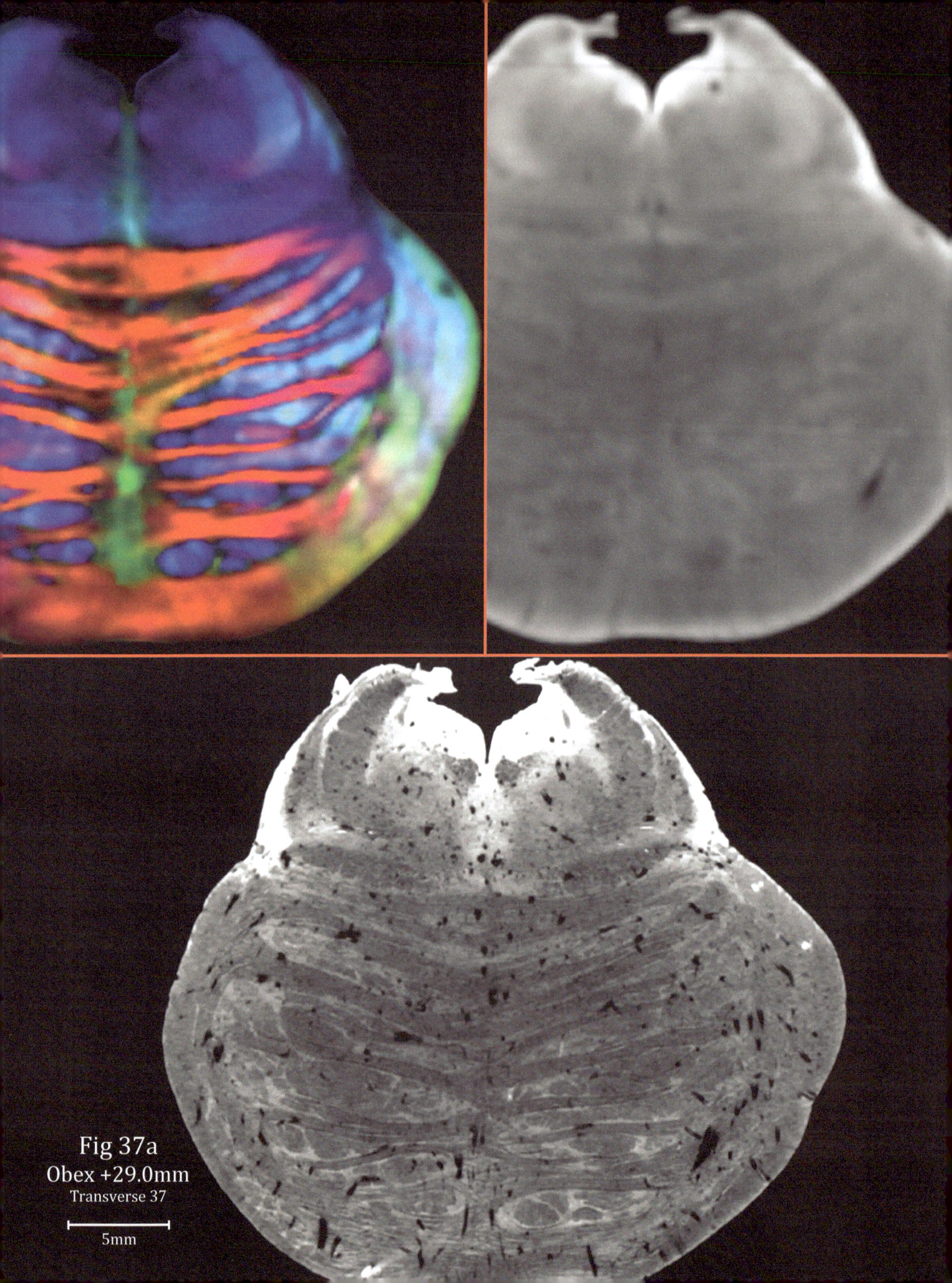

Fig 37a
Obex +29.0mm
Transverse 37

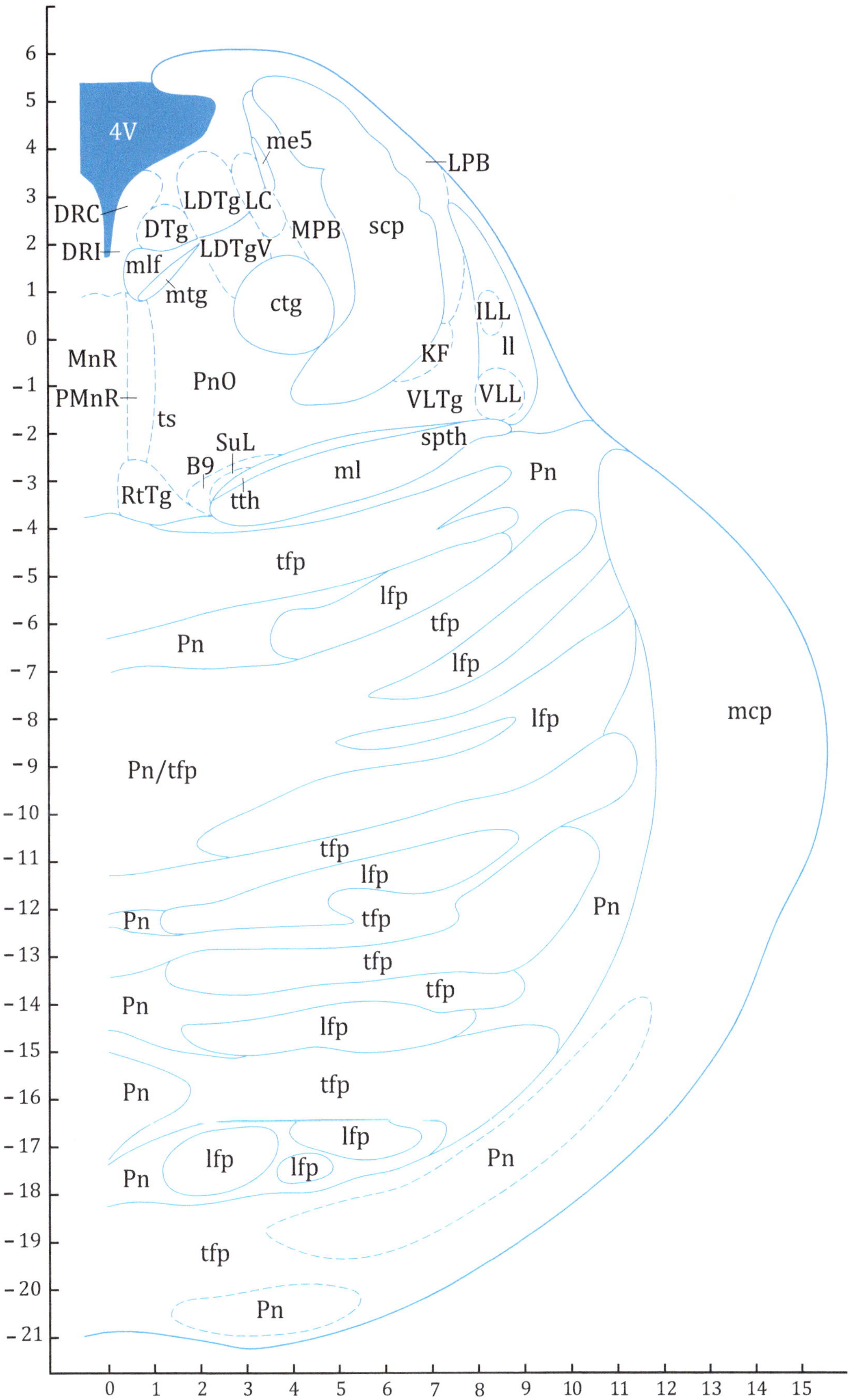

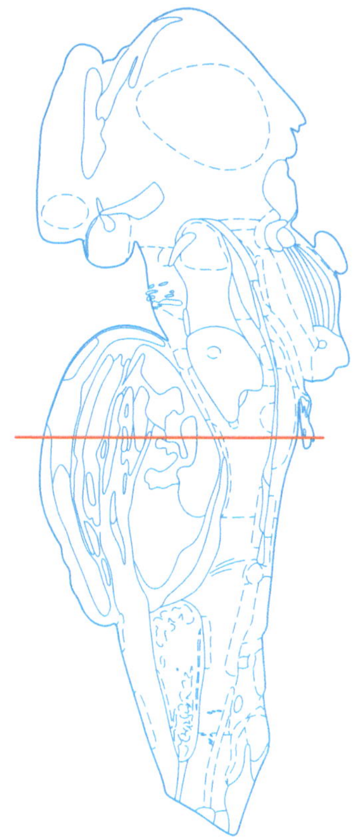

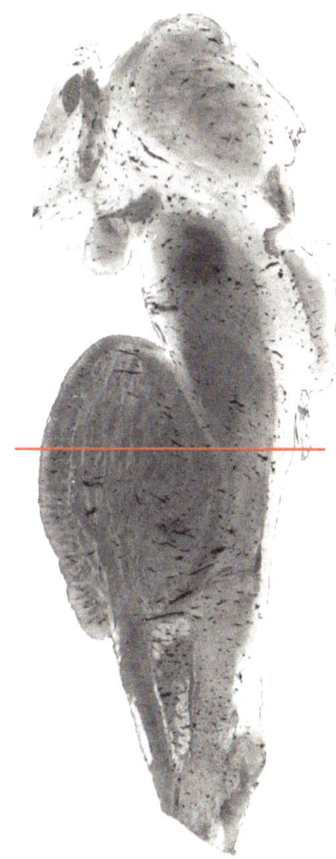

4V 4th ventricle
B9 B9 serotonin cells
ctg central tegmental tract
DRC dorsal raphe nu, caudal
DRI dorsal raphe, interfascicular
DTg dorsal tegmental nu
ILL intermed nu lat lemniscus
KF Kölliker-Fuse nu
LC locus coeruleus
LDTg laterodor tegmental nu
LDTgV laterodor tegmental nu, vent
lfp longitudinal fasciculus pons
ll lateral lemniscus
LPB lateral parabrachial nu
mcp middle cerebellar peduncle
me5 mesencephalic trigem tract
ml medial lemniscus
mlf med longitudinal fasciculus
MnR median raphe nu
MPB medial parabrachial nu
mtg mammillotegmental tract
PMnR paramedian raphe nu
Pn pontine nuclei
Pn/tfp pontine nu/transv fibres pons
PnO pontine reticular nu, oral
RtTg reticulotegmental nu pons
scp superior Cb peduncle
spth spinothalamic tract
SuL supralemniscal nucleus
tfp transverse fibers pons
ts tectospinal tr
tth trigeminothalamic tr
VLL ventral nu lat lemniscus
VLTg ventrolateral tegmental area

Fig 37b
Obex +29.0mm
Transverse 37

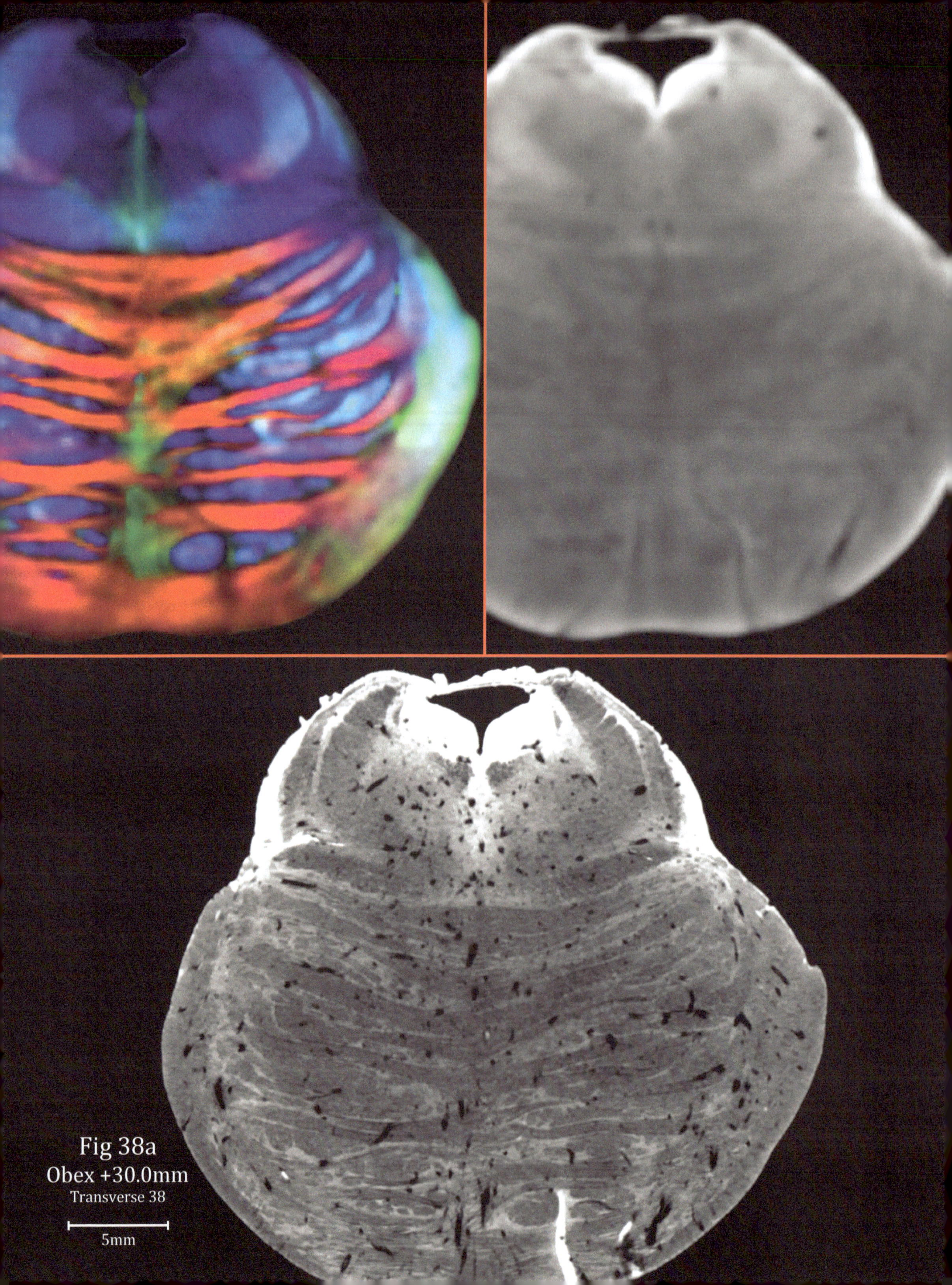
Fig 38a
Obex +30.0mm
Transverse 38
5mm

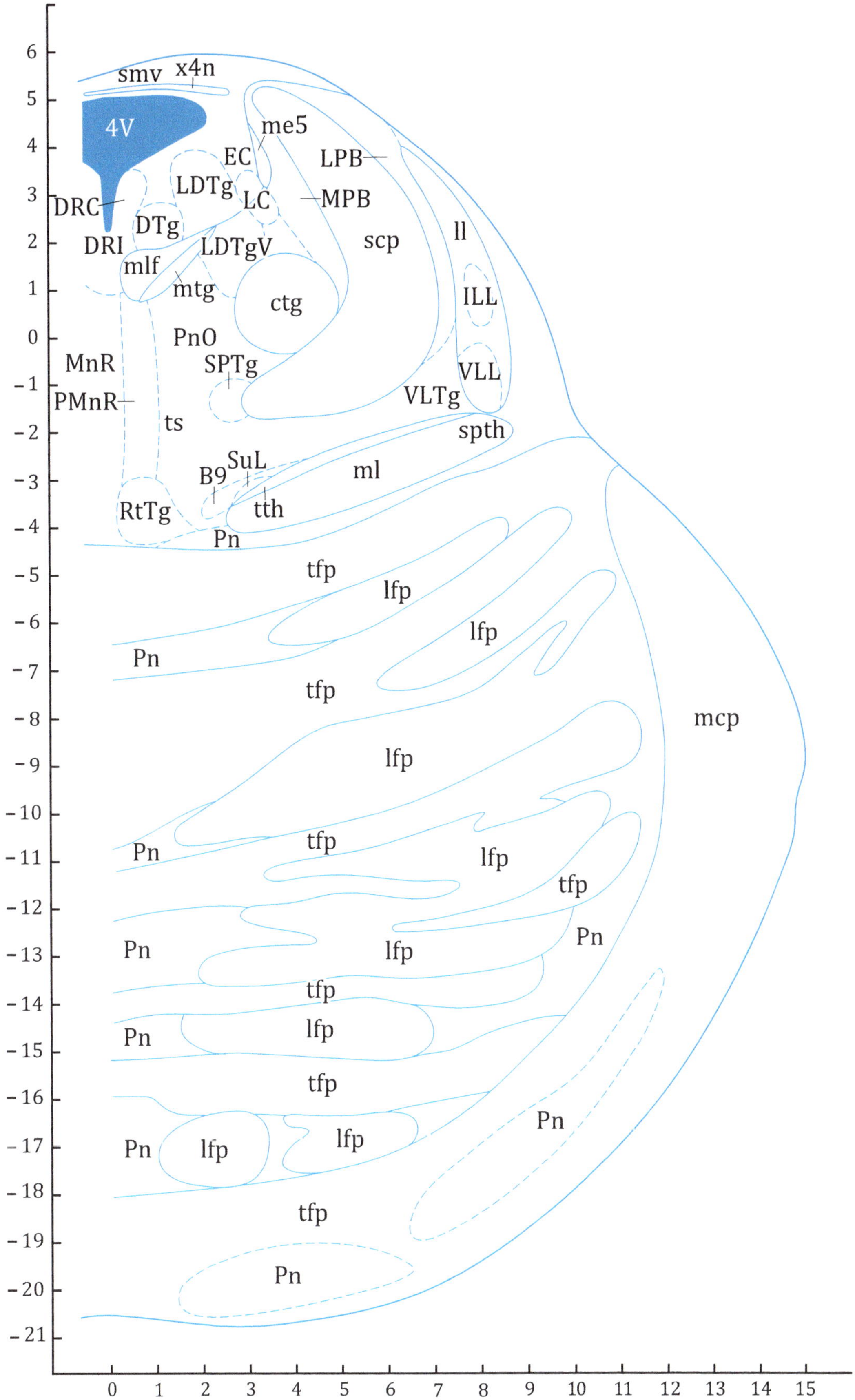

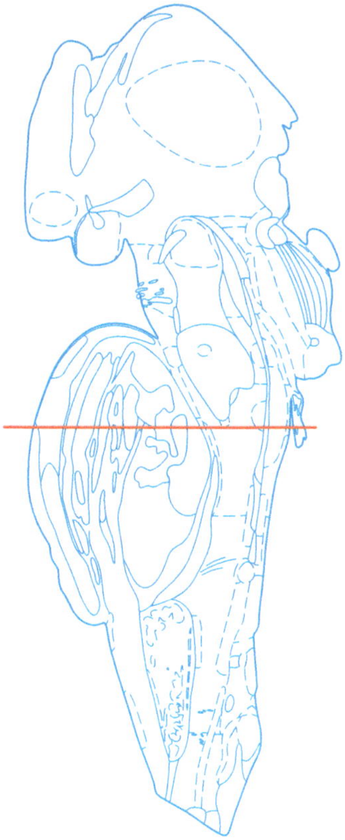

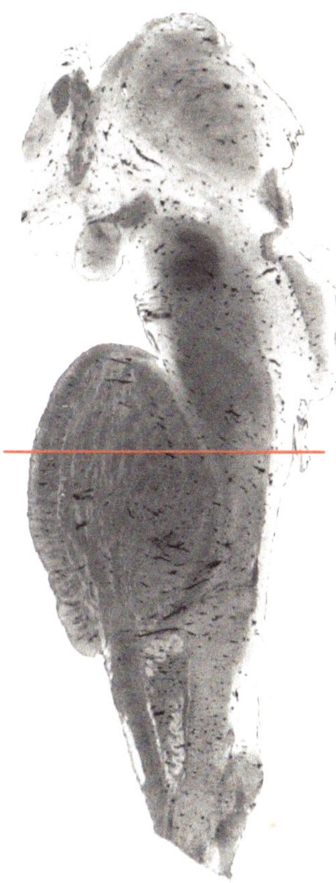

4V 4th ventricle
B9 B9 serotonin cells
ctg central tegmental tract
DRC dorsal raphe nu, caudal
DRI dorsal raphe, interfascicular
DTg dorsal tegmental nu
EC epicoeruleus nucleus
ILL intermed nu lat lemniscus
LC locus coeruleus
LDTg laterodor tegmental nu
LDTgV laterodor tegmental nu, vent
lfp longitudinal fasciculus pons
ll lateral lemniscus
LPB lateral parabrachial nu
mcp middle cerebellar peduncle
me5 mesencephalic trigem tract
ml medial lemniscus
mlf med longitudinal fasciculus
MnR median raphe nu
MPB medial parabrachial nu
mtg mammillotegmental tract
PMnR paramedian raphe nu
Pn pontine nuclei
PnO pontine reticular nu, oral
RtTg reticulotegmental nu pons
scp superior Cb peduncle
smv superior medullary velum
SPTg subpeduncular tegmental nu
spth spinothalamic tract
SuL supralemniscal nucleus
tfp transverse fibers pons
ts tectospinal tr
tth trigeminothalamic tr
VLL ventral nu lat lemniscus
VLTg ventrolateral tegmental area
x4n decussation of the trochlear nerve

Fig 38b
Obex +30.0mm
Transverse 38

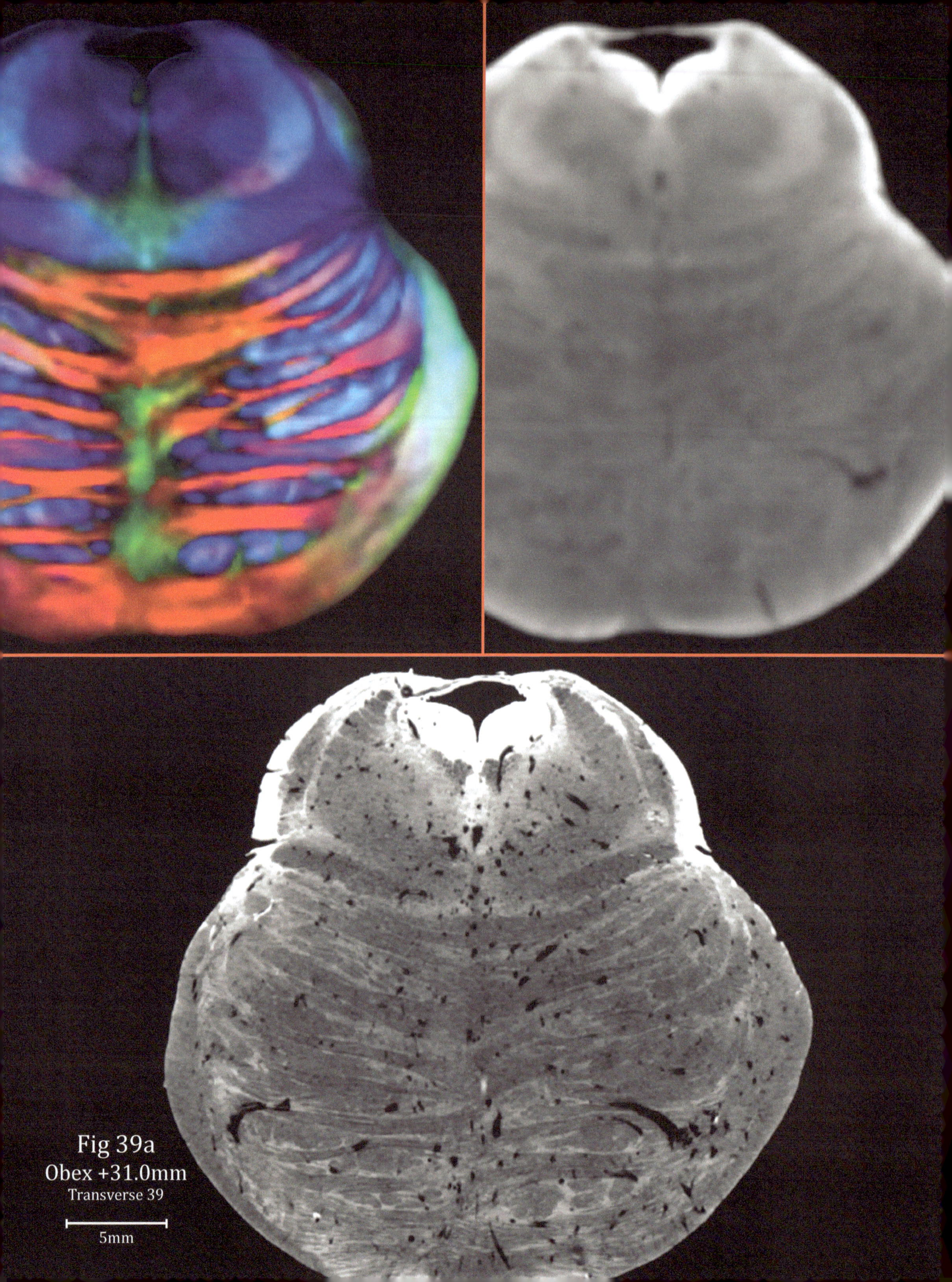

Fig 39a
Obex +31.0mm
Transverse 39

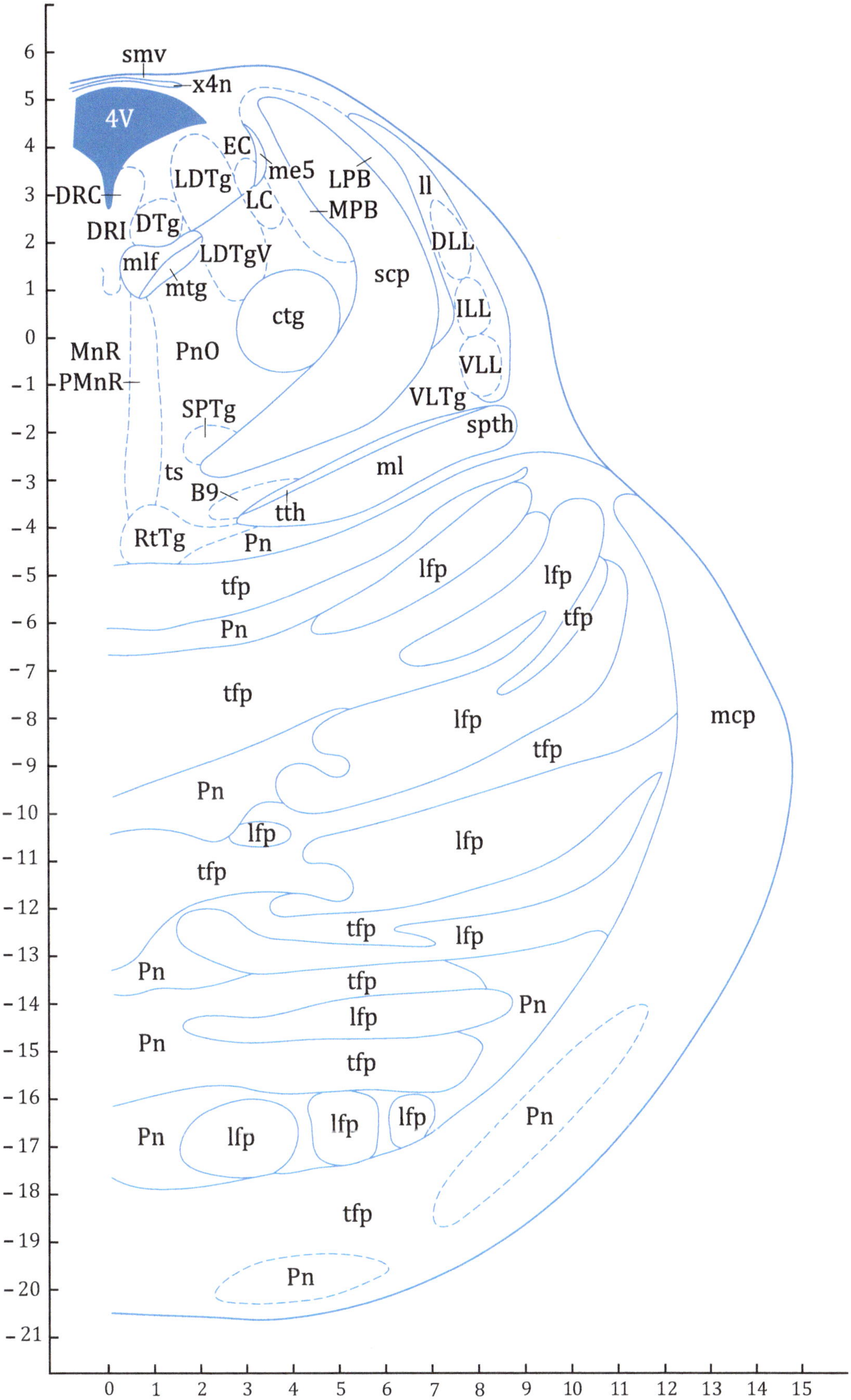

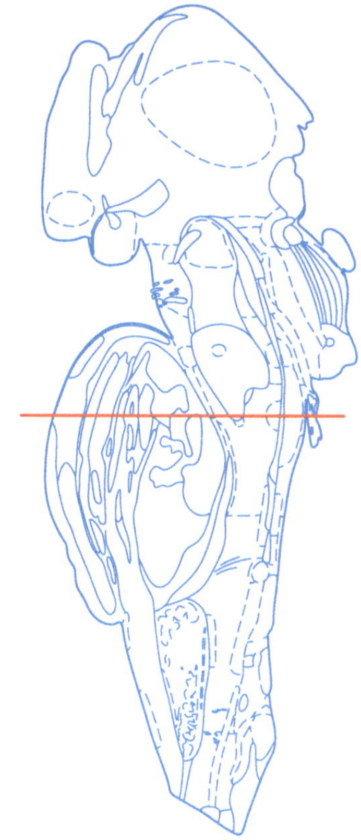

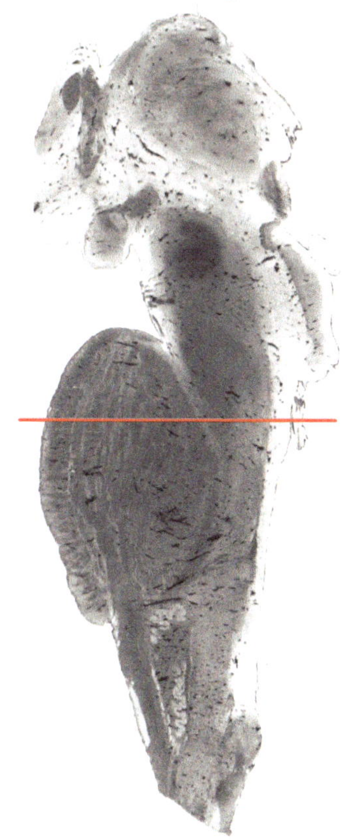

4V 4th ventricle
B9 B9 serotonin cells
ctg central tegmental tract
DLL dorsal nu lateral lemniscus
DRC dorsal raphe nu, caudal
DRI dorsal raphe, interfascicular
DTg dorsal tegmental nu
EC epicoeruleus nucleus
ILL intermed nu lat lemniscus
LC locus coeruleus
LDTg laterodor tegmental nu
LDTgV laterodor tegmental nu, vent
lfp longitudinal fasciculus pons
ll lateral lemniscus
LPB lateral parabrachial nu
mcp middle cerebellar peduncle
me5 mesencephalic trigem tract
ml medial lemniscus
mlf med longitudinal fasciculus
MnR median raphe nu
MPB medial parabrachial nu
mtg mammillotegmental tract
PMnR paramedian raphe nu
Pn pontine nuclei
PnO pontine reticular nu, oral
RtTg reticulotegmental nu pons
scp superior Cb peduncle
smv superior medullary velum
SPTg subpeduncular tegmental nu
spth spinothalamic tract
tfp transverse fibers pons
ts tectospinal tr
tth trigeminothalamic tr
VLL ventral nu lat lemniscus
VLTg ventrolateral tegmental area
x4n decussation of the trochlear nerve

Fig 39b
Obex +31.0mm
Transverse 39

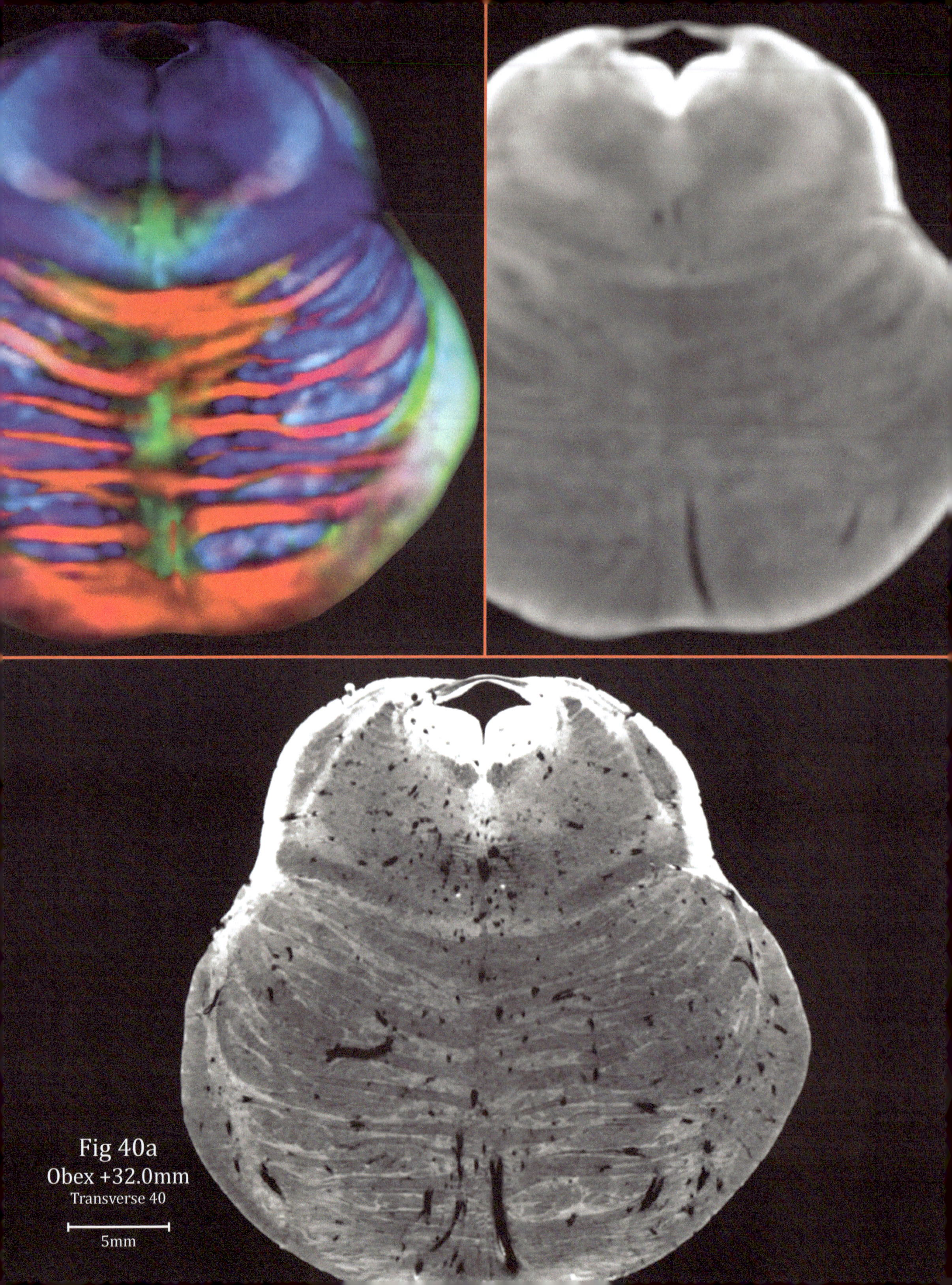

Fig 40a
Obex +32.0mm
Transverse 40

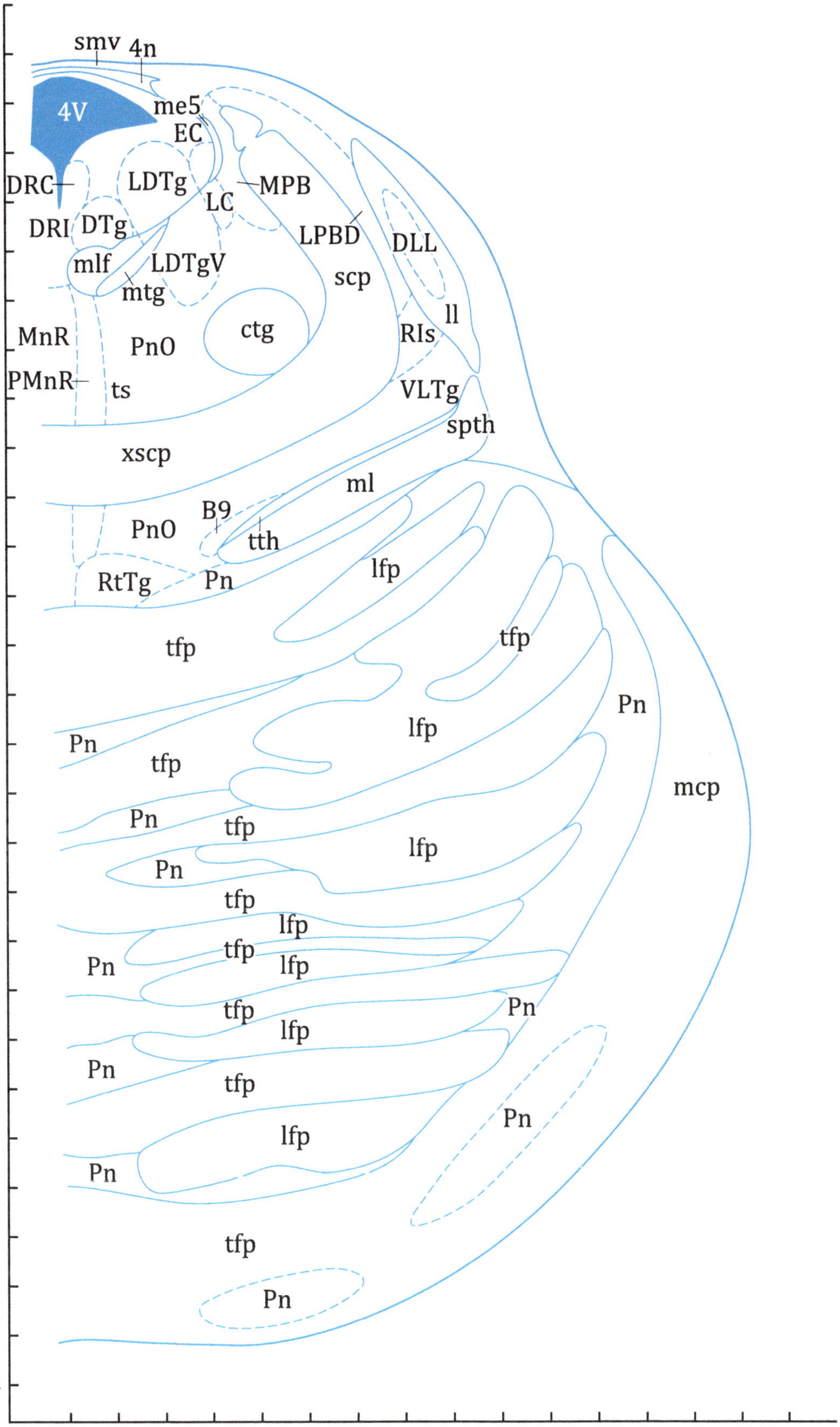

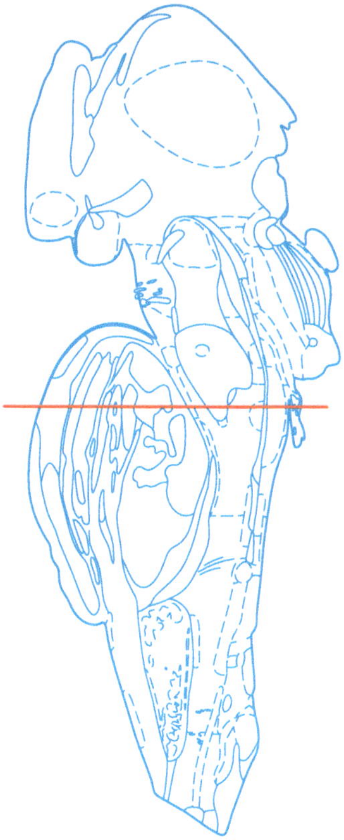

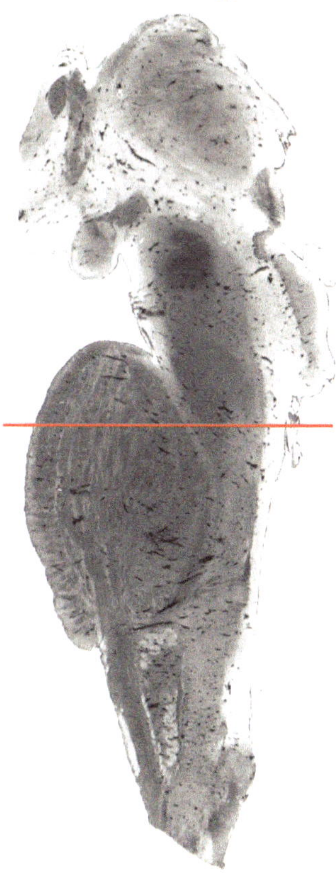

4n trochlear nerve
4V 4th ventricle
B9 B9 serotonin cells
ctg central tegmental tract
DLL dorsal nu lateral lemniscus
DRC dorsal raphe nu, caudal
DRI dorsal raphe, interfascicular
DTg dorsal tegmental nu
EC epicoeruleus nucleus
LC locus coeruleus
LDTg laterodor tegmental nu
LDTgV laterodor tegmental nu, vent
lfp longitudinal fasciculus pons
ll lateral lemniscus
LPBD lat parabrachial nu, dors
mcp middle cerebellar peduncle
me5 mesencephalic trigem tract
ml medial lemniscus
mlf med longitudinal fasciculus
MnR median raphe nu
MPB medial parabrachial nu
mtg mammillotegmental tract
PMnR paramedian raphe nu
Pn pontine nuclei
PnO pontine reticular nu, oral
RIs retroisthmic nucleus
RtTg reticulotegmental nu pons
scp superior Cb peduncle
smv superior medullary velum
spth spinothalamic tract
tfp transverse fibers pons
ts tectospinal tr
tth trigeminothalamic tr
VLTg ventrolateral tegmental area
xscp decussation sup cereb ped

Fig 40b
Obex +32.0mm
Transverse 40

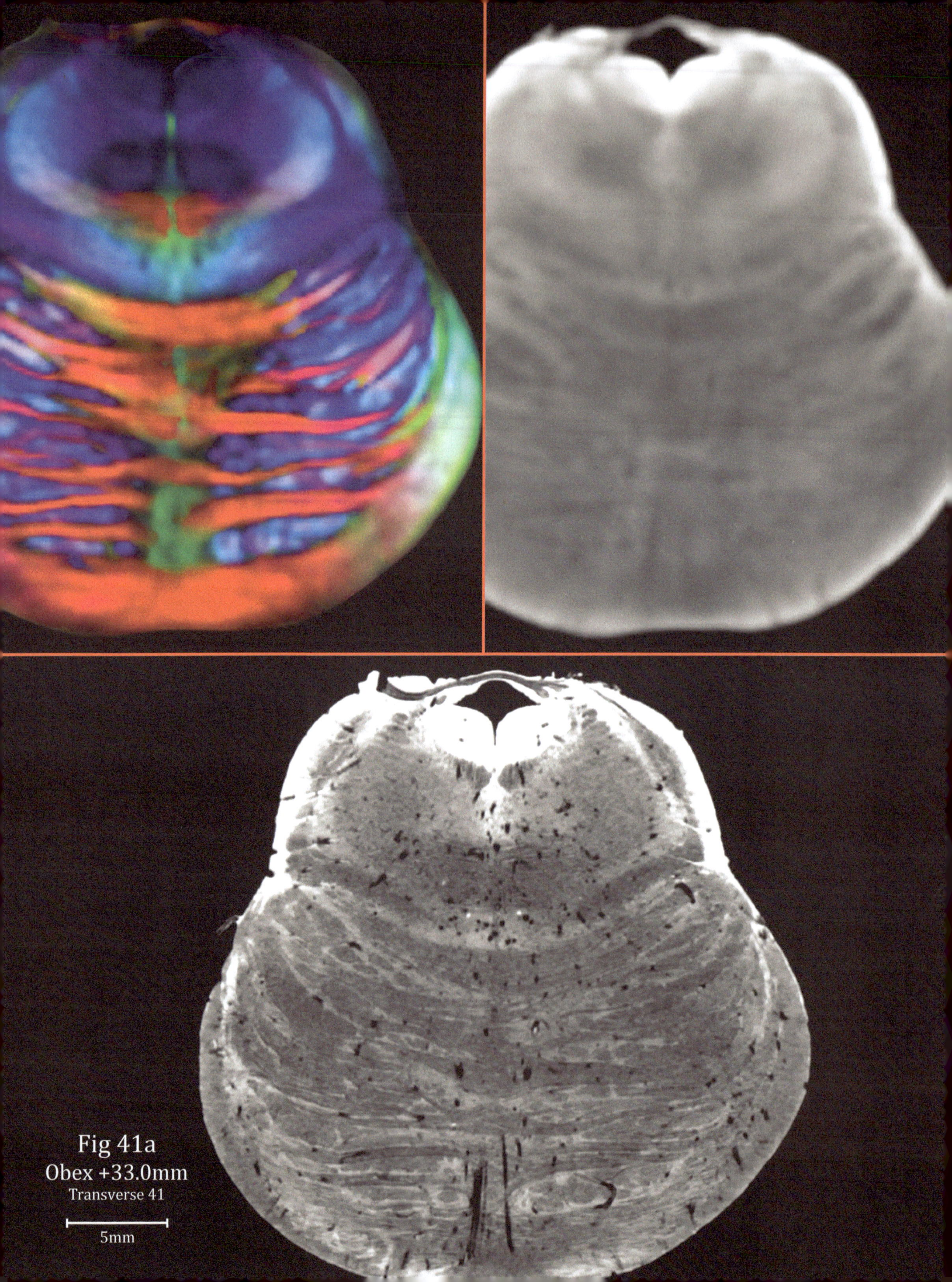

Fig 41a
Obex +33.0mm
Transverse 41
5mm

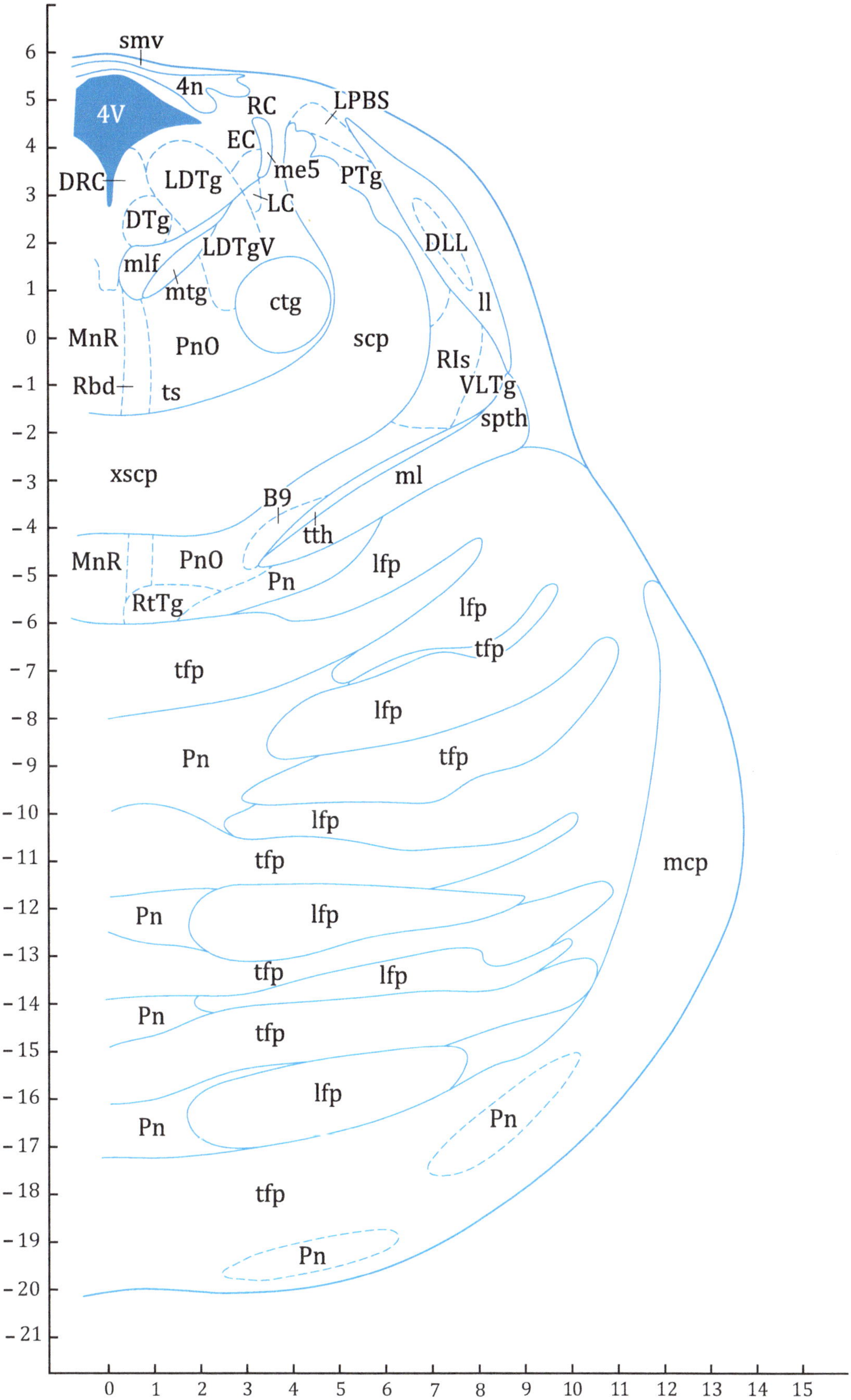

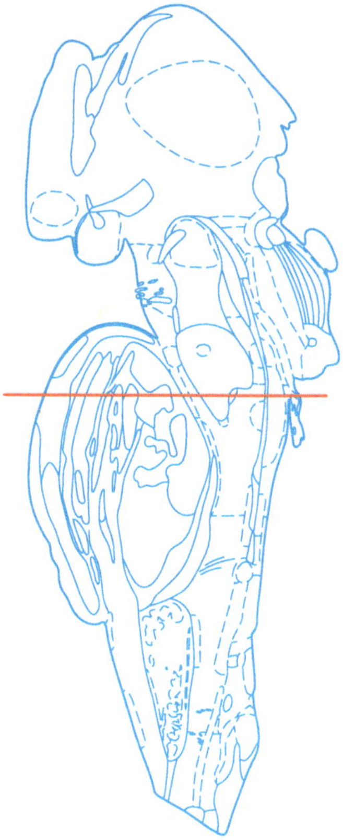

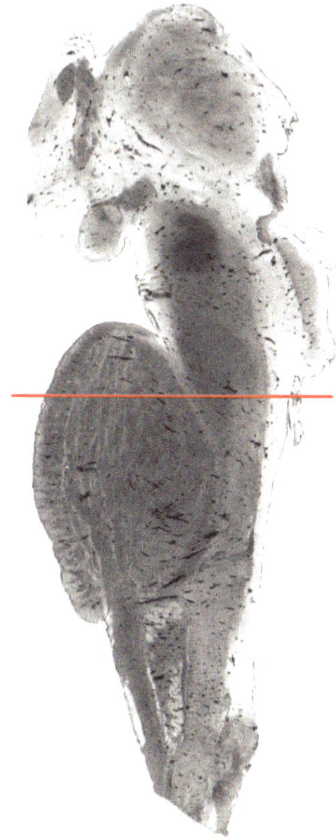

4n trochlear nerve
4V 4th ventricle
B9 B9 serotonin cells
ctg central tegmental tract
DLL dorsal nu lateral lemniscus
DRC dorsal raphe nu, caudal
DTg dorsal tegmental nu
EC epicoeruleus nucleus
LC locus coeruleus
LDTg laterodor tegmental nu
LDTgV laterodor tegmental nu, vent
lfp longitudinal fasciculus pons
ll lateral lemniscus

LPBS lat parabrachial nu, sup
mcp middle cerebellar peduncle
me5 mesencephalic trigem tract
ml medial lemniscus
mlf med longitudinal fasciculus
MnR median raphe nu
mtg mammillotegmental tract
Pn pontine nuclei
PnO pontine reticular nu, oral
PTg pedunculopontine tegmental
Rbd rhabdoid nu
RC raphe cap
RIs retroisthic nucleus

RtTg reticulotegmental nu pons
scp superior Cb peduncle
smv superior medullary velum
spth spinothalamic tract
tfp transverse fibers pons
ts tectospinal tr
tth trigeminothalamic tr
VLTg ventrolateral tegmental area
xscp decussation sup cereb ped

Fig 41b
Obex +33.0mm
Transverse 41

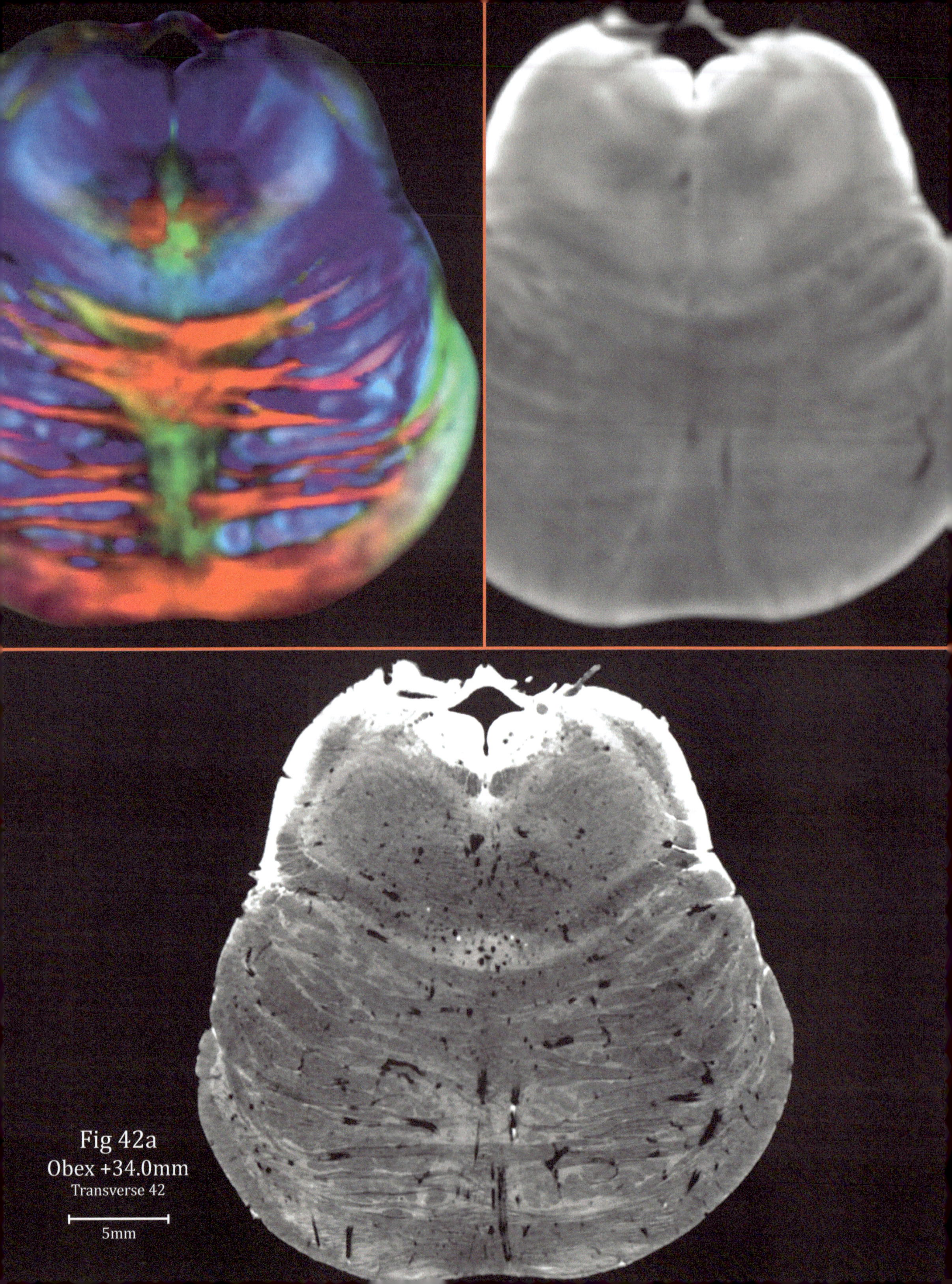

Fig 42a
Obex +34.0mm
Transverse 42

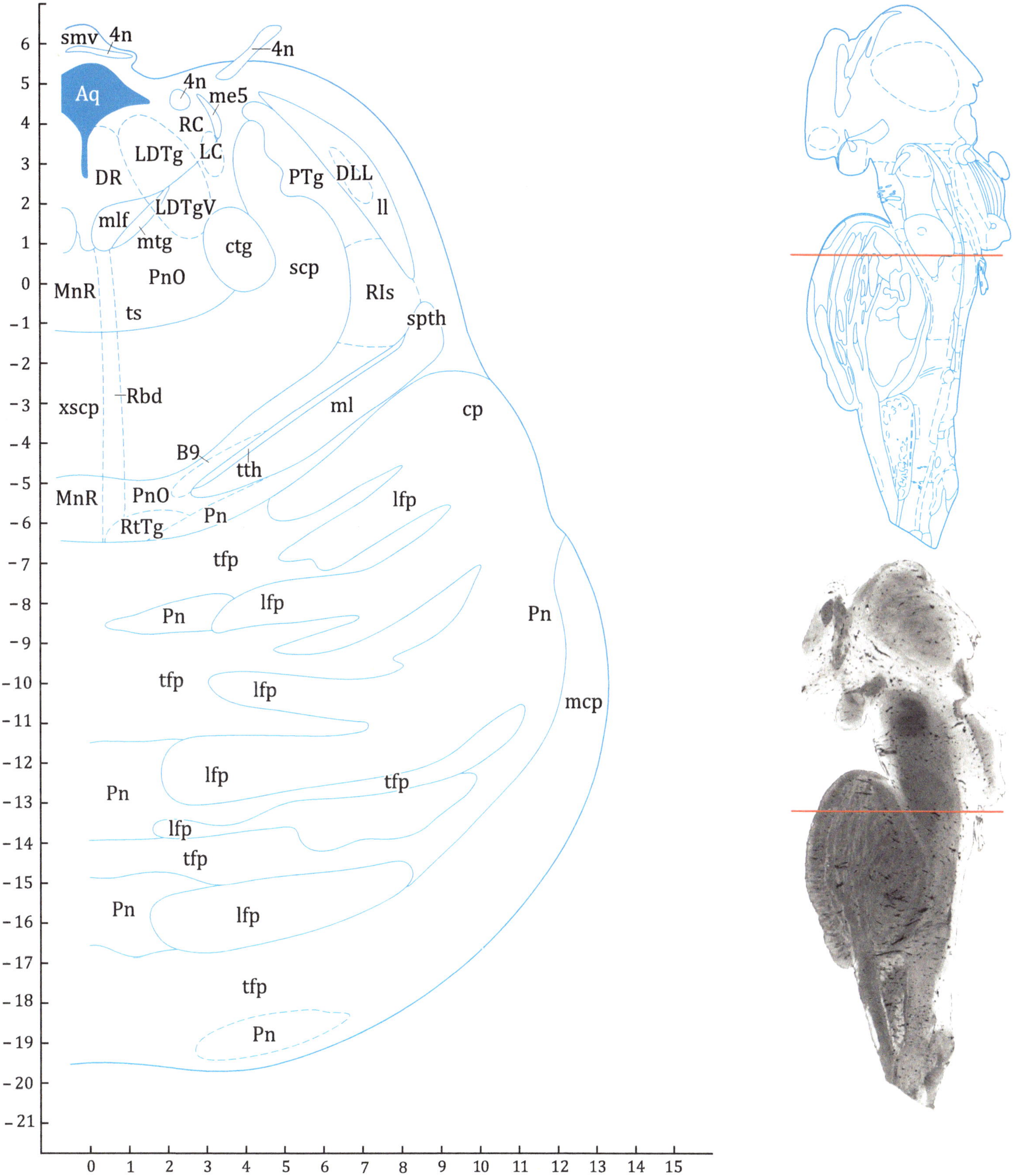

4n trochlear nerve
Aq aqueduct
B9 B9 serotonin cells
cp cerebral peduncle
ctg central tegmental tract
DLL dorsal nu lateral lemniscus
DR dorsal raphe nu
LC locus coeruleus
LDTg laterodor tegmental nu
LDTgV laterodor tegmental nu, vent
lfp longitudinal fasciculus pons
ll lateral lemniscus
mcp middle cerebellar peduncle
me5 mesencephalic trigem tract
ml medial lemniscus
mlf med longitudinal fasciculus
MnR median raphe nu
mtg mammillotegmental tract
Pn pontine nuclei
PnO pontine reticular nu, oral
PTg pedunculopontine tegmental
Rbd rhabdoid nu
RC raphe cap
RIs retroisthic nucleus
RtTg reticulotegmental nu pons
scp superior Cb peduncle
smv superior medullary velum
spth spinothalamic tract
tfp transverse fibers pons
ts tectospinal tr
tth trigeminothalamic tr
xscp decussation sup cereb ped

Fig 42b
Obex +34.0mm
Transverse 42

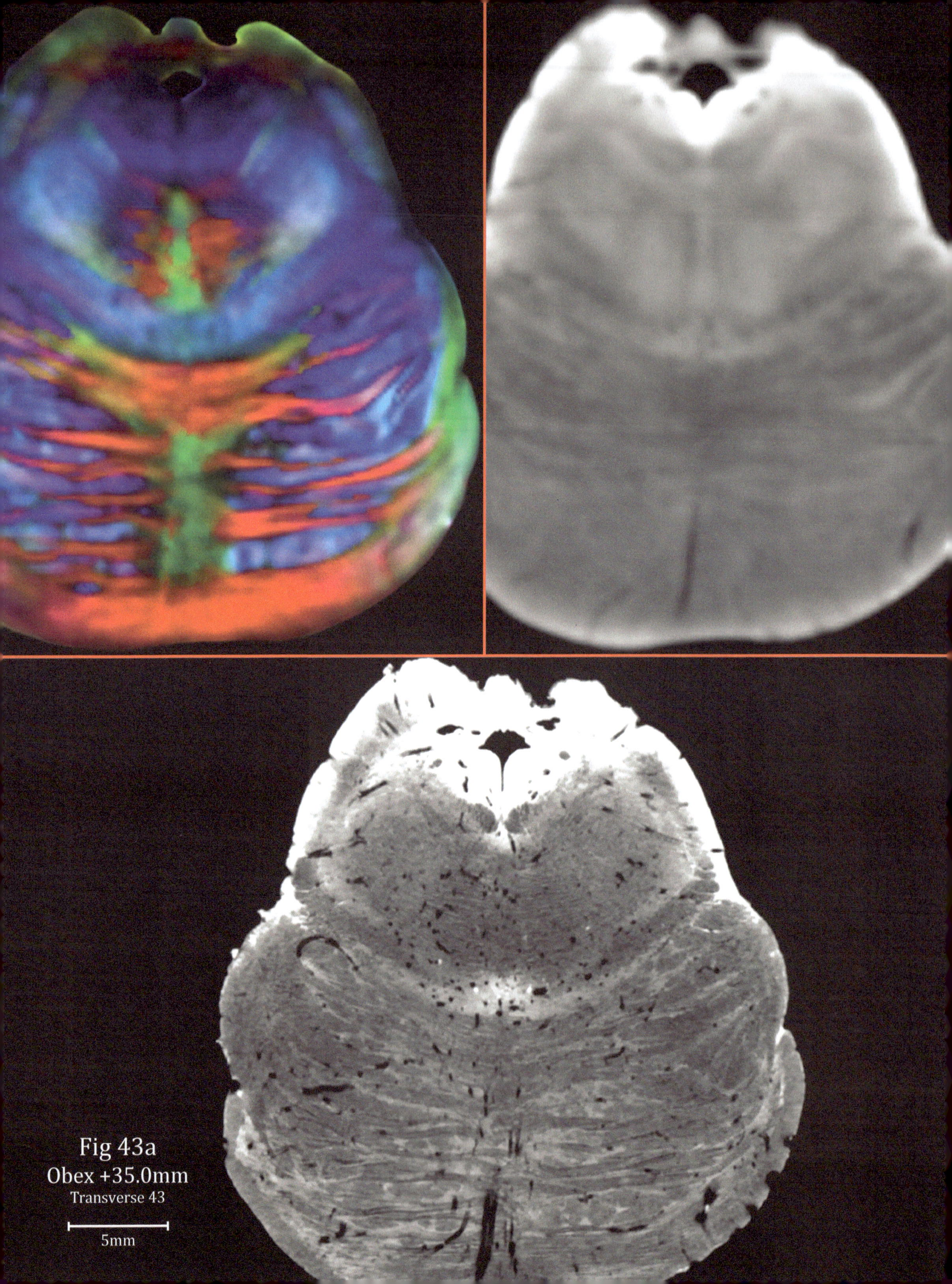

Fig 43a
Obex +35.0mm
Transverse 43

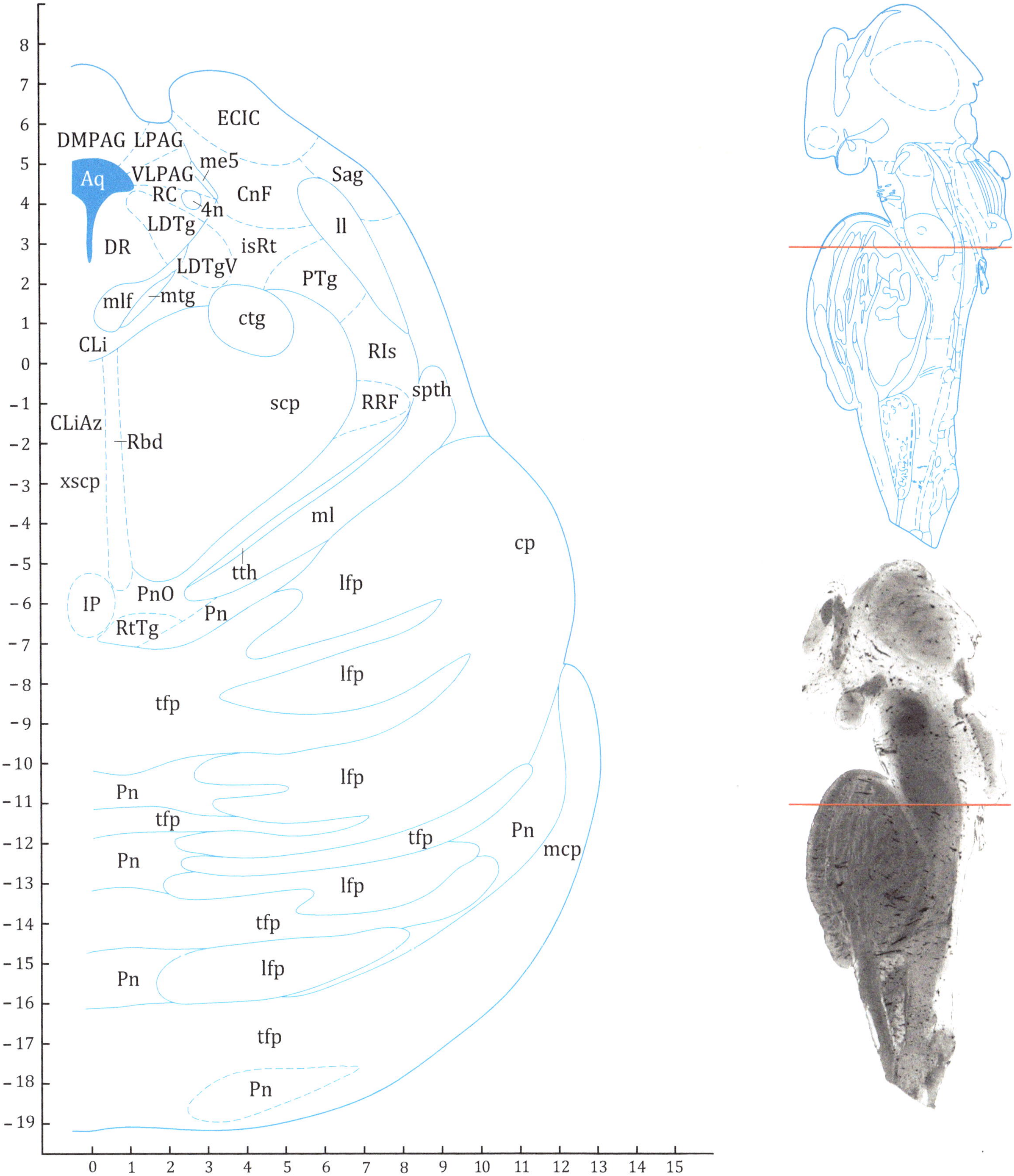

4n trochlear nerve
Aq aqueduct
CLi caudal linear nu raphe
CLiAz caudal linear nu of raphe, azygos
CnF cuneiform nu
cp cerebral peduncle
ctg central tegmental tract
DMPAG dorsomed periaqueductl gray
DR dorsal raphe nu
ECIC external cx inferior coll
IP interpedunc nu
isRt isthmic reticular formation
LDTg laterodor tegmental nu
LDTgV laterodor tegmental nu, vent
lfp longitudinal fasciculus pons
ll lateral lemniscus
LPAG lat periaqueductal gray
mcp middle cerebellar peduncle
me5 mesencephalic trigem tract
ml medial lemniscus
mlf med longitudinal fasciculus
mtg mammillotegmental tract
Pn pontine nuclei
PnO pontine reticular nu, oral
PTg pedunculopontine tegmental
Rbd rhabdoid nu
RC raphe cap
RIs retroisthic nucleus
RRF retrorubral field
RtTg reticulotegmental nu pons
Sag sagulum nu
scp superior Cb peduncle
spth spinothalamic tract
tfp transverse fibers pons
tth trigeminothalamic tr
VLPAG ventlat periaqueductal gray
xscp decussation sup cereb ped

Fig 43b
Obex +35.0mm
Transverse 43

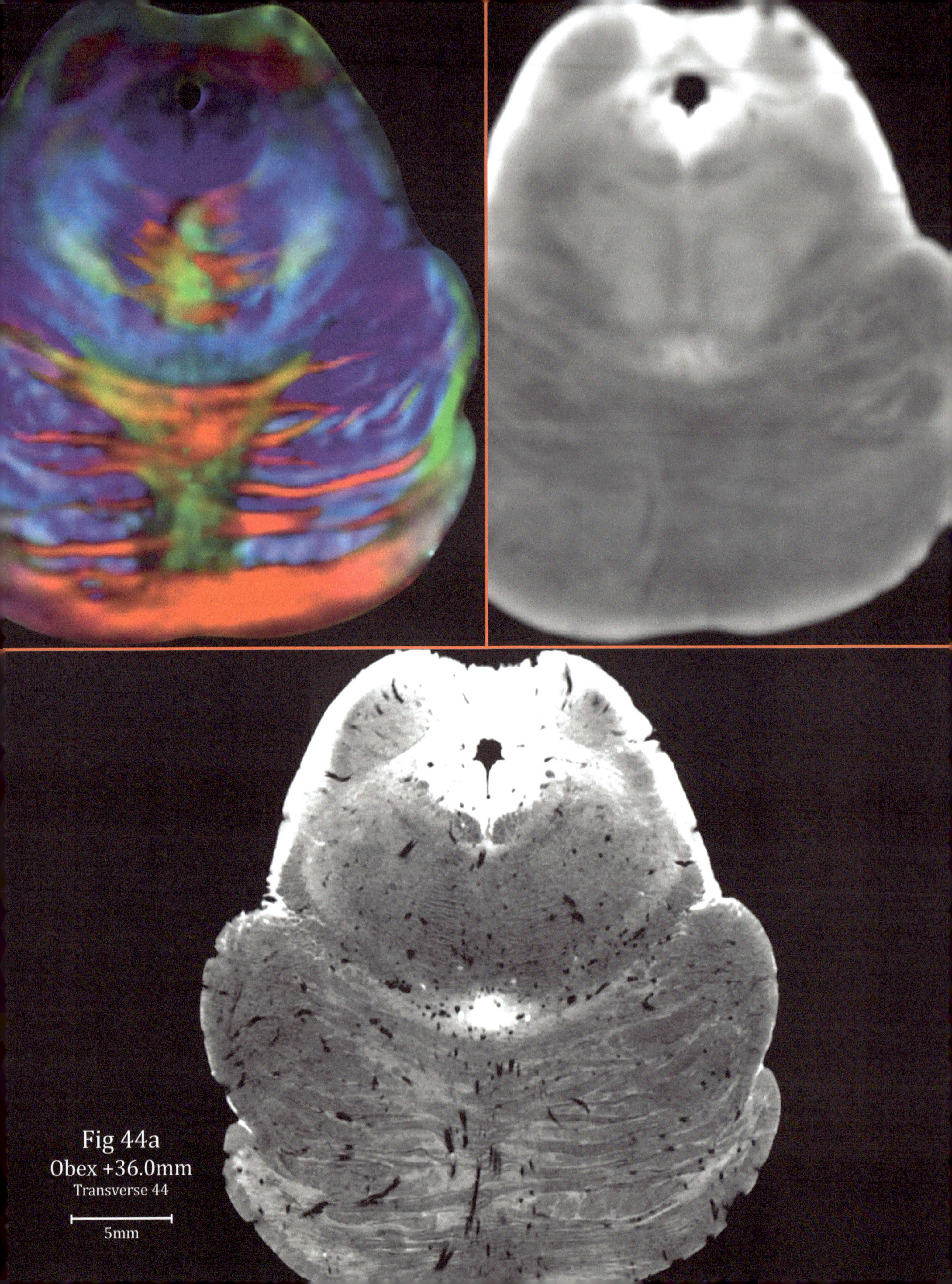

Fig 44a
Obex +36.0mm
Transverse 44

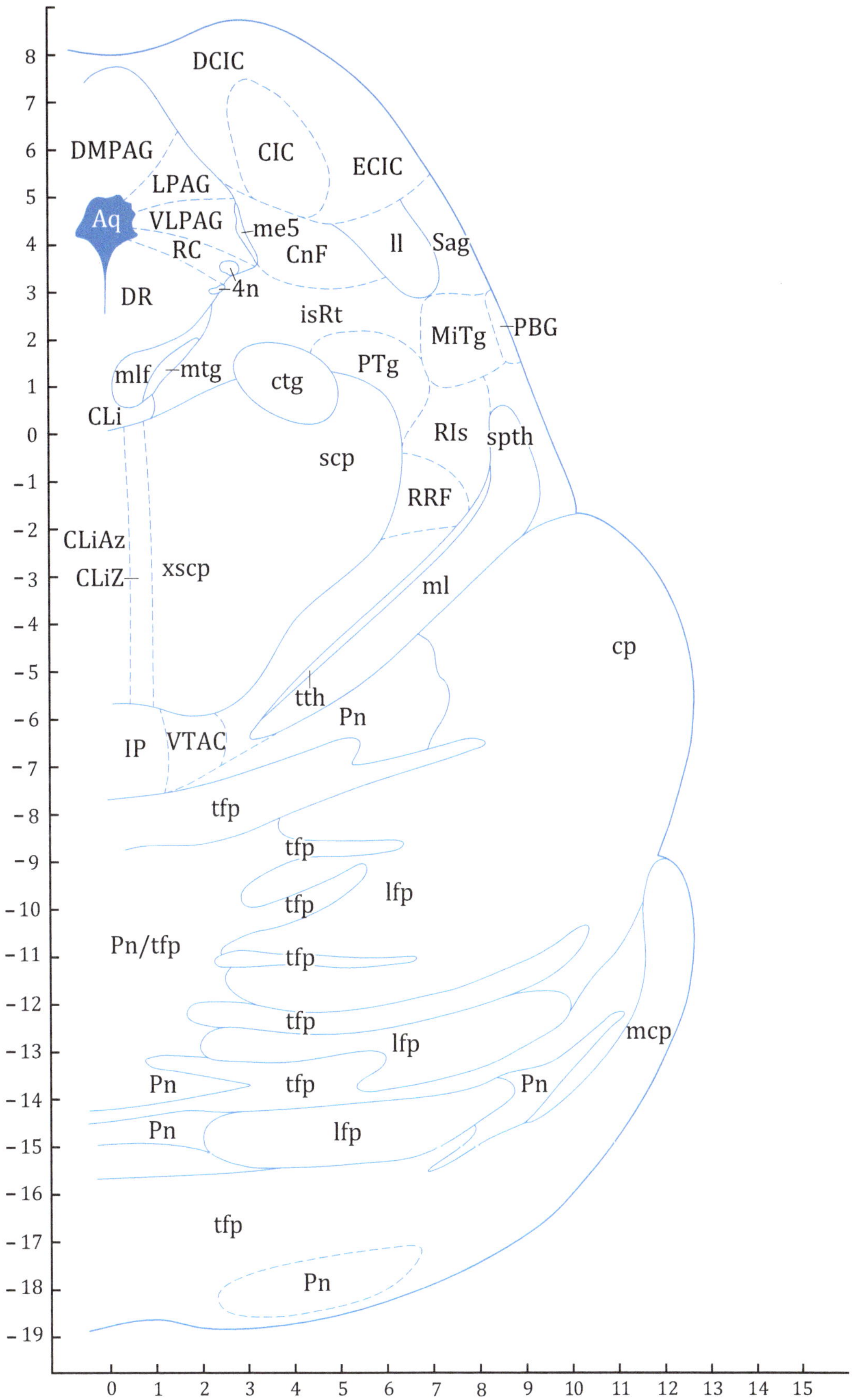

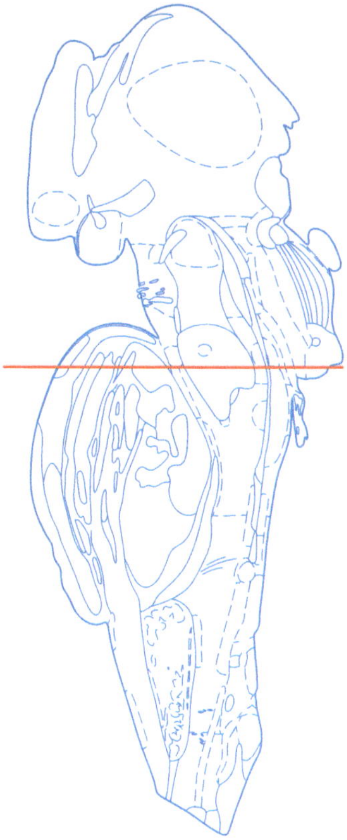

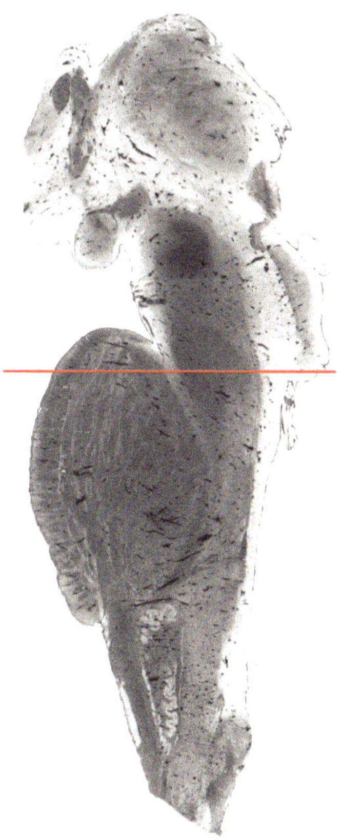

4n trochlear nerve
Aq aqueduct
CIC central nu inf colliculus
CLi caudal linear nu raphe
CLiAz caudal linear nu of raphe, azygos
CLiZ caudal linear nu of raphe, zygos
CnF cuneiform nu
cp cerebral peduncle
ctg central tegmental tract
DCIC dors cx inf colliculus
DMPAG dorsomed periaqueductl gray
DR dorsal raphe nu
ECIC external cx inferior coll
IP interpedunc nu
isRt isthmic reticular formation
lfp longitudinal fasciculus pons
ll lateral lemniscus
LPAG lat periaqueductal gray
mcp middle cerebellar peduncle
me5 mesencephalic trigem tract
MiTg microcell tegmental nu
ml medial lemniscus
mlf med longitudinal fasciculus
mtg mammillotegmental tract
PBG parabigeminal nu
Pn pontine nuclei
Pn/tfp pontine nu/transv fibres pons
PTg pedunculopontine tegmental
RC raphe cap
RIs retroisthmic nucleus
RRF retrorubral field
Sag sagulum nu
scp superior Cb peduncle
spth spinothalamic tract
tfp transverse fibers pons
tth trigeminothalamic tr
VLPAG ventlat periaqueductal gray
VTAC vent tegmental area, caudal
xscp decussation sup cereb ped

Fig 44b
Obex +36.0mm
Transverse 44

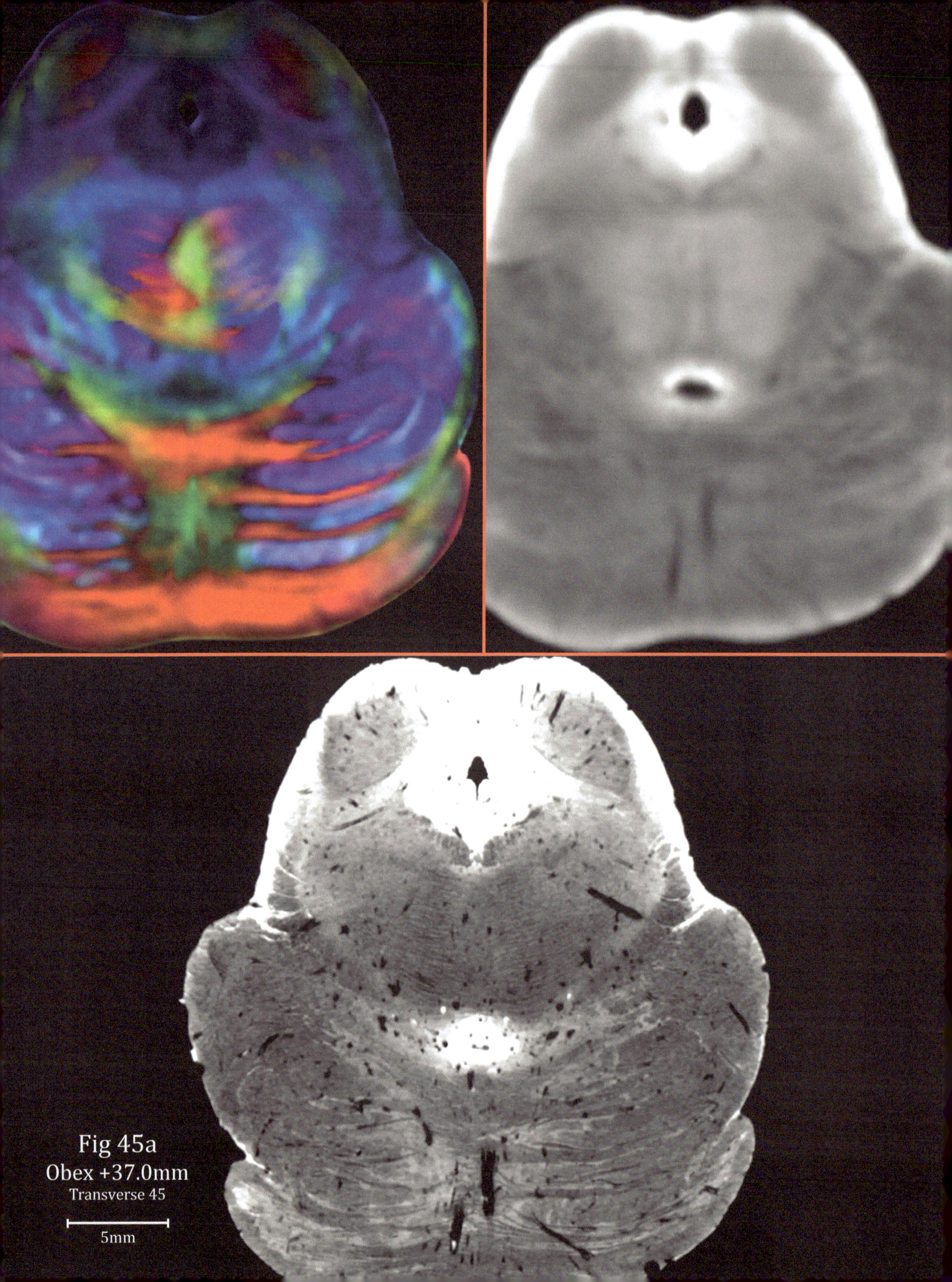

Fig 45a
Obex +37.0mm
Transverse 45
5mm

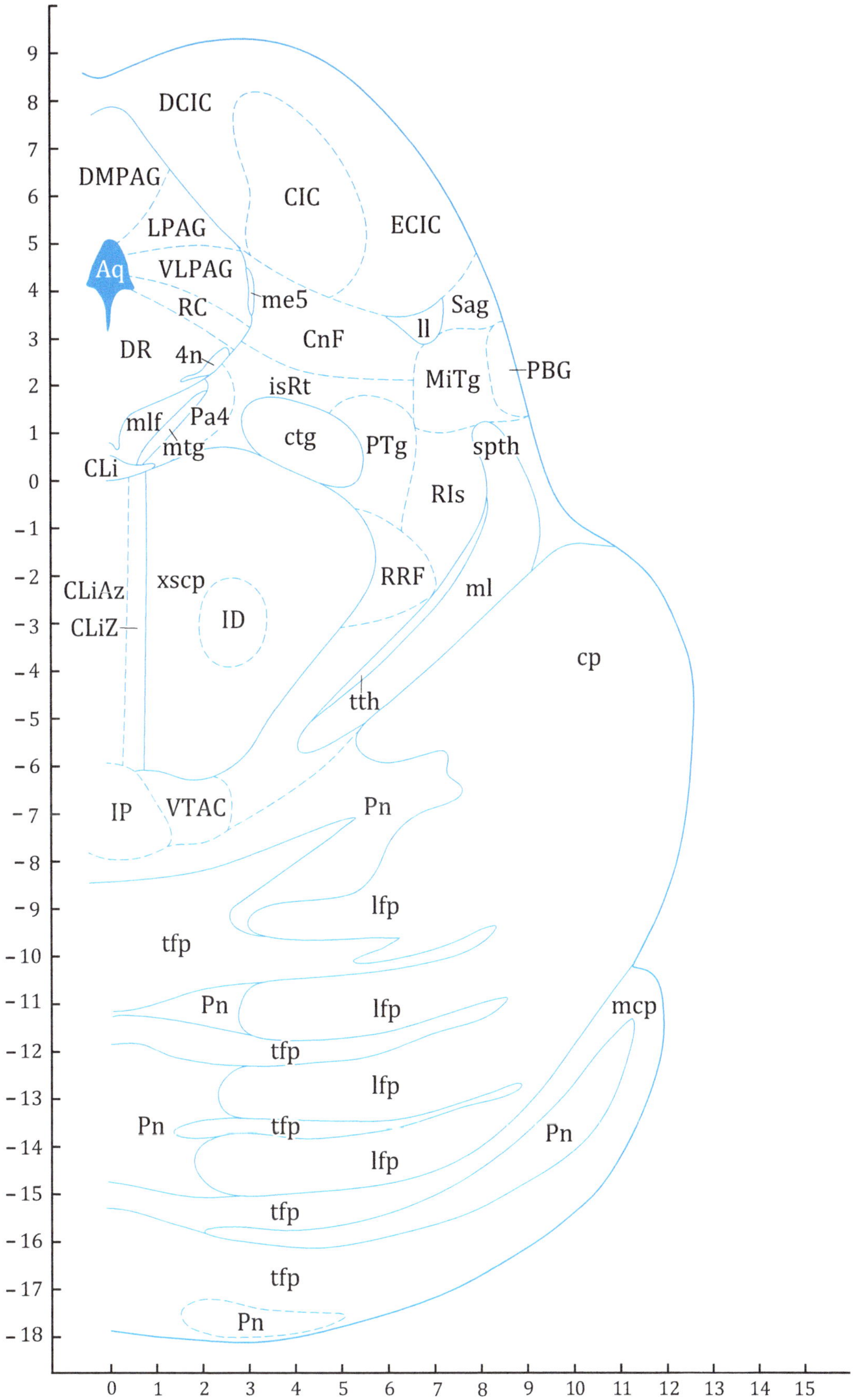

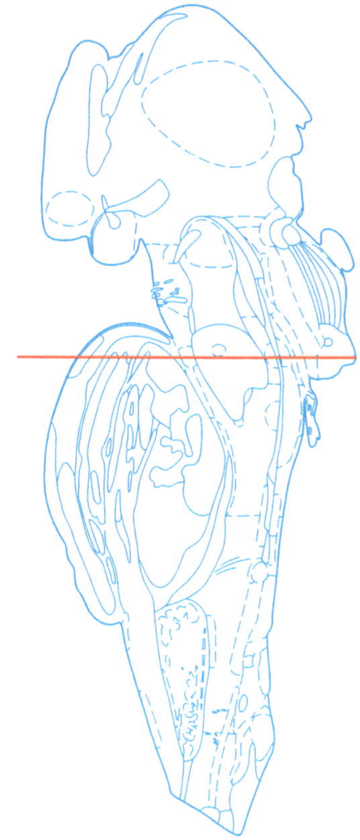

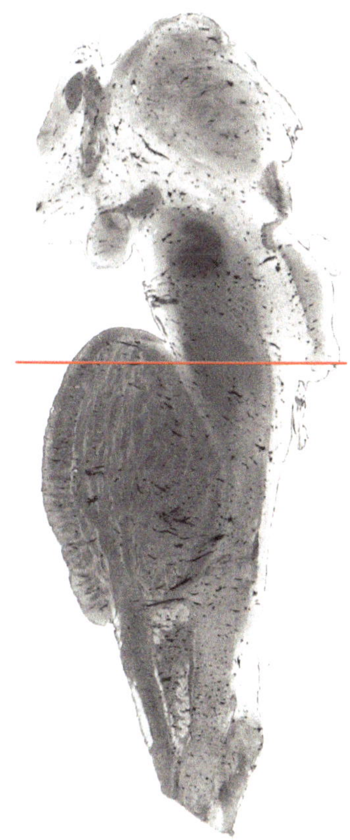

4n trochlear nerve
Aq aqueduct
CIC central nu inf colliculus
CLi caudal linear nu raphe
CLiAz caudal linear nu of raphe, azygos
CLiZ caudal linear nu of raphe, zygos
CnF cuneiform nu
cp cerebral peduncle
ctg central tegmental tract
DCIC dors cx inf colliculus
DMPAG dorsomed periaqueductl gray
DR dorsal raphe nu
ECIC external cx inferior coll
ID intradecussational nu scp
IP interpedunc nu
isRt isthmic reticular formation
lfp longitudinal fasciculus pons
ll lateral lemniscus
LPAG lat periaqueductal gray
mcp middle cerebellar peduncle
me5 mesencephalic trigem tract
MiTg microcell tegmental nu
ml medial lemniscus
mlf med longitudinal fasciculus
mtg mammillotegmental tract
Pa4 paratrochlear nu
PBG parabigeminal nu
Pn pontine nuclei
PTg pedunculopontine tegmental
RC raphe cap
RIs retroisthic nucleus
RRF retrorubral field
Sag sagulum nu
spth spinothalamic tract
tfp transverse fibers pons
tth trigeminothalamic tr
VLPAG ventlat periaqueductal gray
VTAC vent tegmental area, caudal
xscp decussation sup cereb ped

Fig 45b
Obex +37.0mm
Transverse 45

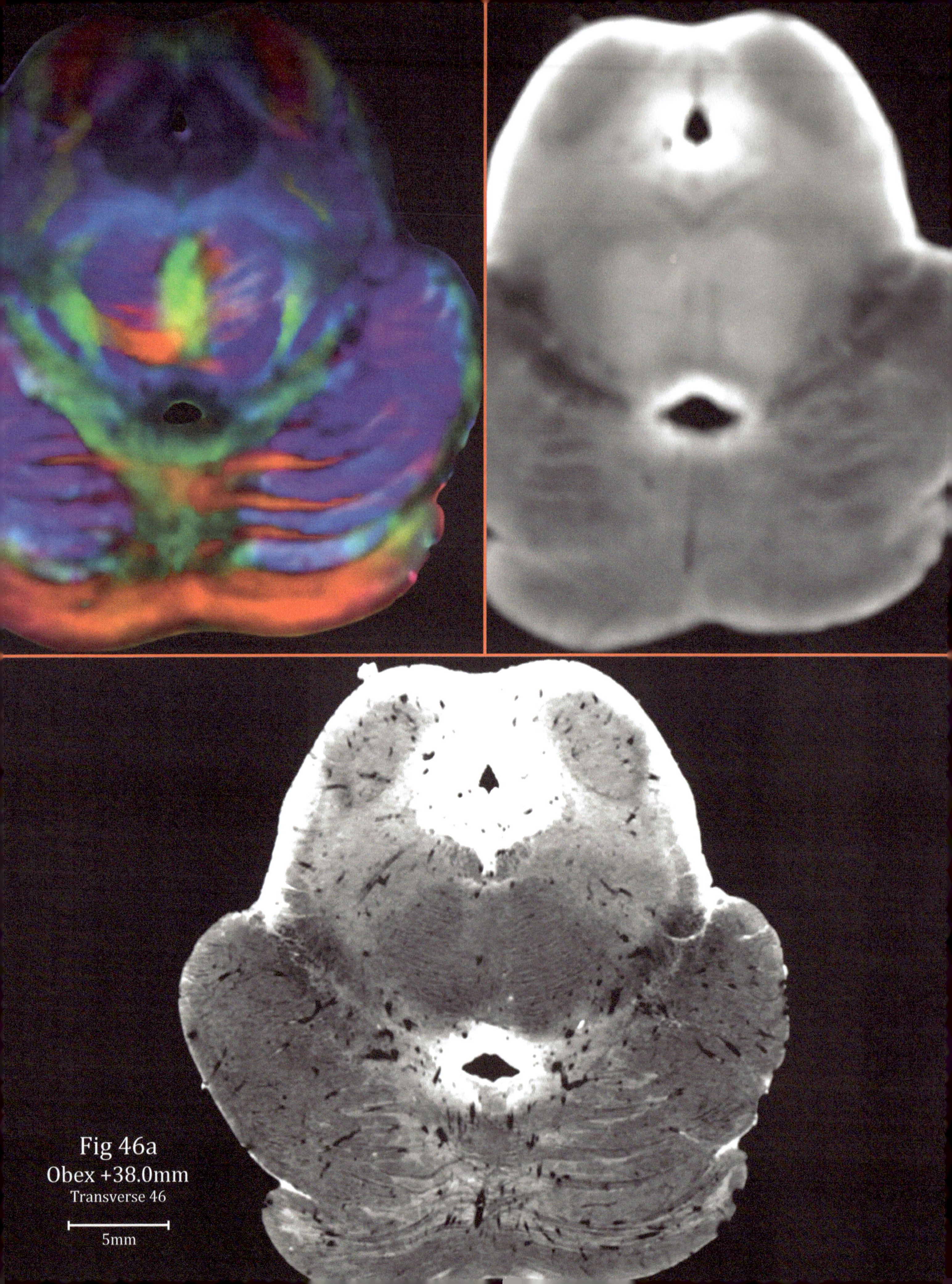

Fig 46a
Obex +38.0mm
Transverse 46

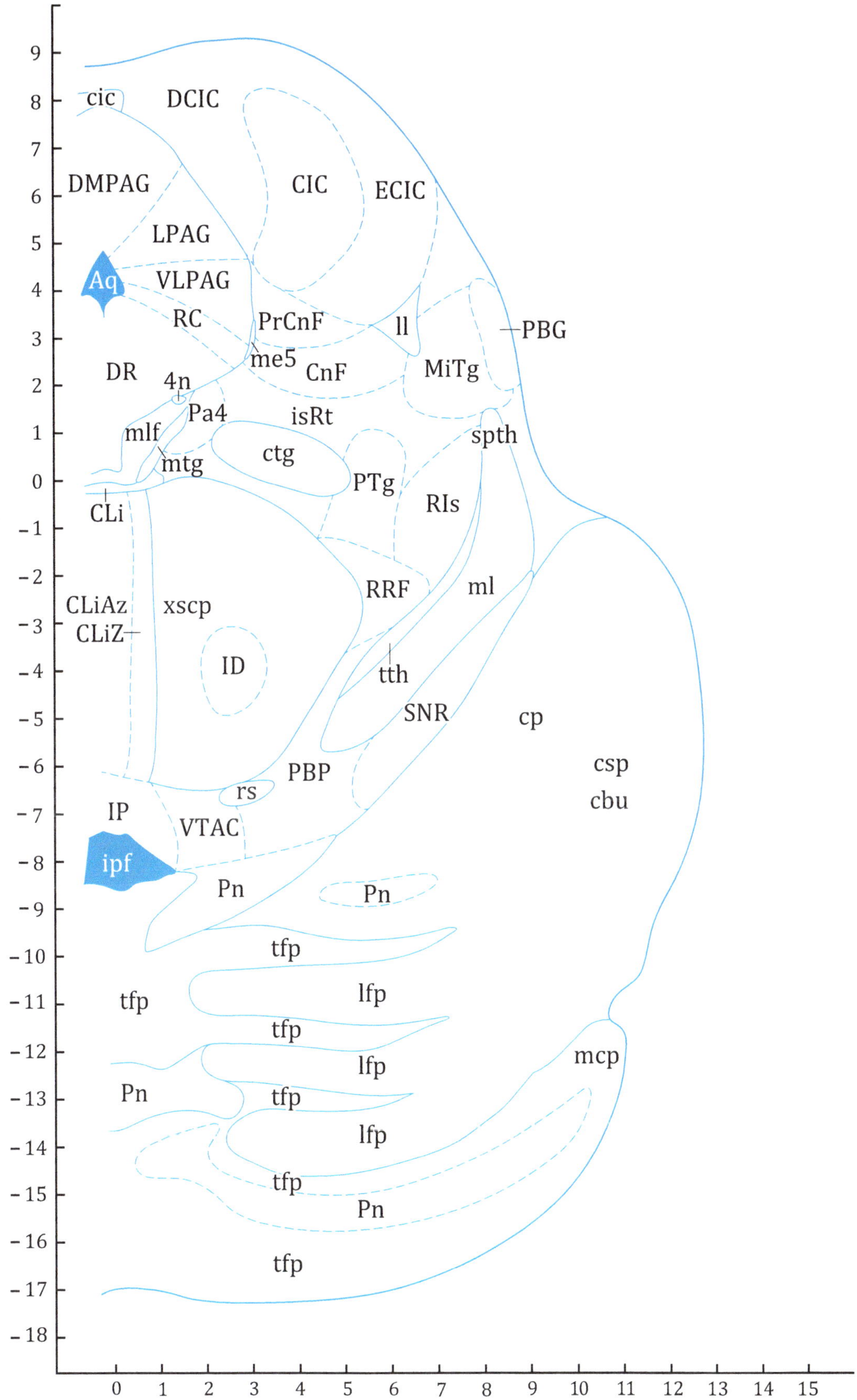

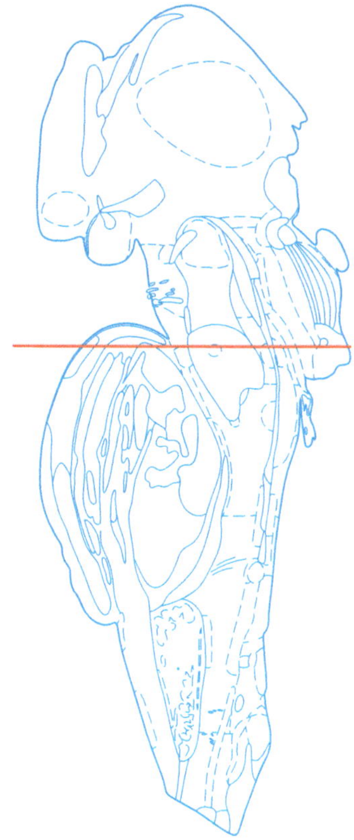

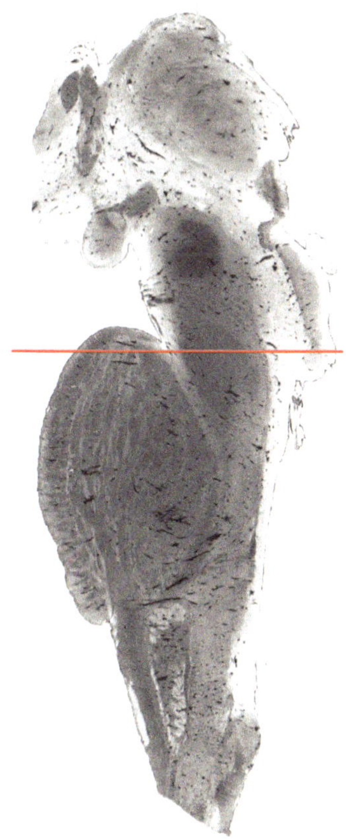

4n trochlear nerve
Aq aqueduct
cbu corticobulbar tract
cic commissure inf colliculus
CIC central nu inf colliculus
CLi caudal linear nu raphe
CLiAz caudal linear nu of raphe, azygos
CLiZ caudal linear nu of raphe, zygos
CnF cuneiform nu
cp cerebral peduncle
csp corticospinal tract
ctg central tegmental tract
DCIC dors cx inf colliculus
DMPAG dorsomed periaqueductl gray
DR dorsal raphe nu
ECIC external cx inferior coll
ID intradecussational nu scp
IP interpedunc nu
ipf interpedunc fossa
isRt isthmic reticular formation
lfp longitudinal fasciculus pons
ll lateral lemniscus
LPAG lat periaqueductal gray
mcp middle cerebellar peduncle
me5 mesencephalic trigem tract
MiTg microcell tegmental nu
ml medial lemniscus
mlf med longitudinal fasciculus
mtg mammillotegmental tract
Pa4 paratrochlear nu
PBG parabigeminal nu
PBP parabrachial pigmented nu
Pn pontine nuclei
PrCnF precuneiform area
PTg pedunculopontine tegmental
RC raphe cap
RIs retroisthic nucleus
RRF retrorubral field
rs rubrospinal tract
SNR s nigra, reticular
spth spinothalamic tract
tfp transverse fibers pons
tth trigeminothalamic tr
VLPAG ventlat periaqueductal gray
VTAC vent tegmental area, caudal
xscp decussation sup cereb ped

Fig 46b
Obex +38.0mm
Transverse 46

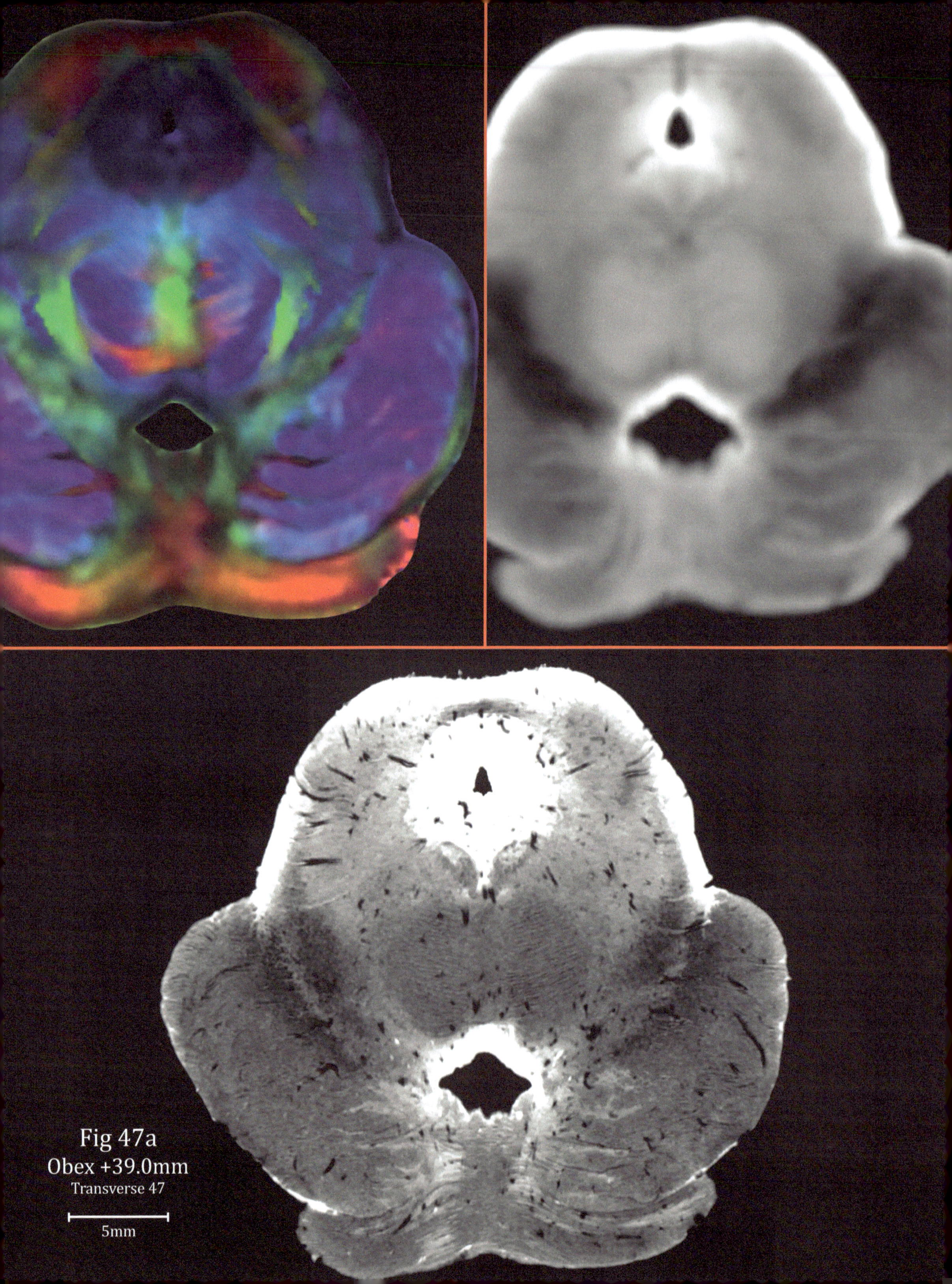

Fig 47a
Obex +39.0mm
Transverse 47

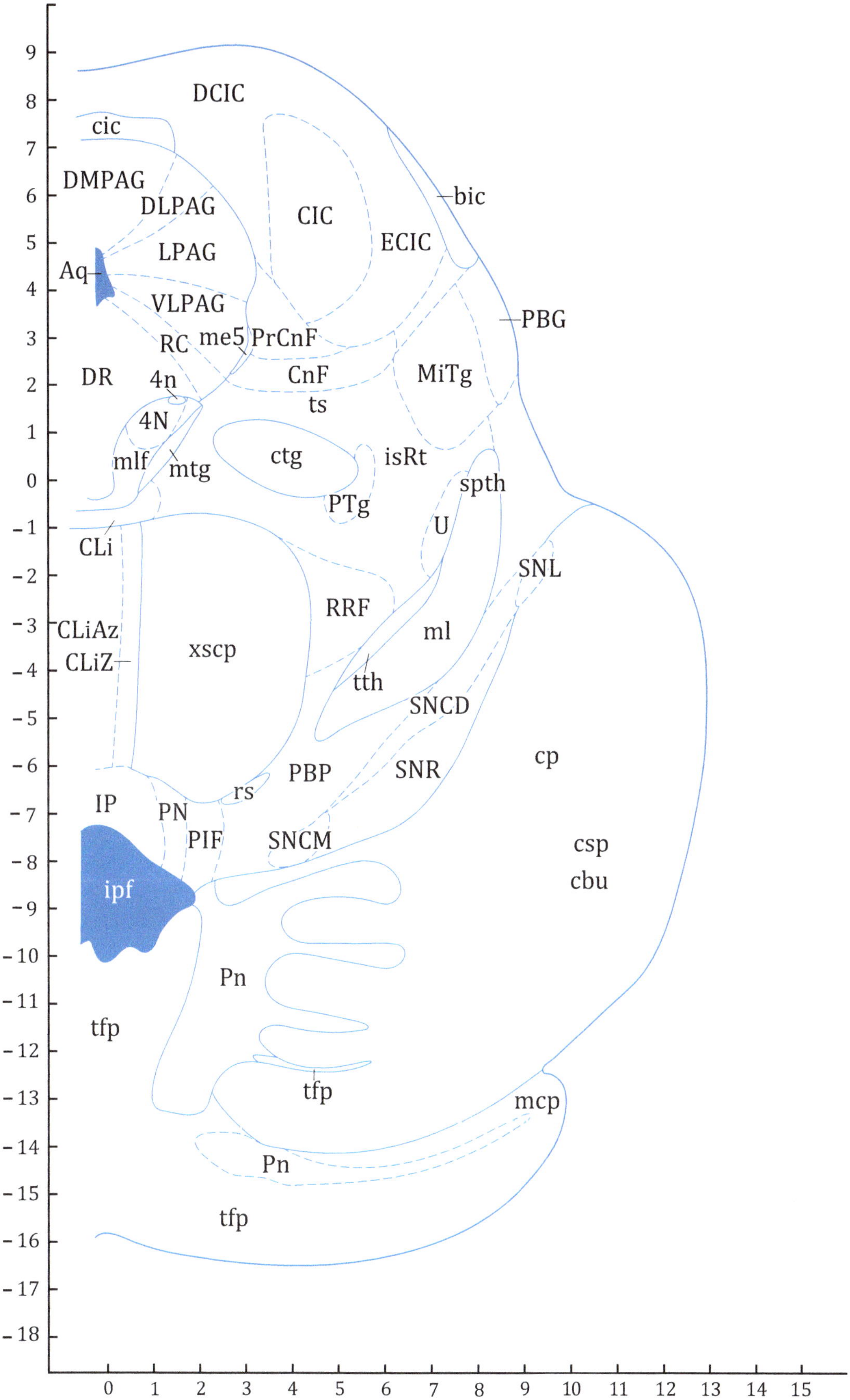

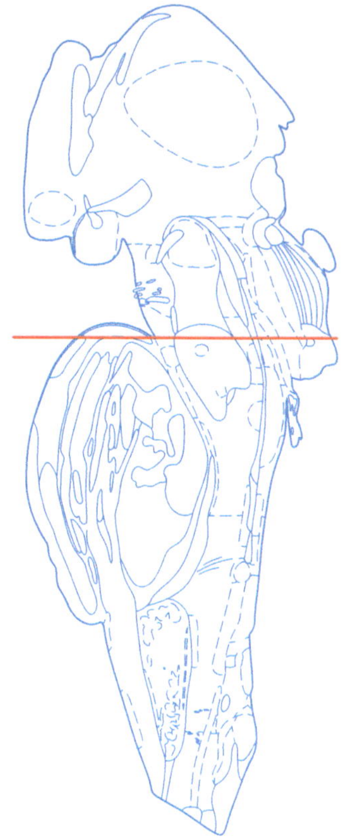

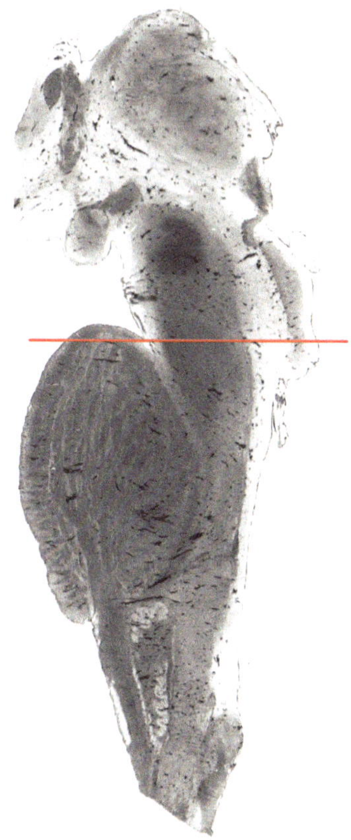

4n trochlear nerve
4N trochlear nu
Aq aqueduct
bic brachium inf colliculus
cbu corticobulbar tract
cic commissure inf colliculus
CIC central nu inf colliculus
CLi caudal linear nu raphe
CLiAz caudal linear nu of raphe, azygos
CLiZ caudal linear nu of raphe, zygos
CnF cuneiform nu
cp cerebral peduncle
csp corticospinal tract
ctg central tegmental tract
DCIC dors cx inf colliculus
DLPAG dorsolat periaqueductal gray
DMPAG dorsomed periaqueductl gray
DR dorsal raphe nu
ECIC external cx inferior coll
IP interpedunc nu
ipf interpedunc fossa
isRt isthmic reticular formation
LPAG lat periaqueductal gray
mcp middle cerebellar peduncle
me5 mesencephalic trigem tract
MiTg microcell tegmental nu
ml medial lemniscus
mlf med longitudinal fasciculus
mtg mammillotegmental tract
PBG parabigeminal nu
PBP parabrachial pigmented nu
PIF parainterfascicular nu VTA
PN paranigral nu of the VTA
Pn pontine nuclei
PrCnF precuneiform area
PTg pedunculopontine tegmental
RC raphe cap
RRF retrorubral field
rs rubrospinal tract
SNCD s nigra, compact, dors tier
SNCM s nigra, compact, med tier
SNL s nigra, lateral
SNR s nigra, reticular
spth spinothalamic tract
tfp transverse fibers pons
ts tectospinal tr
tth trigeminothalamic tr
U nucleus U
VLPAG ventlat periaqueductal gray
xscp decussation sup cereb ped

Fig 47b
Obex +39.0mm
Transverse 47

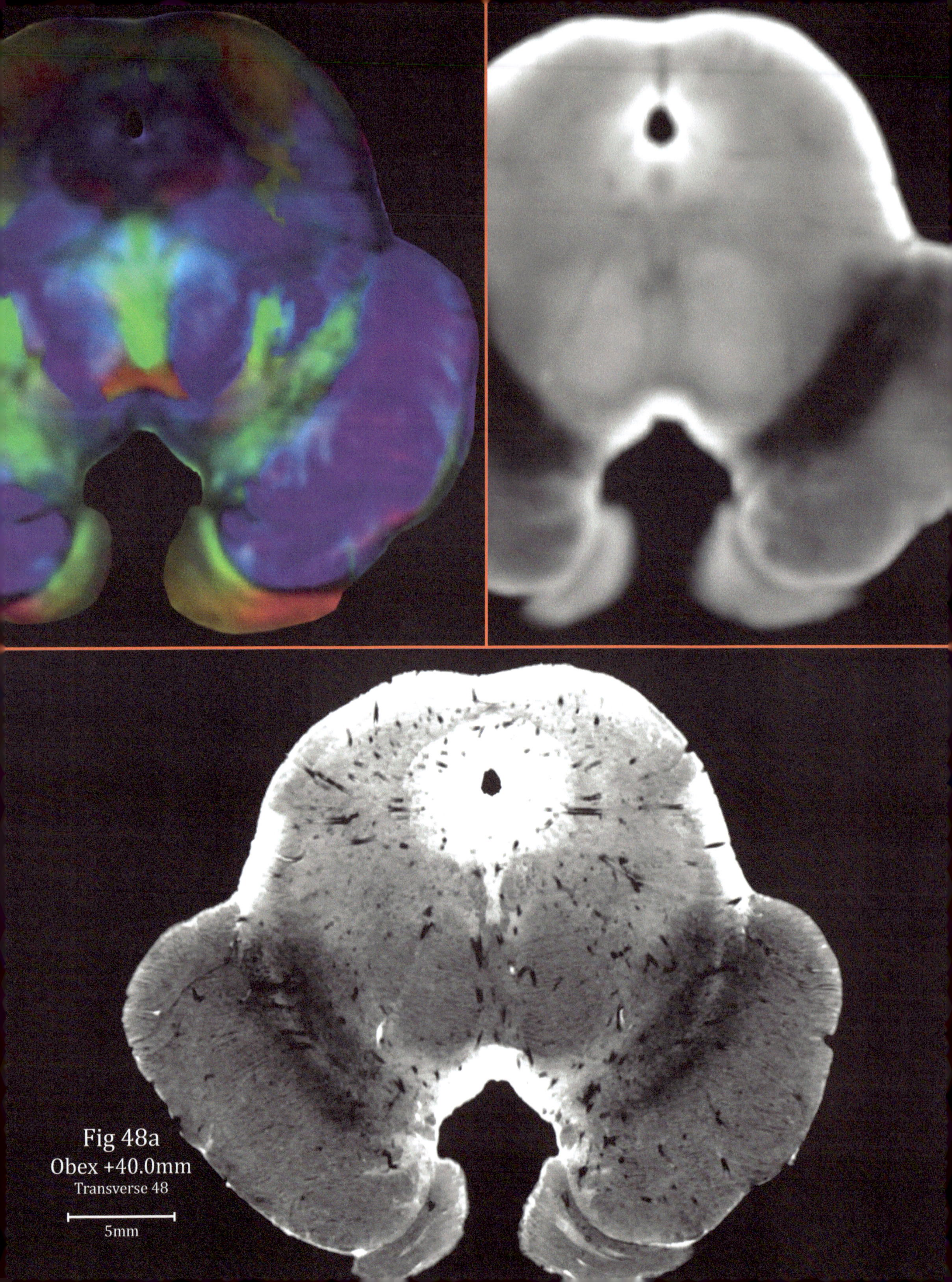

Fig 48a
Obex +40.0mm
Transverse 48

5mm

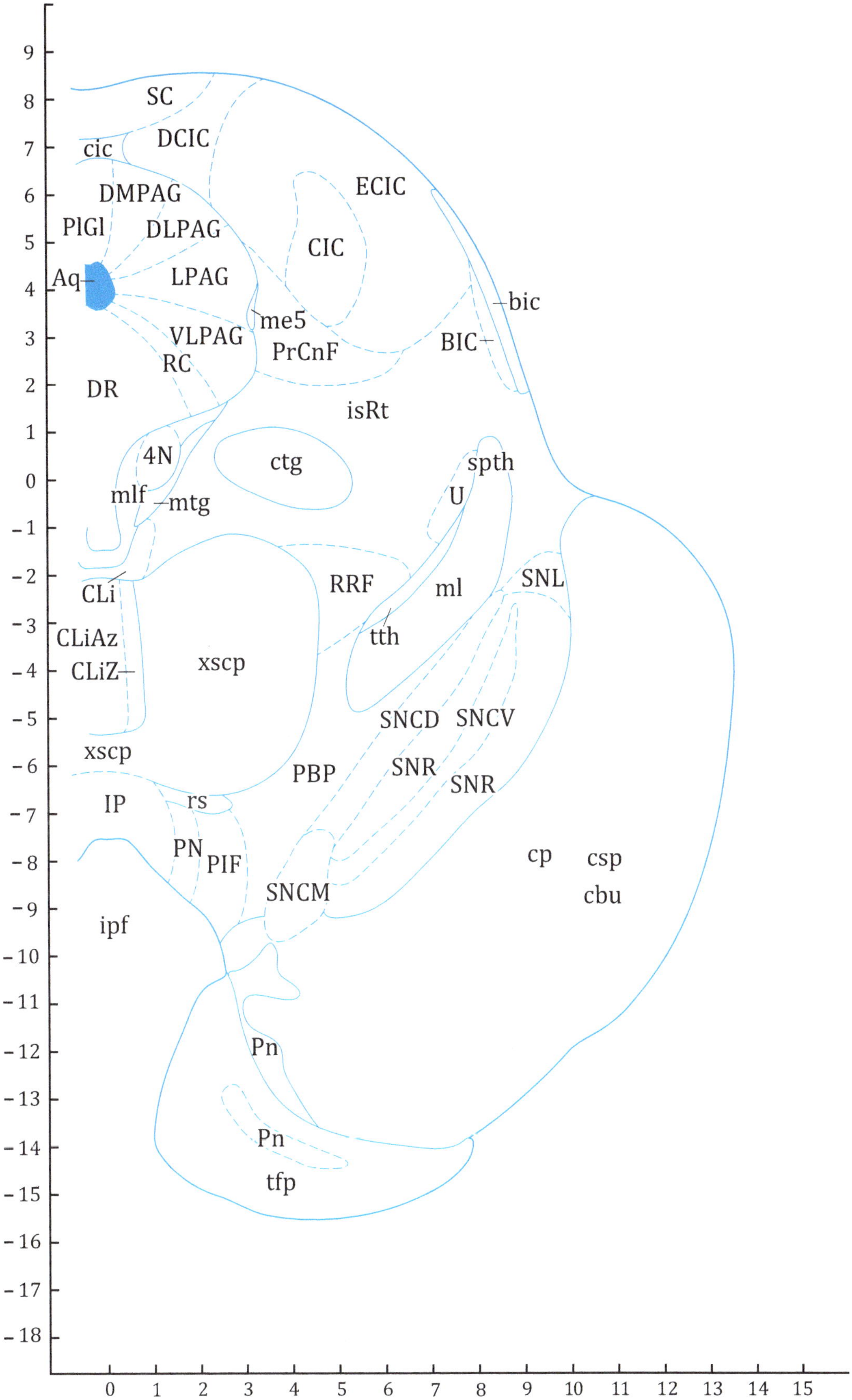

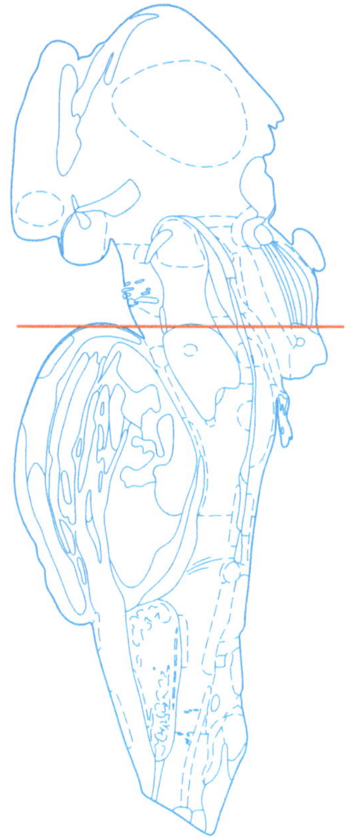

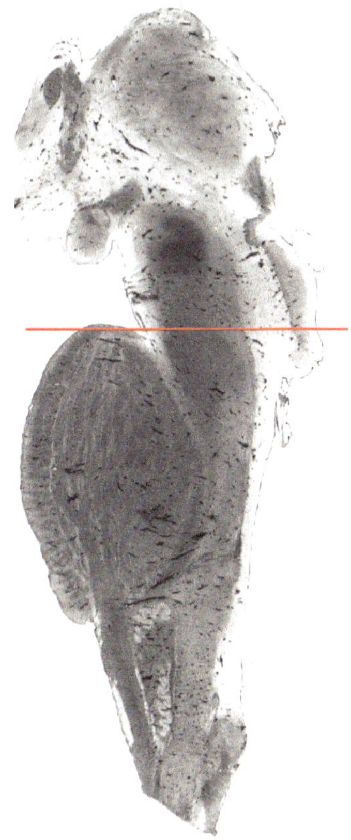

4N trochlear nu
Aq aqueduct
bic brachium inf colliculus
BIC nu brachium inf colliculus
cbu corticobulbar tract
cic commissure inf colliculus
CIC central nu inf colliculus
CLi caudal linear nu raphe
CLiAz caudal linear nu of raphe, azygos
CLiZ caudal linear nu of raphe, zygos
cp cerebral peduncle
csp corticospinal tract
ctg central tegmental tract
DCIC dors cx inf colliculus
DLPAG dorsolat periaqueductal gray
DMPAG dorsomed periaqueductl gray
DR dorsal raphe nu
ECIC external cx inferior coll
IP interpedunc nu
ipf interpedunc fossa
isRt isthmic reticular formation
LPAG lat periaqueductal gray
me5 mesencephalic trigem tract
ml medial lemniscus
mlf med longitudinal fasciculus
mtg mammillotegmental tract
PBP parabrachial pigmented nu
PIF parainterfascicular nu VTA
PlGl pleioglia periaqueductal gray
Pn pontine nuclei
PN paranigral nu of the VTA
PrCnF precuneiform area
RC raphe cap
RRF retrorubral field
rs rubrospinal tract
SC superior colliculus
SNCD s nigra, compact, dors tier
SNCM s nigra, compact, med tier
SNCV s nigra, compact, vent tier
SNL s nigra, lateral
SNR s nigra, reticular
spth spinothalamic tract
tfp transverse fibers pons
tth trigeminothalamic tr
U nucleus U
VLPAG ventlat periaqueductal gray
xscp decussation sup cereb ped

Fig 48b
Obex +40.0mm
Transverse 48

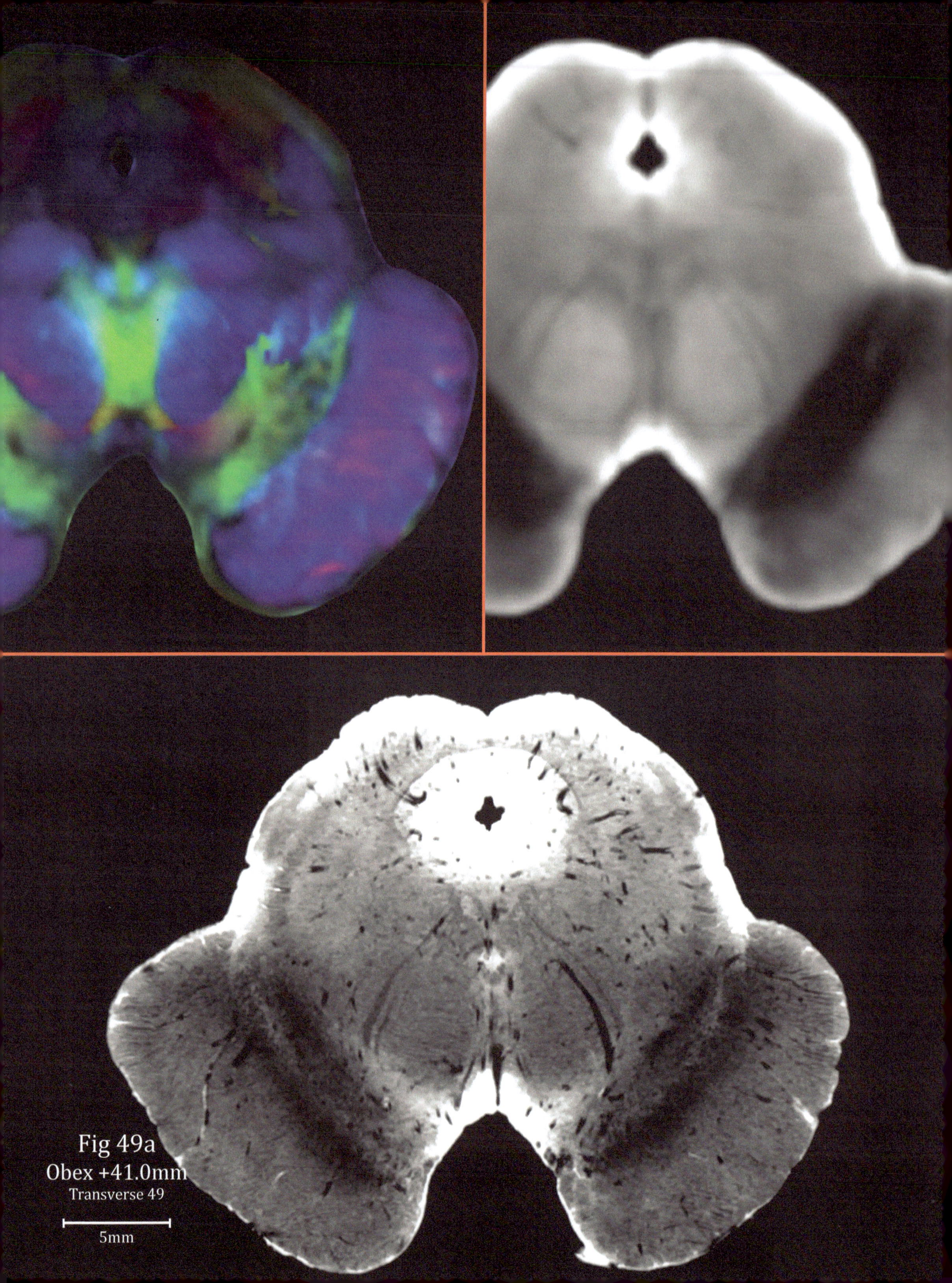

Fig 49a
Obex +41.0mm
Transverse 49
5mm

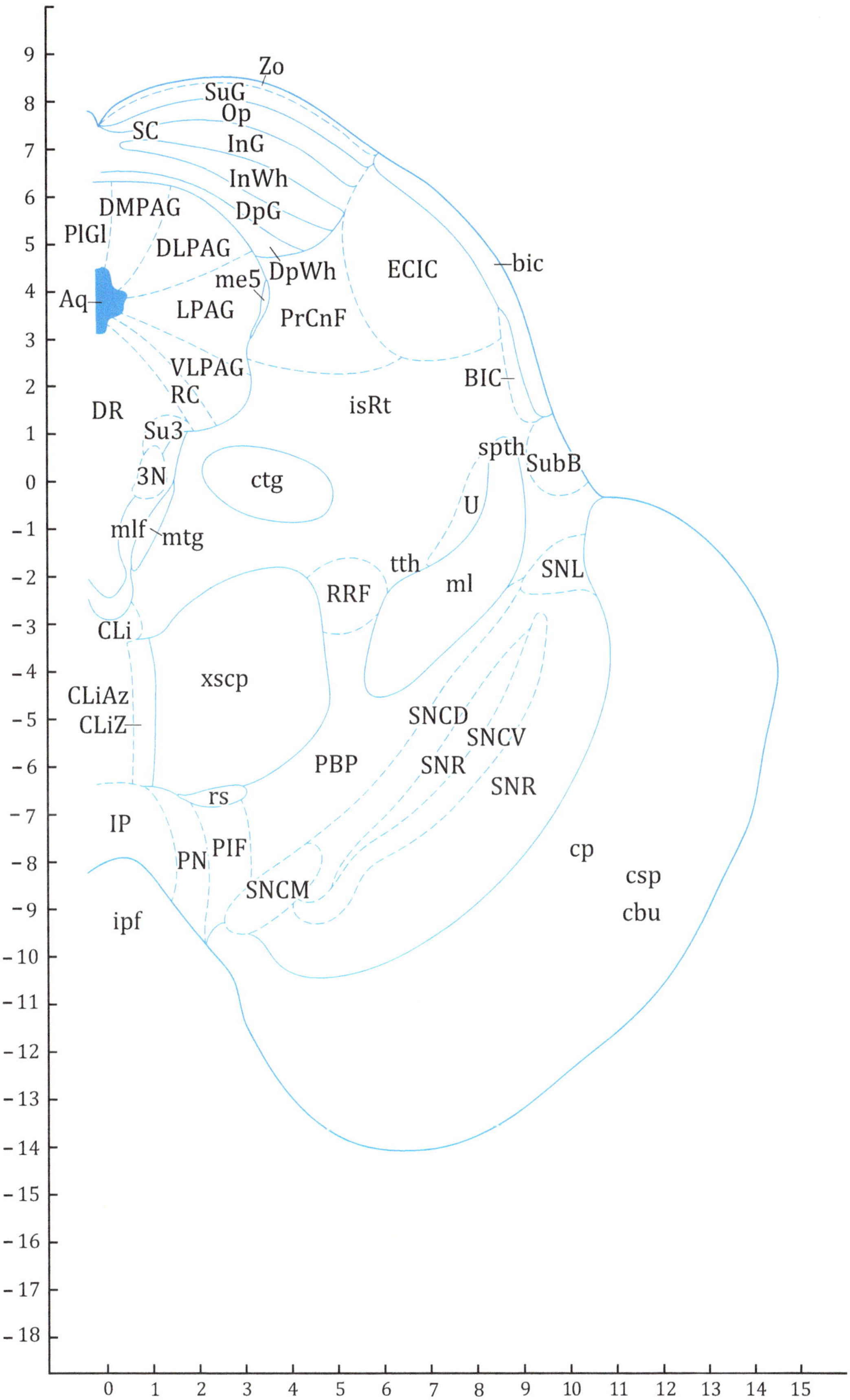

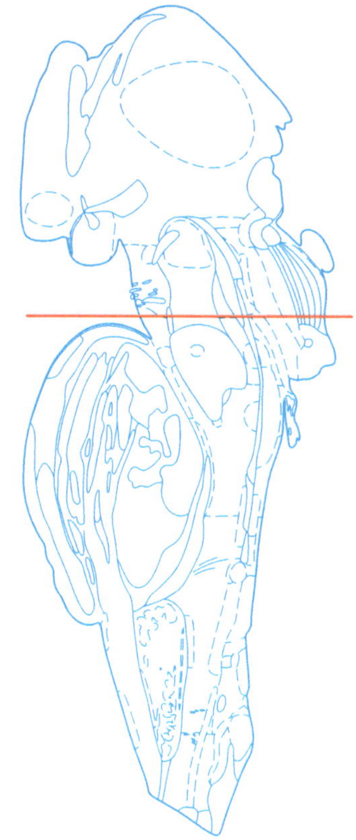

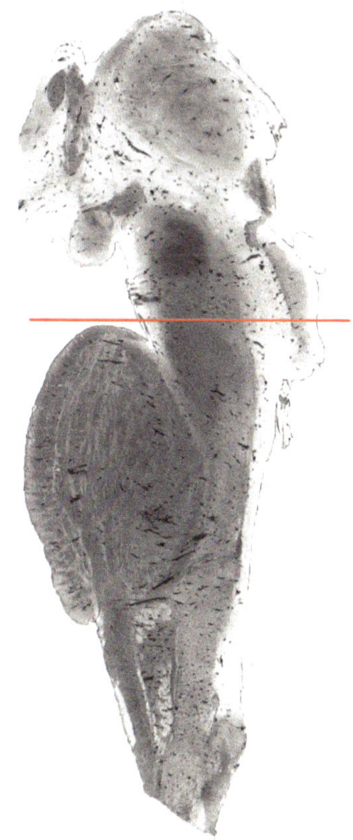

3N oculomotor nu
Aq aqueduct
bic brachium inf colliculus
BIC nu brachium inf colliculus
cbu corticobulbar tract
CLi caudal linear nu raphe
CLiAz caudal linear nu of raphe, azygos
CLiZ caudal linear nu of raphe, zygos
cp cerebral peduncle
csp corticospinal tract
ctg central tegmental tract
DLPAG dorsolat periaqueductal gray
DMPAG dorsomed periaqueductl gray
DpG deep gray superior coll
DpWh deep white superior coll
DR dorsal raphe nu
ECIC external cx inferior coll
InG intermed gray layer SC
InWh intermediate white layer SC
IP interpedunc nu
ipf interpedunc fossa
isRt isthmic reticular formation
LPAG lat periaqueductal gray
me5 mesencephalic trigem tract
ml medial lemniscus
mlf med longitudinal fasciculus
mtg mammillotegmental tract
Op optic nerve layer sup coll
PBP parabrachial pigmented nu
PIF parainterfascicular nu VTA
PlGl pleioglia periaqueductal gray
PN paranigral nu of the VTA
PrCnF precuneiform area
RC raphe cap
RRF retrorubral field
rs rubrospinal tract
SC superior colliculus
SNCD s nigra, compact, dors tier
SNCM s nigra, compact, med tier
SNCV s nigra, compact, vent tier
SNL s nigra, lateral
SNR s nigra, reticular
spth spinothalamic tract
Su3 supraoculomotor PAG
SubB subbrachial nu
SuG superficial gray sup coll
tth trigeminothalamic tr
U nucleus U
VLPAG ventlat periaqueductal gray
xscp decussation sup cereb ped
Zo zona layer sup coll

Fig 49b
Obex +41.0mm
Transverse 49

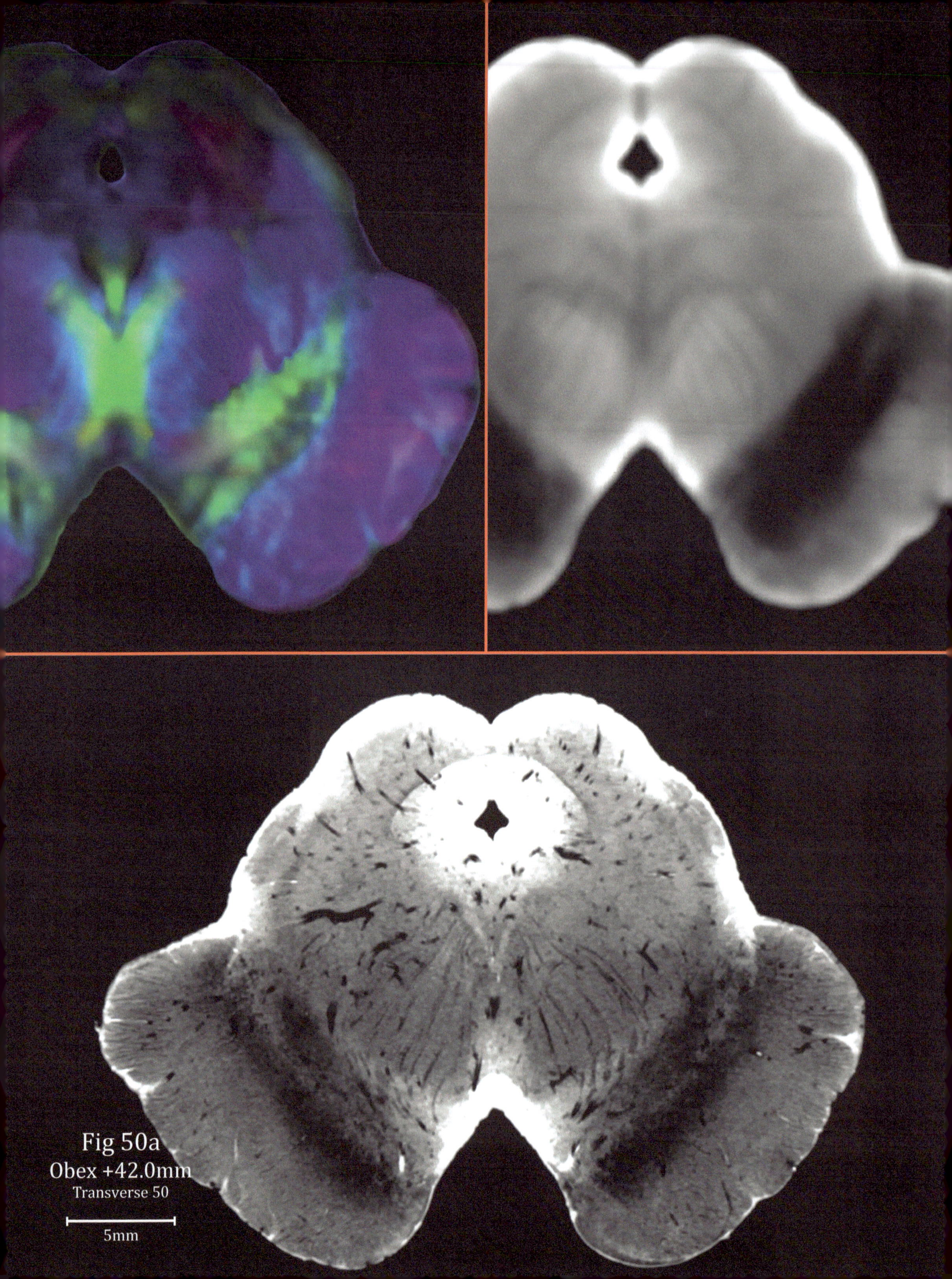

Fig 50a
Obex +42.0mm
Transverse 50

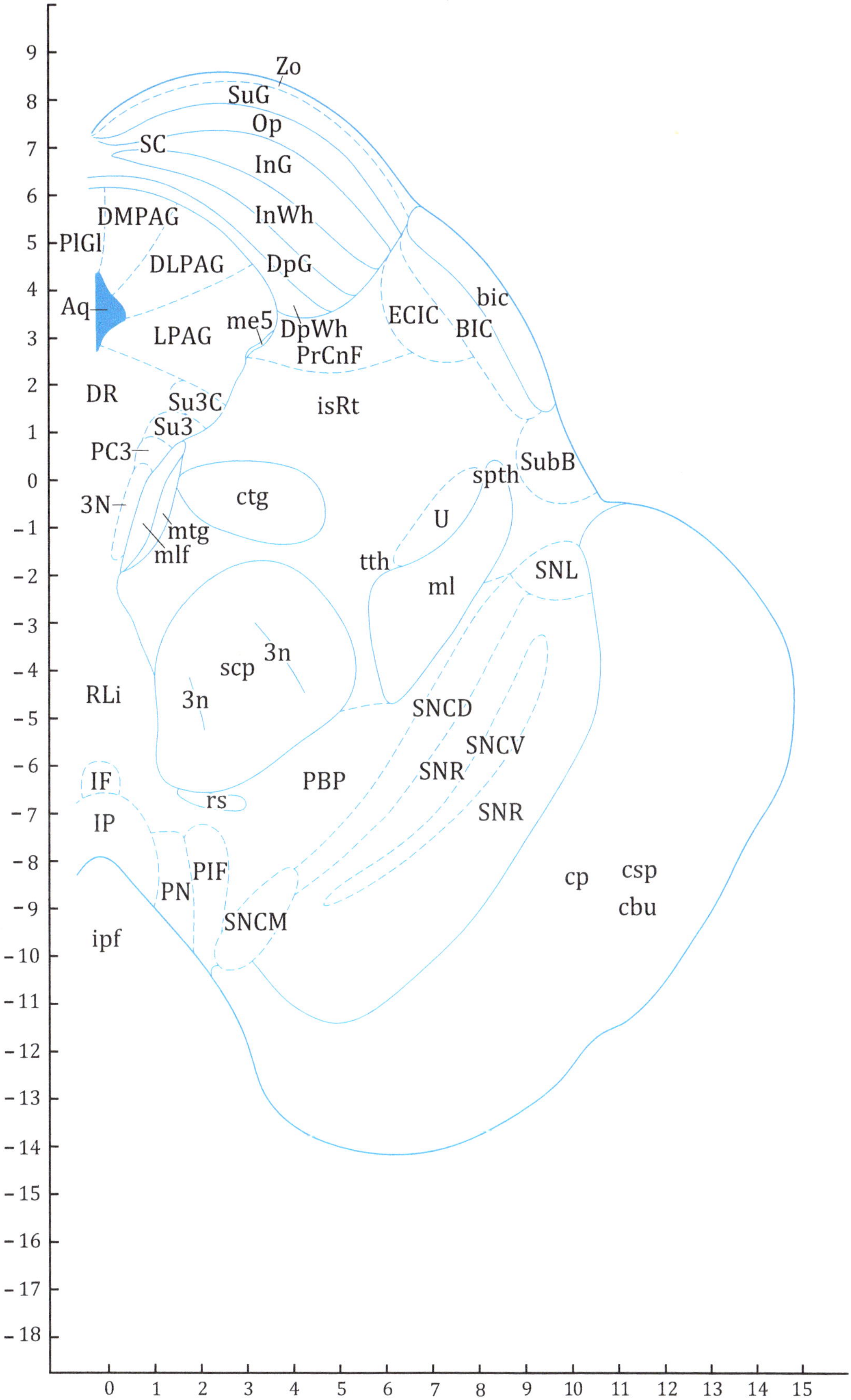

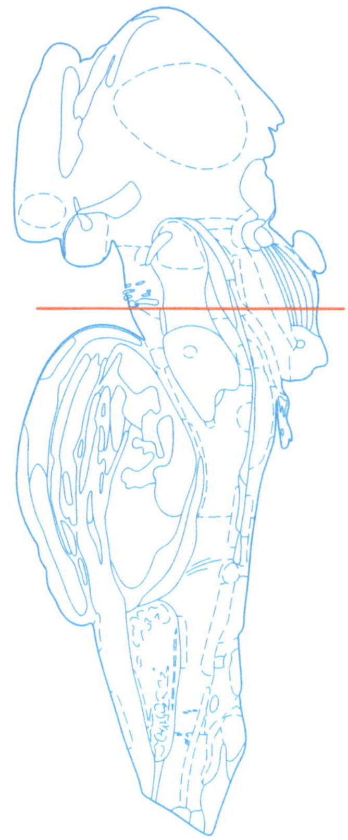

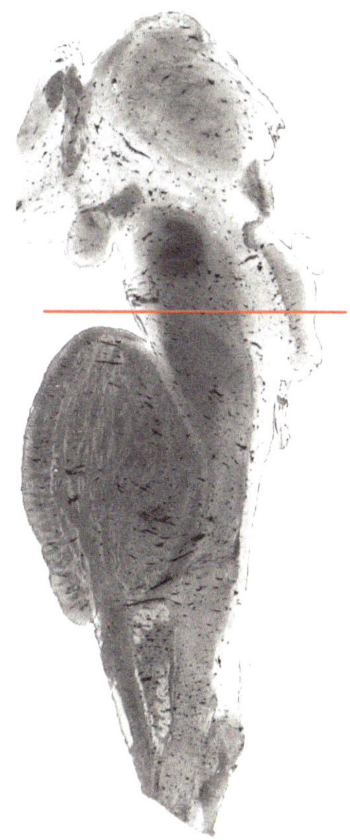

3n oculomotor nerve
3N oculomotor nu
Aq aqueduct
BIC nu brachium inf colliculus
bic brachium inf colliculus
cbu corticobulbar tract
cp cerebral peduncle
csp corticospinal tract
ctg central tegmental tract
DLPAG dorsolat periaqueductal gray
DMPAG dorsomed periaqueductl gray
DpG deep gray superior coll
DpWh deep white superior coll
DR dorsal raphe nu
ECIC external cx inferior coll
IF interfascicular nu
InG intermed gray layer SC
InWh intermediate white layer SC
IP interpedunc nu
ipf interpedunc fossa
isRt isthmic reticular formation
LPAG lat periaqueductal gray
me5 mesencephalic trigem tract
ml medial lemniscus
mlf med longitudinal fasciculus
mtg mammillotegmental tract
Op optic nerve layer sup coll
PBP parabrachial pigmented nu
PC3 parvicellular oculomotor nucleus
PIF parainterfascicular nu VTA
PlGl pleioglia periaqueductal gray
PN paranigral nu of the VTA
PrCnF precuneiform area
RLi rostral linear nu raphe
rs rubrospinal tract
SC superior colliculus
scp superior Cb peduncle
SNCD s nigra, compact, dors tier
SNCM s nigra, compact, med tier
SNCV s nigra, compact, vent tier
SNL s nigra, lateral
SNR s nigra, reticular
spth spinothalamic tract
Su3 supraoculomotor PAG
Su3C supraoculomotor cap
SubB subbrachial nu
SuG superficial gray sup coll
tth trigeminothalamic tr
U nucleus U
Zo zona layer sup coll

Fig 50b
Obex +42.0mm
Transverse 50

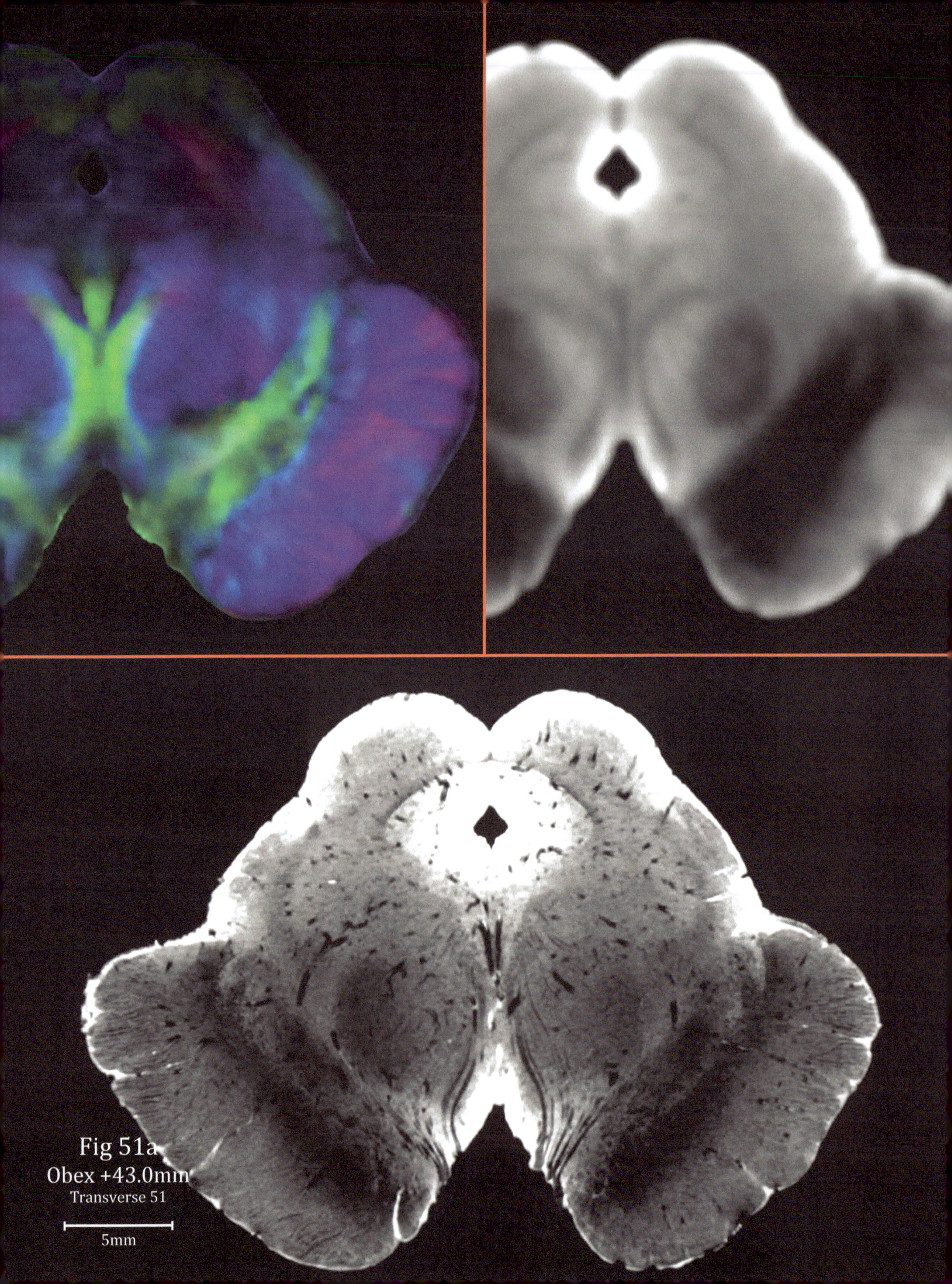

Fig 51a
Obex +43.0mm
Transverse 51

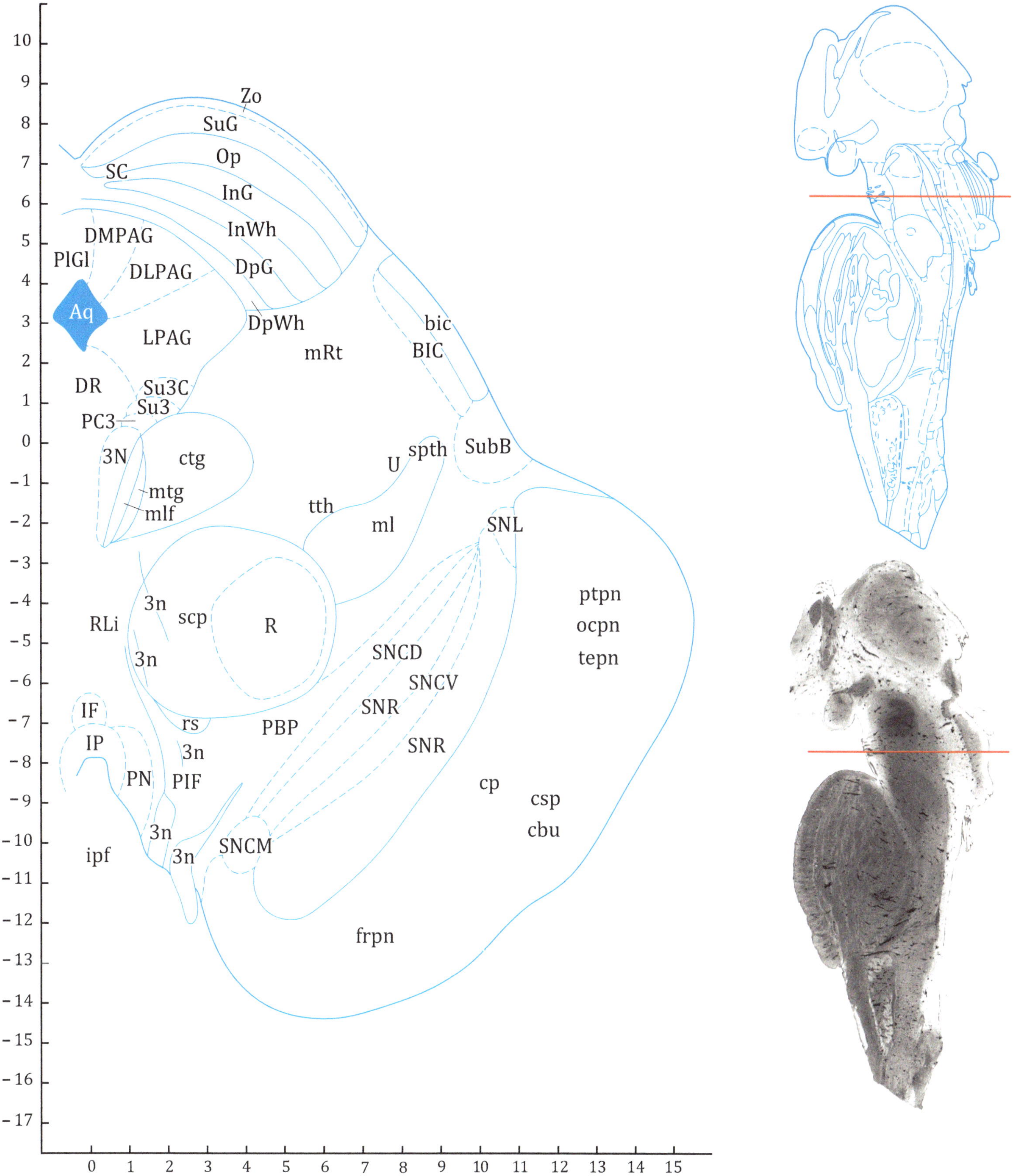

3n oculomotor nerve
3N oculomotor nu
Aq aqueduct
bic brachium inf colliculus
BIC nu brachium inf colliculus
cbu corticobulbar tract
cp cerebral peduncle
csp corticospinal tract
ctg central tegmental tract
DLPAG dorsolat periaqueductal gray
DMPAG dorsomed periaqueductl gray
DpG deep gray superior coll
DpWh deep white superior coll
DR dorsal raphe nu
frpn frontopontine fibers
IF interfascicular nu
InG intermed gray layer SC
InWh intermediate white layer SC
IP interpedunc nu
ipf interpedunc fossa
LPAG lat periaqueductal gray
ml medial lemniscus
mlf med longitudinal fasciculus
mRt mesencephalic retic form
mtg mammillotegmental tract
ocpn occipitopontine fibers
Op optic nerve layer sup coll
PBP parabrachial pigmented nu
PC3 parvicellular oculomotor nucleus
PIF parainterfascicular nu VTA
PlGl pleioglia periaqueductal gray
PN paranigral nu of the VTA
ptpn parietal pontine fibers
R red nu
RLi rostral linear nu raphe
rs rubrospinal tract
SC superior colliculus
scp superior Cb peduncle
SNCD s nigra, compact, dors tier
SNCM s nigra, compact, med tier
SNCV s nigra, compact, vent tier
SNL s nigra, lateral
SNR s nigra, reticular
spth spinothalamic tract
Su3 supraoculomotor PAG
Su3C supraoculomotor cap
SubB subbrachial nu
SuG superficial gray sup coll
tepn temporopontine fibers
tth trigeminothalamic tr
U nucleus U
Zo zona layer sup coll

Fig 51b
Obex +43.0mm
Transverse 51

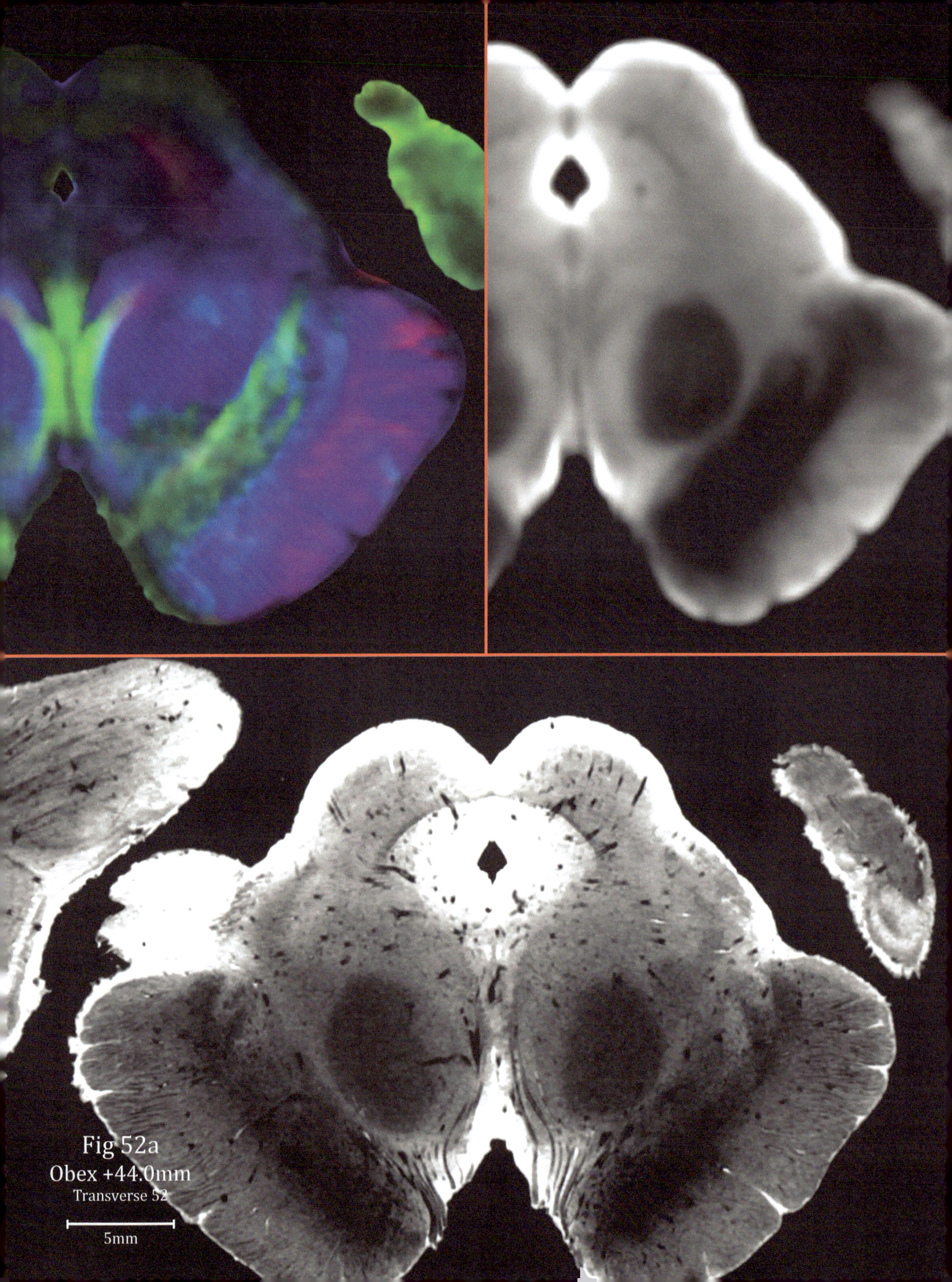

Fig 52a
Obex +44.0mm
Transverse 52

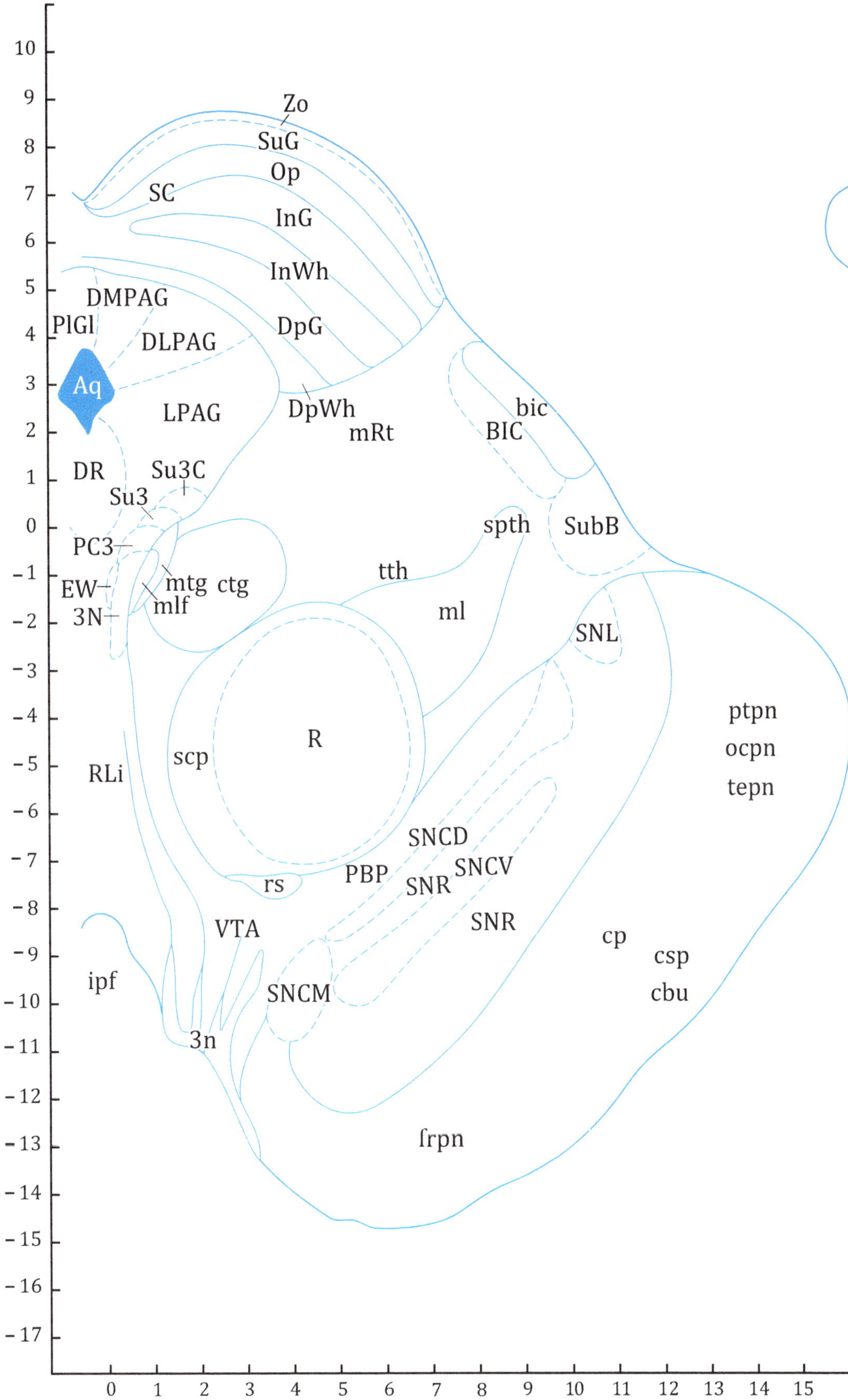

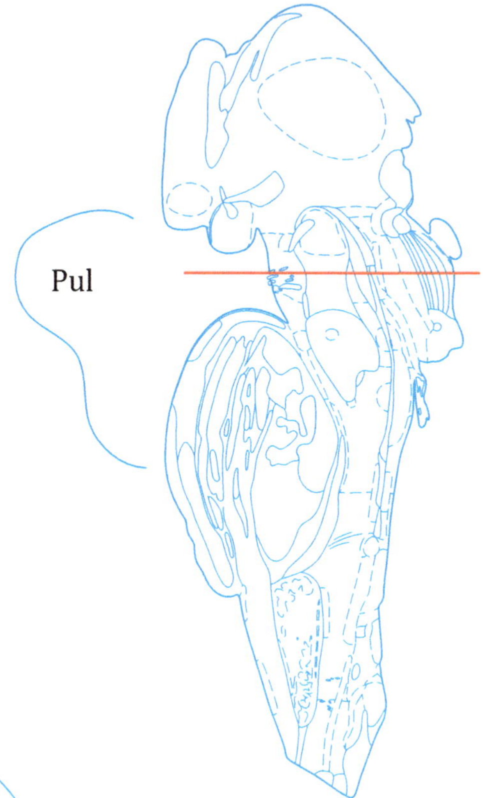

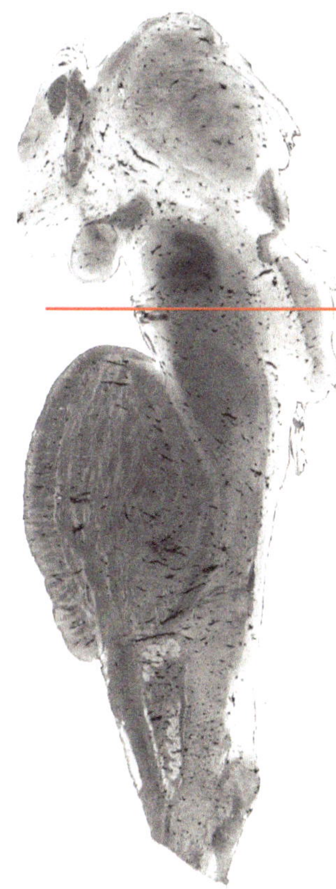

3n oculomotor nerve
3N oculomotor nu
Aq aqueduct
BIC nu brachium inf colliculus
bic brachium inf colliculus
cbu corticobulbar tract
cp cerebral peduncle
csp corticospinal tract
ctg central tegmental tract
DLPAG dorsolat periaqueductal gray
DMPAG dorsomed periaqueductl gray
DpG deep gray superior coll
DpWh deep white superior coll

DR dorsal raphe nu
EW Edinger-Westphal nu
frpn frontopontine fibers
InG intermed gray layer SC
InWh intermediate white layer SC
ipf interpedunc fossa
LPAG lat periaqueductal gray
ml medial lemniscus
mlf med longitudinal fasciculus
mRt mesencephalic retic form
mtg mammillotegmental tract
ocpn occipitopontine fibers
Op optic nerve layer sup coll

PBP parabrachial pigmented nu
PC3 parvicellular oculomotor nucleus
PlGl pleioglia periaqueductal gray
ptpn parietal pontine fibers
Pul pulvinar nuclei
R red nu
RLi rostral linear nu raphe
rs rubrospinal tract
SC superior colliculus
scp superior Cb peduncle
SNCD s nigra, compact, dors tier
SNCM s nigra, compact, med tier
SNCV s nigra, compact, vent tier

SNL s nigra, lateral
SNR s nigra, reticular
spth spinothalamic tract
Su3 supraoculomotor PAG
Su3C supraoculomotor cap
SubB subbrachial nu
SuG superficial gray sup coll
tepn temporopontine fibers
tth trigeminothalamic tr
VTA vent tegmental area
Zo zona layer sup coll

Fig 52b
Obex +44.0mm
Transverse 52

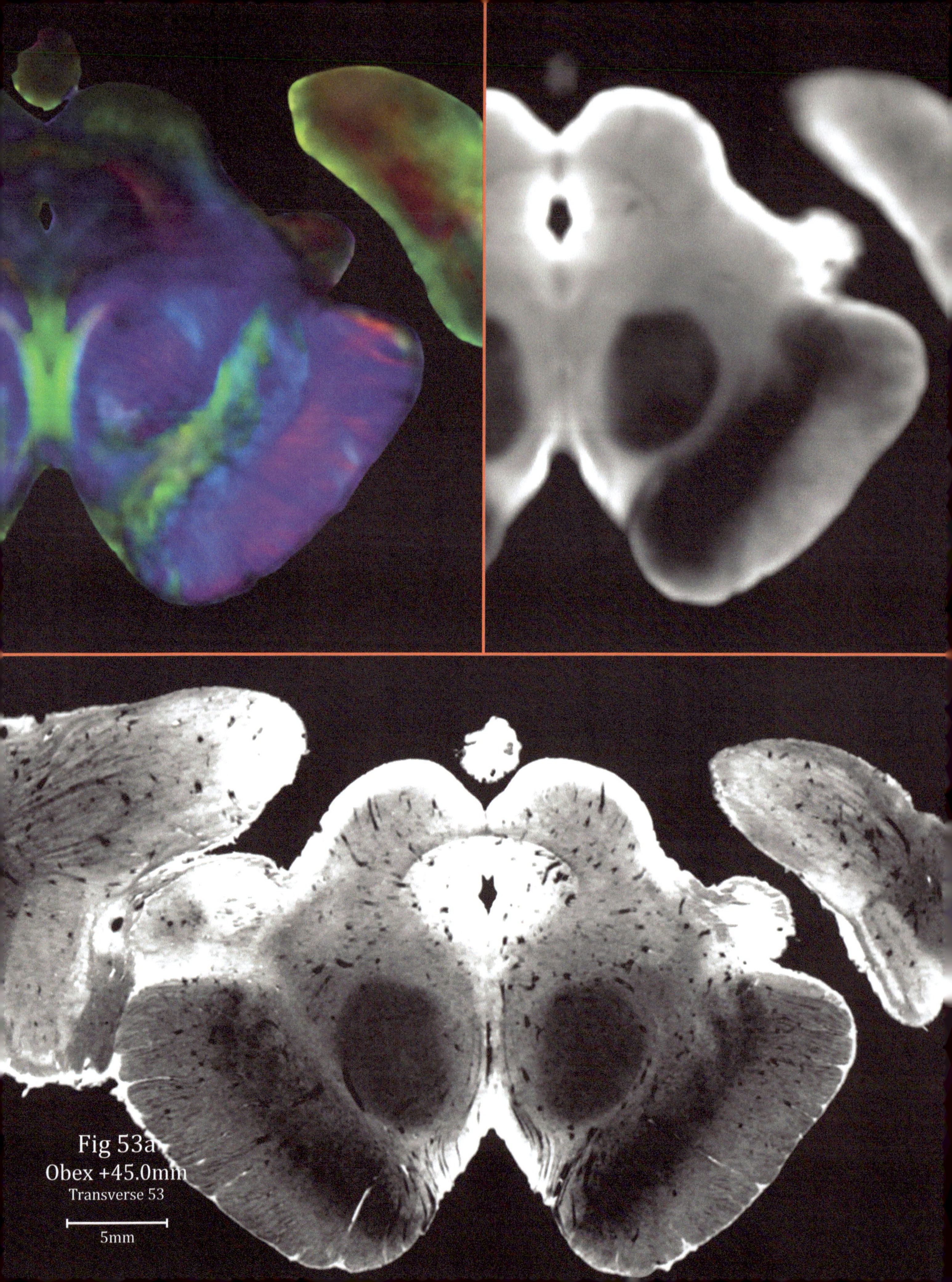
Fig 53a
Obex +45.0mm
Transverse 53
5mm

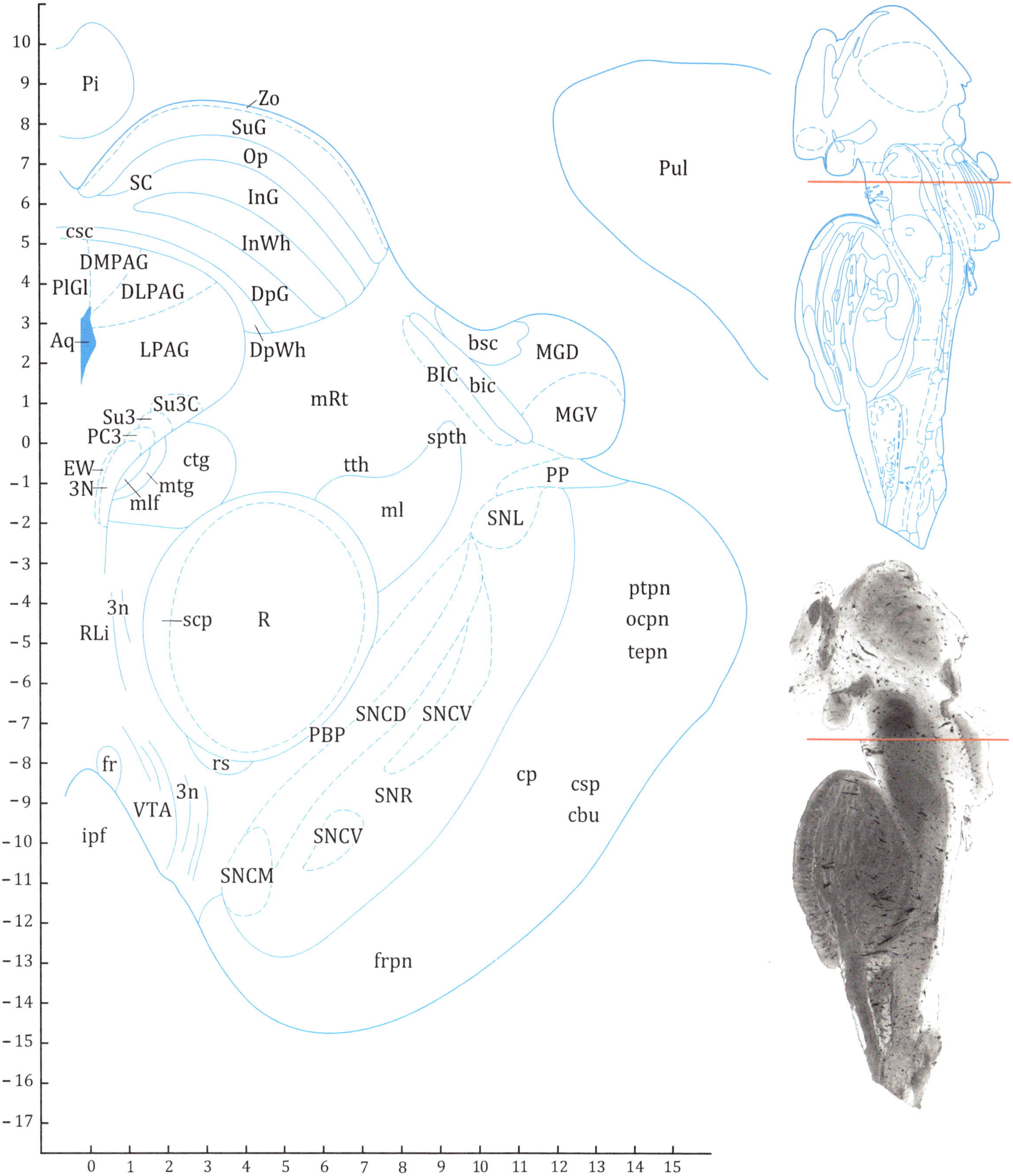

3n oculomotor nerve
3N oculomotor nu
Aq aqueduct
BIC nu brachium inf colliculus
bic brachium inf colliculus
bsc brachium superior colliculus
cbu corticobulbar tract
cp cerebral peduncle
csc comm superior colliculus
csp corticospinal tract
ctg central tegmental tract
DLPAG dorsolat periaqueductal gray
DMPAG dorsomed periaqueductl gray

DpG deep gray superior coll
DpWh deep white superior coll
EW Edinger-Westphal nu
fr fasciculus retroflexus
frpn frontopontine fibers
InG intermed gray layer SC
InWh intermediate white layer SC
ipf interpedunc fossa
LPAG lat periaqueductal gray
MGD med geniculate nu, dorsal
MGV med geniculate nu, ventral
ml medial lemniscus
mlf med longitudinal fasciculus

mRt mesencephalic retic form
mtg mammillotegmental tract
ocpn occipitopontine fibers
Op optic nerve layer sup coll
PBP parabrachial pigmented nu
PC3 parvicellular oculomotor nucleus
Pi pineal gland
PlGl pleioglia periaqueductal gray
PP peripeduncular nu
ptpn parietal pontine fibers
Pul pulvinar nuclei
R red nu
RLi rostral linear nu raphe

rs rubrospinal tract
SC superior colliculus
scp superior Cb peduncle
SNCD s nigra, compact, dors tier
SNCM s nigra, compact, med tier
SNCV s nigra, compact, vent tier
SNL s nigra, lateral
SNR s nigra, reticular
spth spinothalamic tract
Su3 supraoculomotor PAG
Su3C supraoculomotor cap
SuG superficial gray sup coll
tepn temporopontine fibers

Fig 53b
Obex +45.0mm
Transverse 53

tth trigeminothalamic tr
VTA vent tegmental area
Zo zona layer sup coll

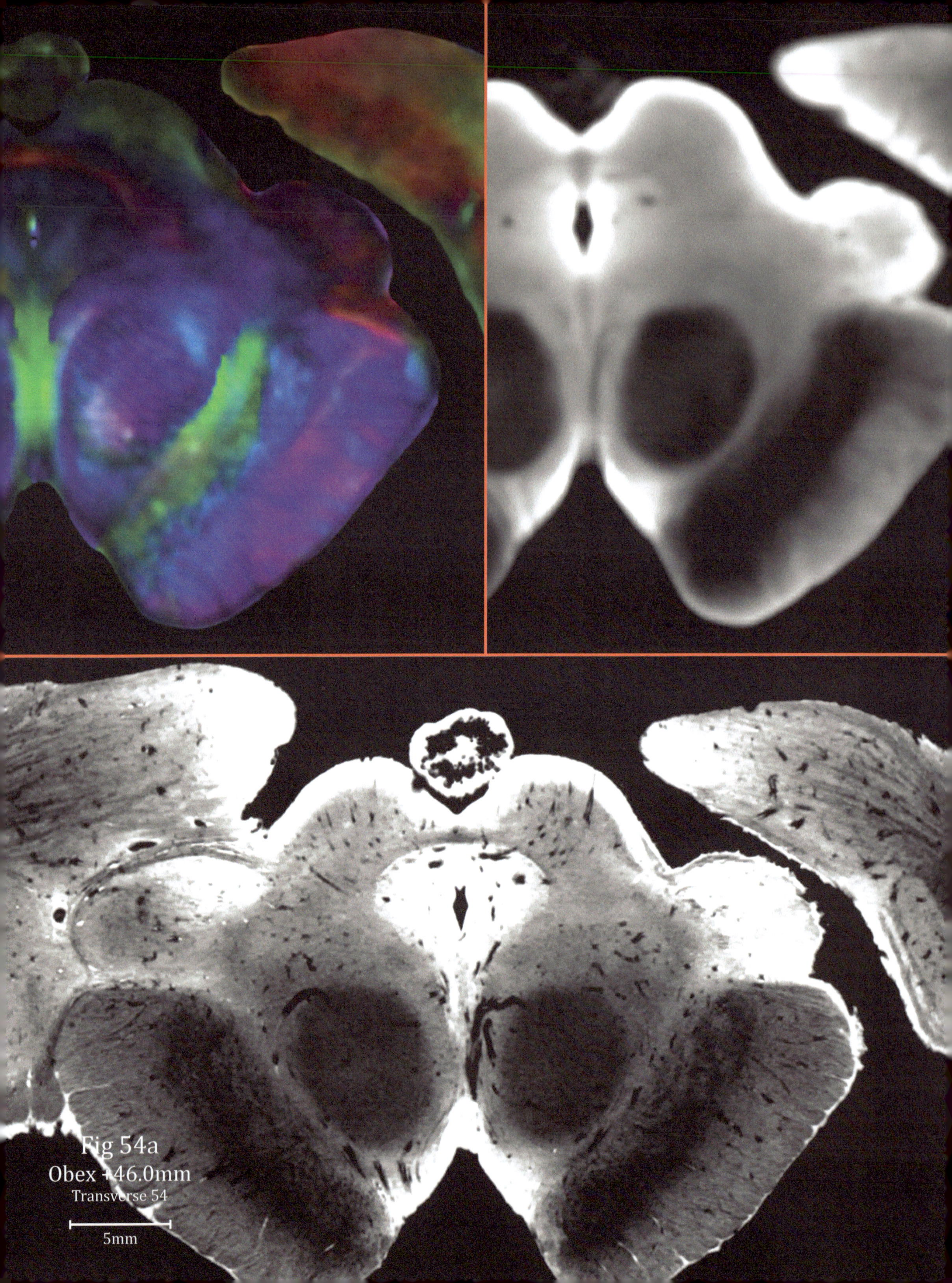

Fig 54a
Obex +46.0mm
Transverse 54
5mm

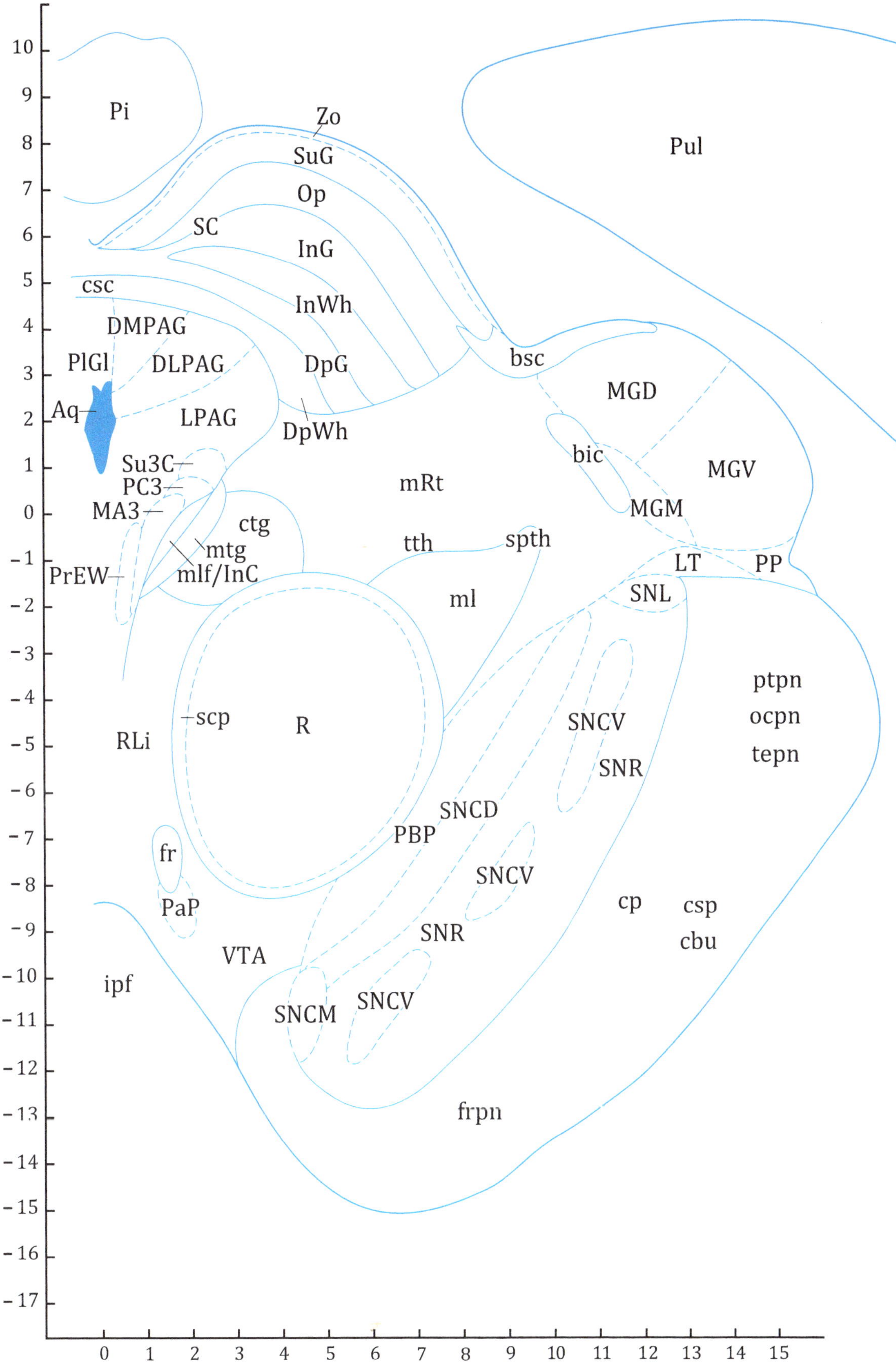

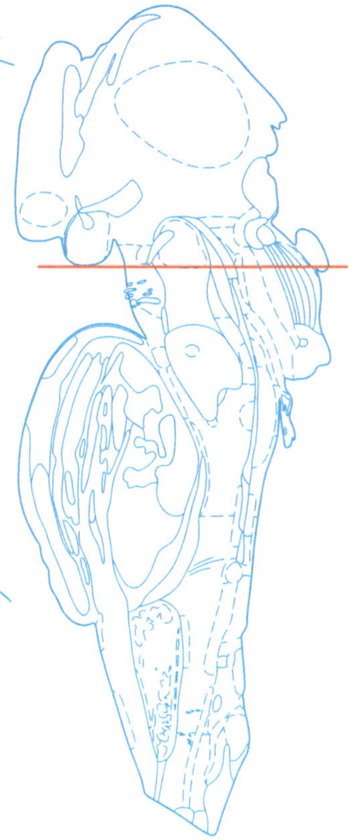

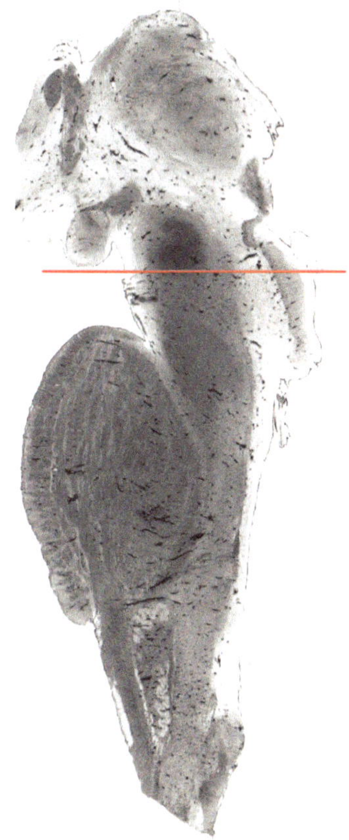

Aq aqueduct
bic brachium inf colliculus
bsc brachium superior colliculus
cbu corticobulbar tract
cp cerebral peduncle
csc comm superior colliculus
csp corticospinal tract
ctg central tegmental tract
DLPAG dorsolat periaqueductal gray
DMPAG dorsomed periaqueductl gray
DpG deep gray superior coll
DpWh deep white superior coll
fr fasciculus retroflexus
frpn frontopontine fibers
InG intermed gray layer SC
InWh intermediate white layer SC
ipf interpedunc fossa
LPAG lat periaqueductal gray
LT lat terminal nu acc optic tr
MA3 med access oculomotor nu
MGD med geniculate nu, dorsal
MGM med geniculate nu, medial
MGV med geniculate nu, ventral
ml medial lemniscus
mlf/InC medial long fas/interstitial nu
mRt mesencephalic retic form
mtg mammillotegmental tract
ocpn occipitopontine fibers
Op optic nerve layer sup coll
PaP parapeduncular nucleus
PBP parabrachial pigmented nu
PC3 parvicellular oculomotor nucleus
Pi pineal gland
PlGl pleioglia periaqueductal gray
PP peripeduncular nu
PrEW pre-Edinger-Westphal nu
ptpn parietal pontine fibers
Pul pulvinar nuclei
R red nu
RLi rostral linear nu raphe
SC superior colliculus
scp superior Cb peduncle
SNCD s nigra, compact, dors tier
SNCM s nigra, compact, med tier
SNCV s nigra, compact, vent tier
SNL s nigra, lateral
SNR s nigra, reticular
spth spinothalamic tract
Su3C supraoculomotor cap
SuG superficial gray sup coll
tepn temporopontine fibers
tth trigeminothalamic tr
VTA vent tegmental area
Zo zona layer sup coll

Fig 54b

Obex +46.0mm

Transverse 54

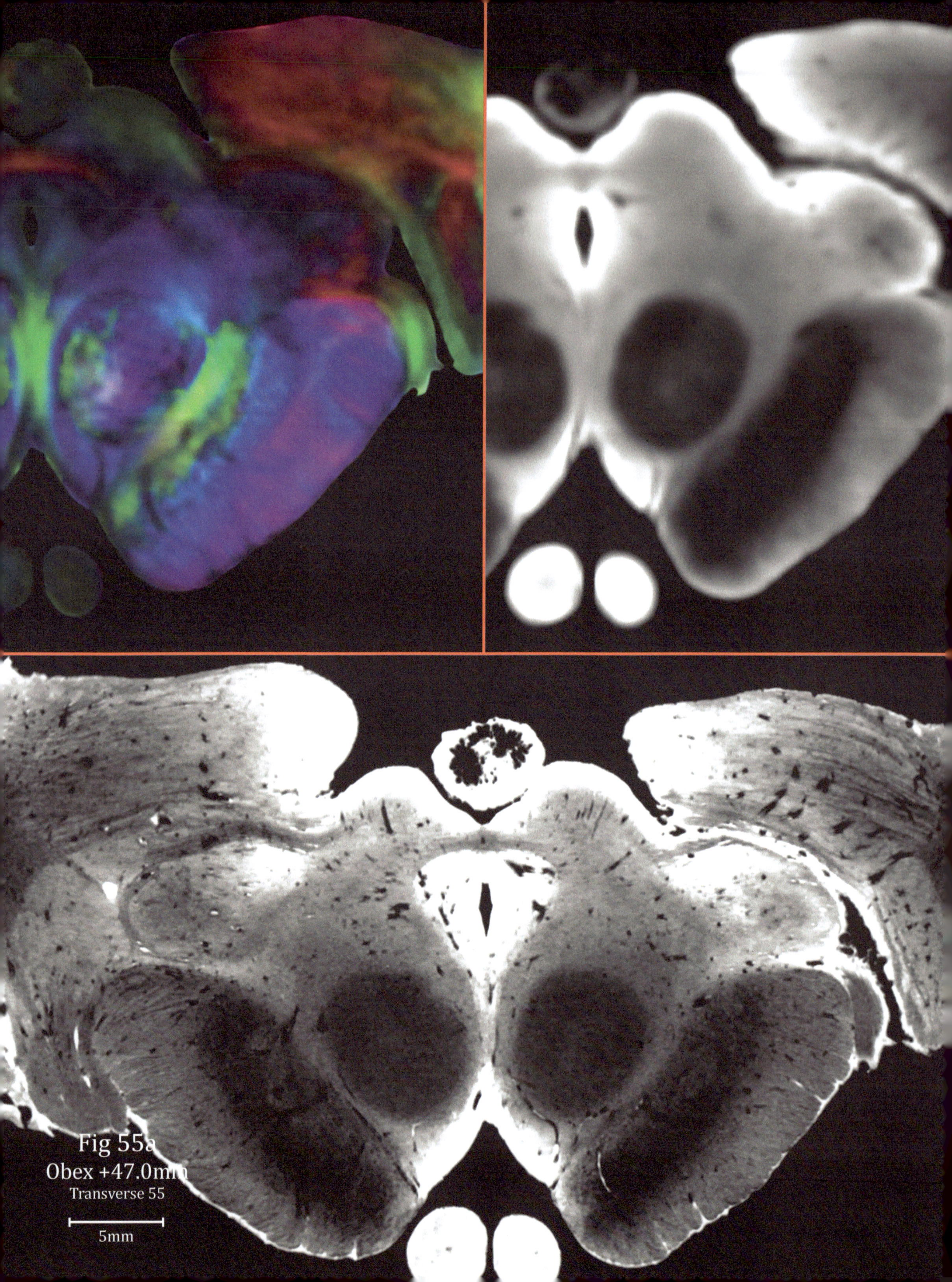

Fig 55a
Obex +47.0mm
Transverse 55
5mm

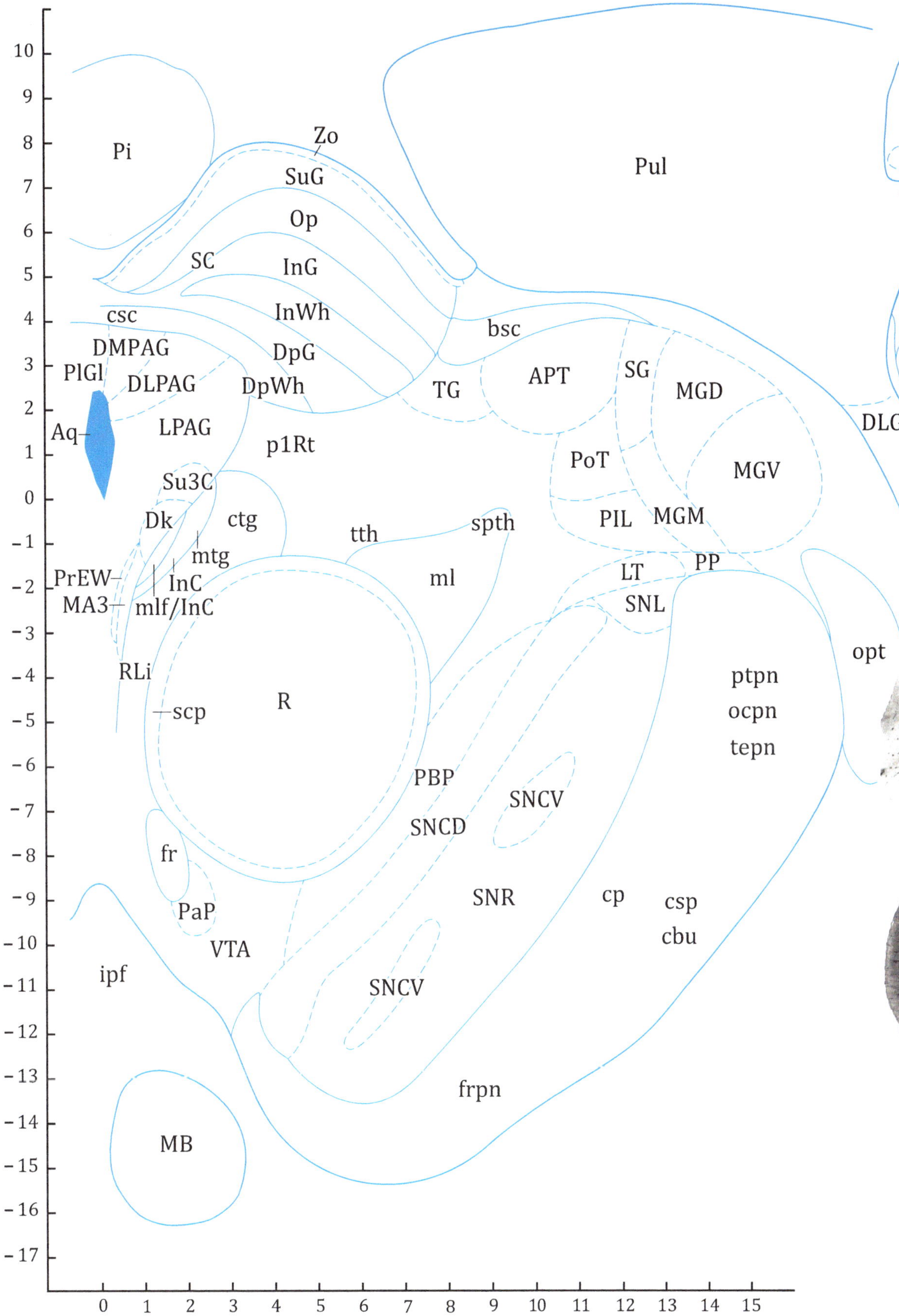

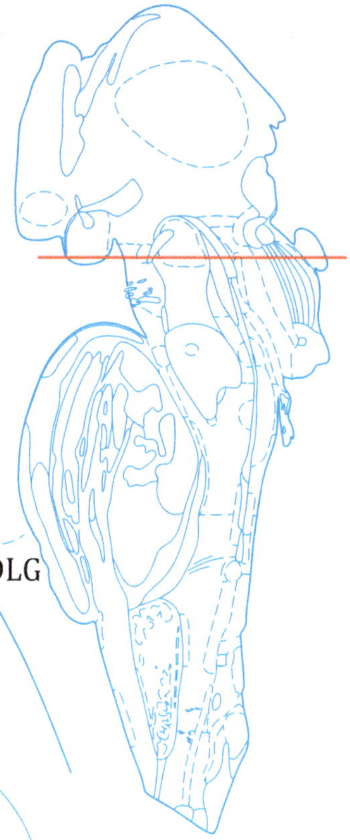

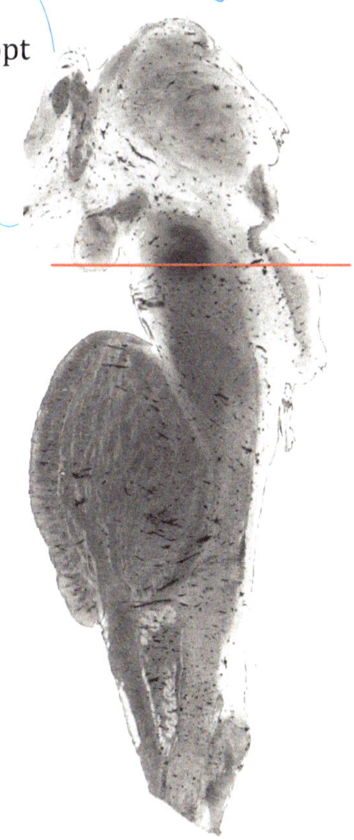

APT anterior pretectal nu
Aq aqueduct
bsc brachium superior colliculus
cbu corticobulbar tract
cp cerebral peduncle
csc comm superior colliculus
csp corticospinal tract
ctg central tegmental tract
Dk nu of Darkschewitsch
DLG dorsal lat geniculate nu
DLPAG dorsolat periaqueductal gray
DMPAG dorsomed periaqueductl gray
DpG deep gray superior coll
DpWh deep white superior coll
fr fasciculus retroflexus
frpn frontopontine fibers
InC interstitial nu of Cajal
InG intermed gray layer SC
InWh intermediate white layer SC
ipf interpedunc fossa
LPAG lat periaqueductal gray
LT lat terminal nu acc optic tr
MA3 med access oculomotor nu
MB mammillary body
MGD med geniculate nu, dorsal
MGM med geniculate nu, medial
MGV med geniculate nu, ventral
ml medial lemniscus
mlf/InC medial long fas/interstitial nu
mtg mammillotegmental tract
ocpn occipitopontine fibers
Op optic nerve layer sup coll
opt optic tract
p1Rt prosomere 1 reticular formation
PaP parapeduncular nucleus
PBP parabrachial pigmented nu
Pi pineal gland
PIL post intralaminar thal nu
PlGl pleioglia periaqueductal gray
PoT posterior thal nu, triangular
PP peripeduncular nu
PrEW pre-Edinger-Westphal nu
ptpn parietal pontine fibers
Pul pulvinar nuclei
R red nu
RLi rostral linear nu raphe
SC superior colliculus
scp superior Cb peduncle
SG suprageniculate thal nu
SNCD s nigra, compact, dors tier
SNCV s nigra, compact, vent tier
SNL s nigra, lateral
SNR s nigra, reticular
spth spinothalamic tract
Su3C supraoculomotor cap
See List of Structures

Fig 55b
Obex +47.0mm
Transverse 55

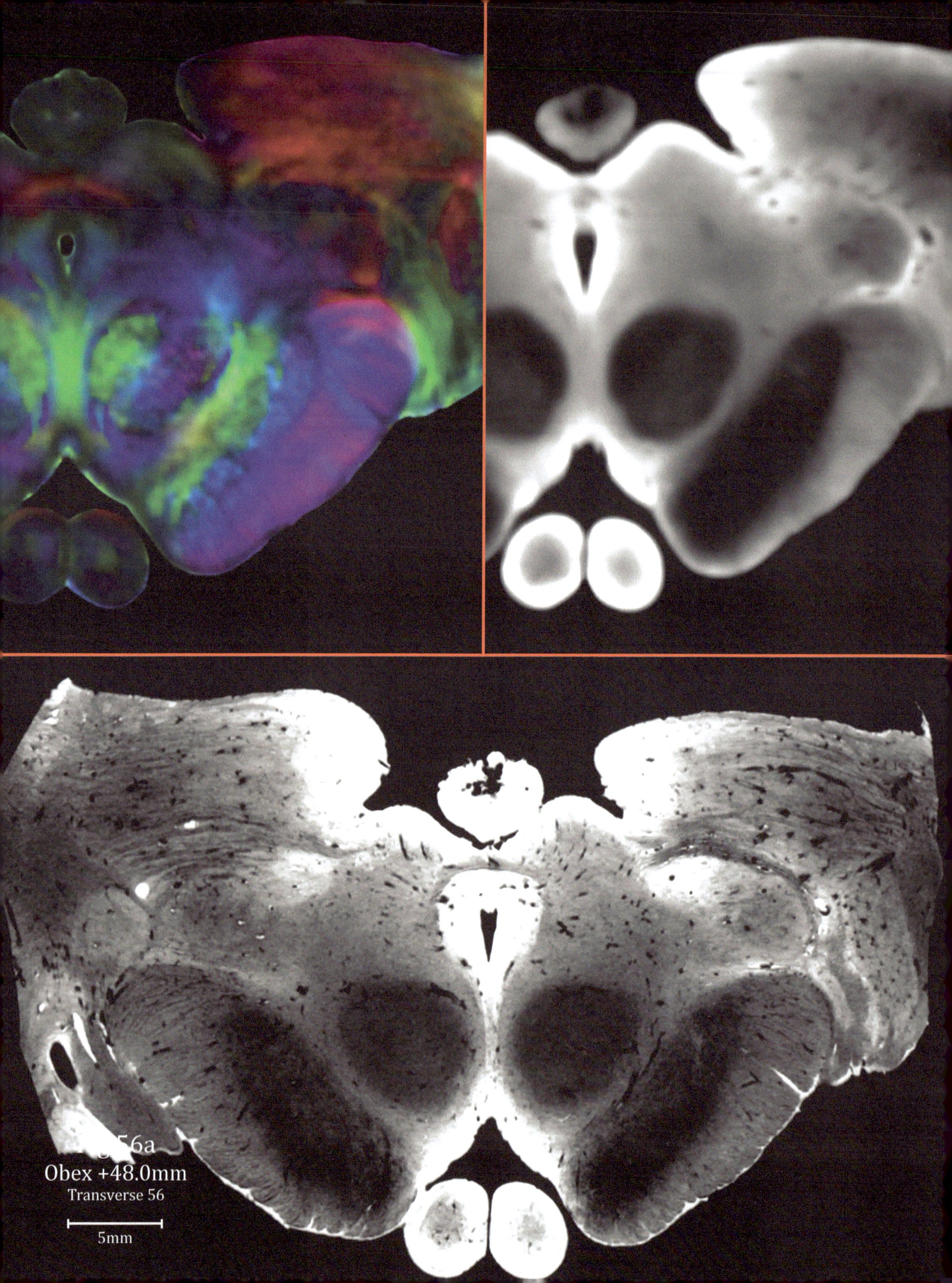
56a
Obex +48.0mm
Transverse 56
5mm

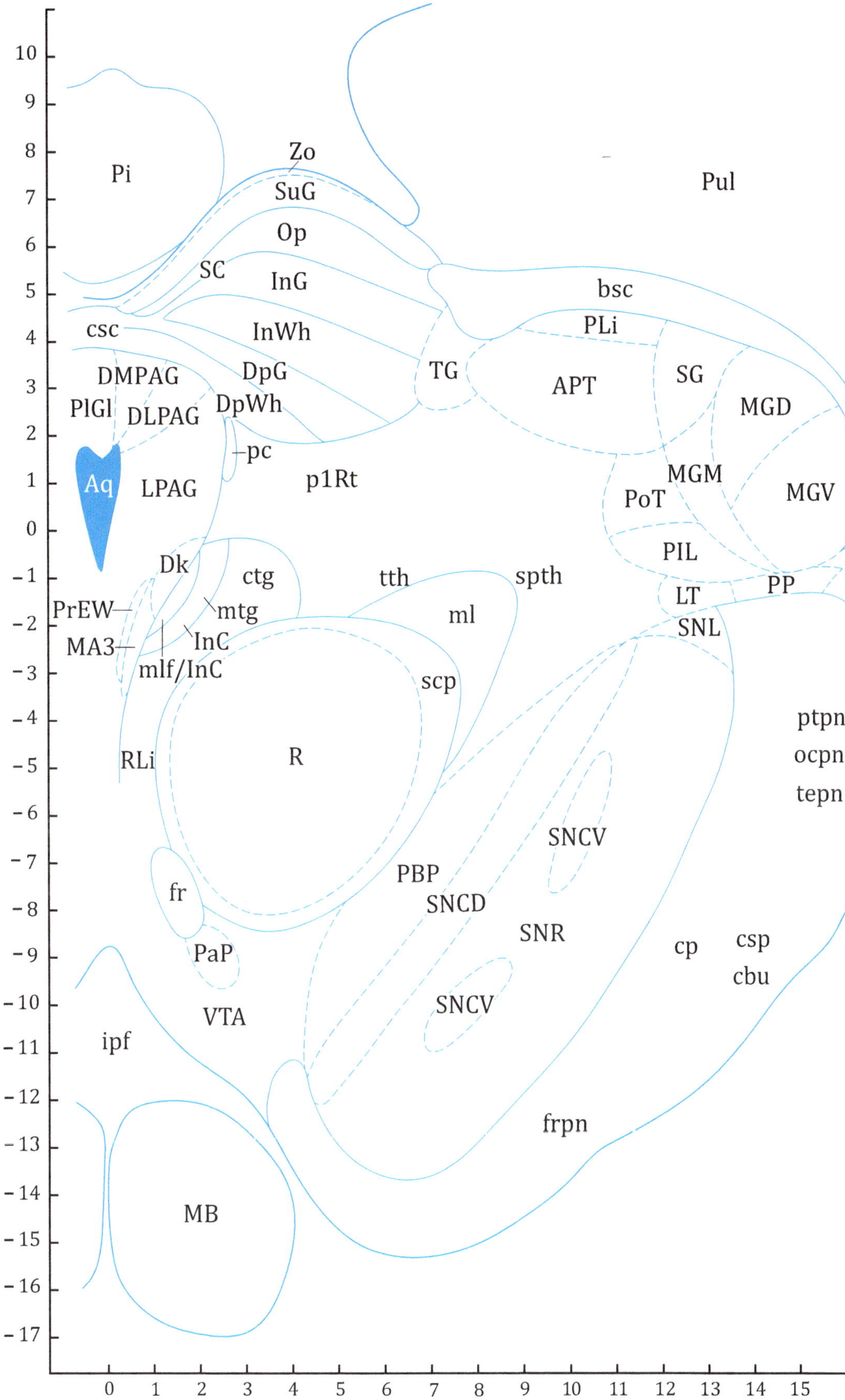

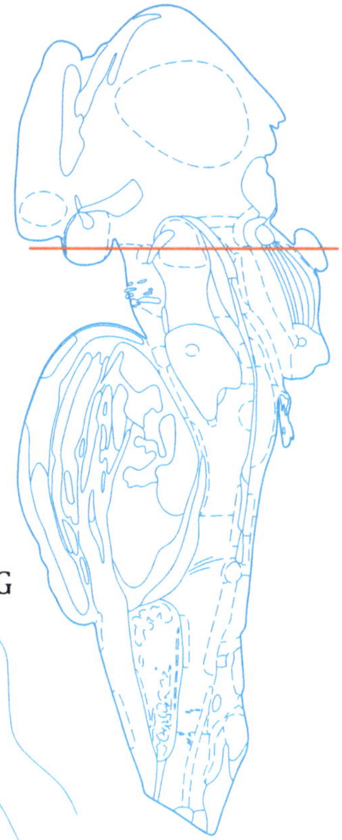

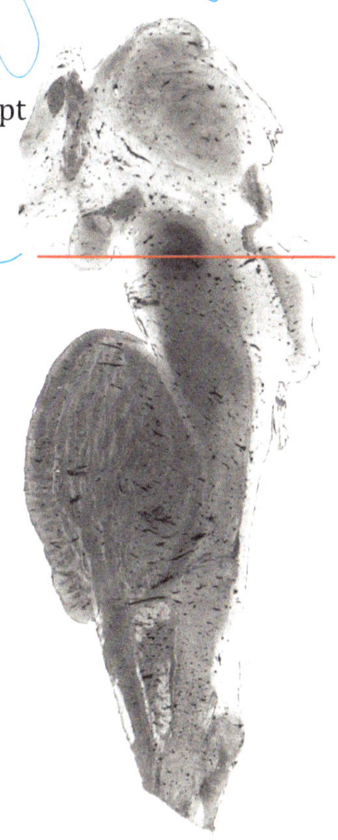

APT anterior pretectal nu
Aq aqueduct
bsc brachium superior colliculus
cbu corticobulbar tract
cp cerebral peduncle
csc comm superior colliculus
csp corticospinal tract
ctg central tegmental tract
Dk nu of Darkschewitsch
DLG dorsal lat geniculate nu
DLPAG dorsolat periaqueductal gray
DMPAG dorsomed periaqueductl gray
DpG deep gray superior coll
DpWh deep white superior coll
fr fasciculus retroflexus
frpn frontopontine fibers
InC interstitial nu of Cajal
InG intermed gray layer SC
InWh intermediate white layer SC
ipf interpedunc fossa
LPAG lat periaqueductal gray
LT lat terminal nu acc optic tr
MA3 med access oculomotor nu
MB mammillary body
MGD med geniculate nu, dorsal
MGM med geniculate nu, medial
MGV med geniculate nu, ventral
ml medial lemniscus
mlf/InC medial long fas/interstitial nu
mtg mammillotegmental tract
ocpn occipitopontine fibers
Op optic nerve layer sup coll
opt optic tract
p1Rt prosomere 1 reticular formation
PaP parapeduncular nucleus
PBP parabrachial pigmented nu
pc posterior comm
Pi pineal gland
PIL post intralaminar thal nu
PlGl pleioglia periaqueductal gray
PLi posterior limitans thal nu
PoT posterior thal nu, triangular
PP peripeduncular nu
PrEW pre-Edinger-Westphal nu
ptpn parietal pontine fibers
Pul pulvinar nuclei
R red nu
RLi rostral linear nu raphe
SC superior colliculus
scp superior Cb peduncle
SG suprageniculate thal nu
SNCD s nigra, compact, dors tier
SNCV s nigra, compact, vent tier
SNL s nigra, lateral
SNR s nigra, reticular
See List of Structures

Fig 56b
Obex +48.0mm
Transverse 56

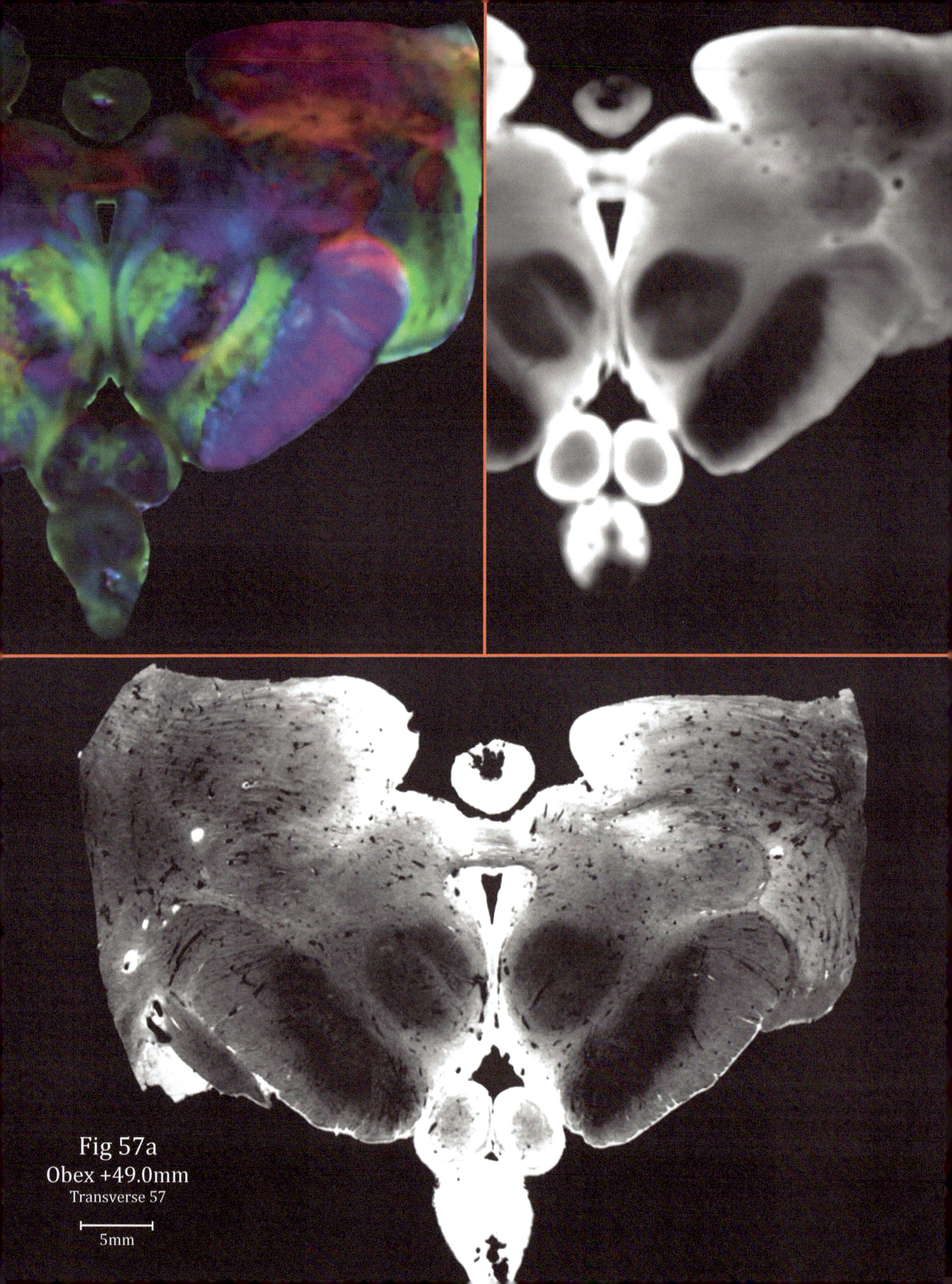

Fig 57a
Obex +49.0mm
Transverse 57

5mm

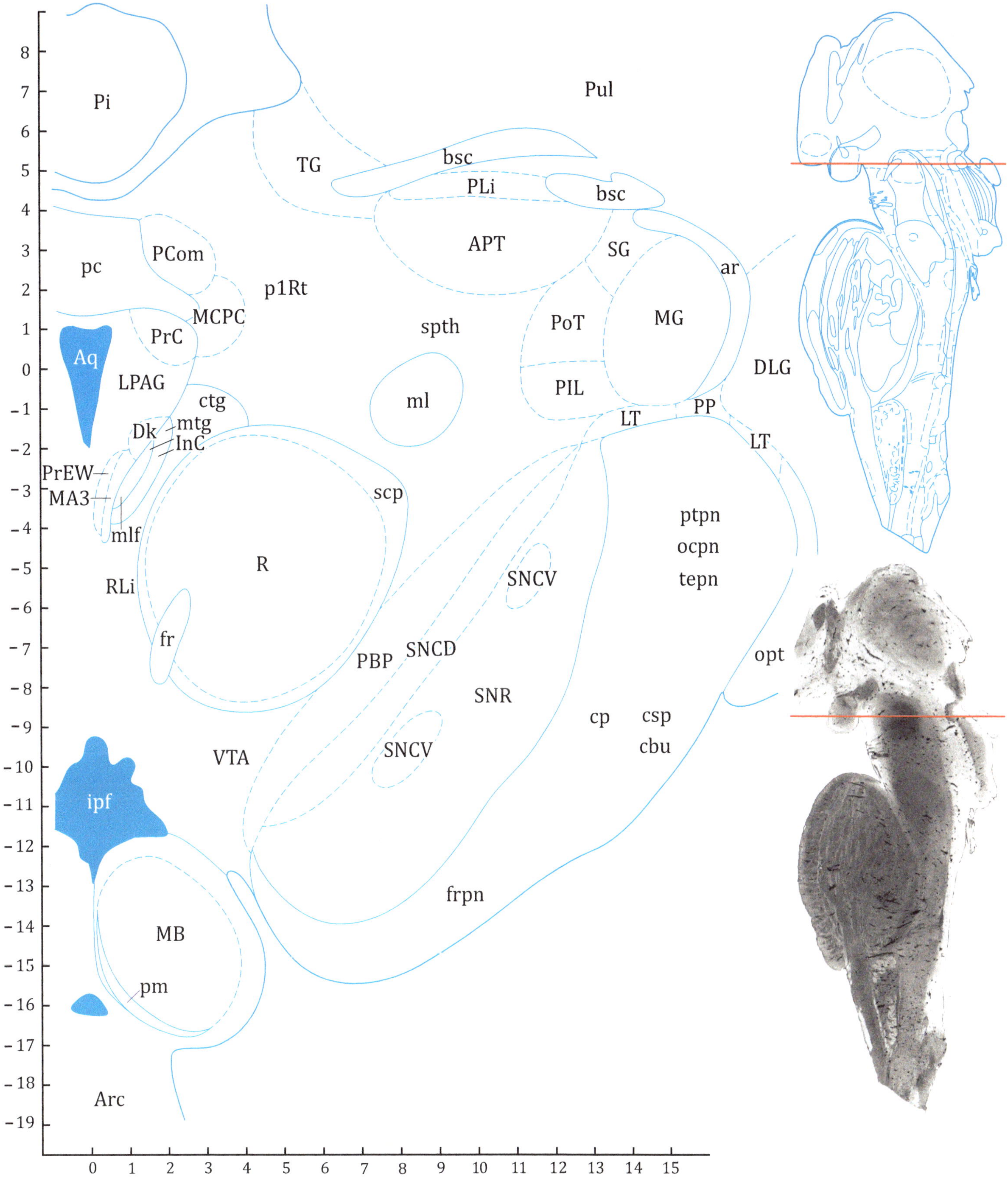

APT anterior pretectal nu
Aq aqueduct
ar acoustic radiation
Arc arcuate hypothal nu
bsc brachium superior colliculus
cbu corticobulbar tract
cp cerebral peduncle
csp corticospinal tract
ctg central tegmental tract
Dk nu of Darkschewitsch
DLG dorsal lat geniculate nu
fr fasciculus retroflexus
frpn frontopontine fibers
InC interstitial nu of Cajal
ipf interpedunc fossa
LPAG lat periaqueductal gray
LT lat terminal nu acc optic tr
MA3 med access oculomotor nu
MB mammillary body
MCPC magnocell nu post comm
MG med geniculate nu
ml medial lemniscus
mlf med longitudinal fasciculus
mtg mammillotegmental tract
ocpn occipitopontine fibers
opt optic tract
p1Rt prosomere 1 reticular formation
PBP parabrachial pigmented nu
pc posterior comm
PCom nu posterior comm
Pi pineal gland
PIL post intralaminar thal nu
PLi posterior limitans thal nu
pm principal mammillary tr
PoT posterior thal nu, triangular
PP peripeduncular nu
PrC precommissural nu
PrEW pre-Edinger-Westphal nu
ptpn parietal pontine fibers
Pul pulvinar nuclei
R red nu
RLi rostral linear nu raphe
scp superior Cb peduncle
SG suprageniculate thal nu
SNCD s nigra, compact, dors tier
SNCV s nigra, compact, vent tier
SNR s nigra, reticular
spth spinothalamic tract
tepn temporopontine fibers
TG tectal gray
VTA vent tegmental area

Fig 57b
Obex +49.0mm
Transverse 57

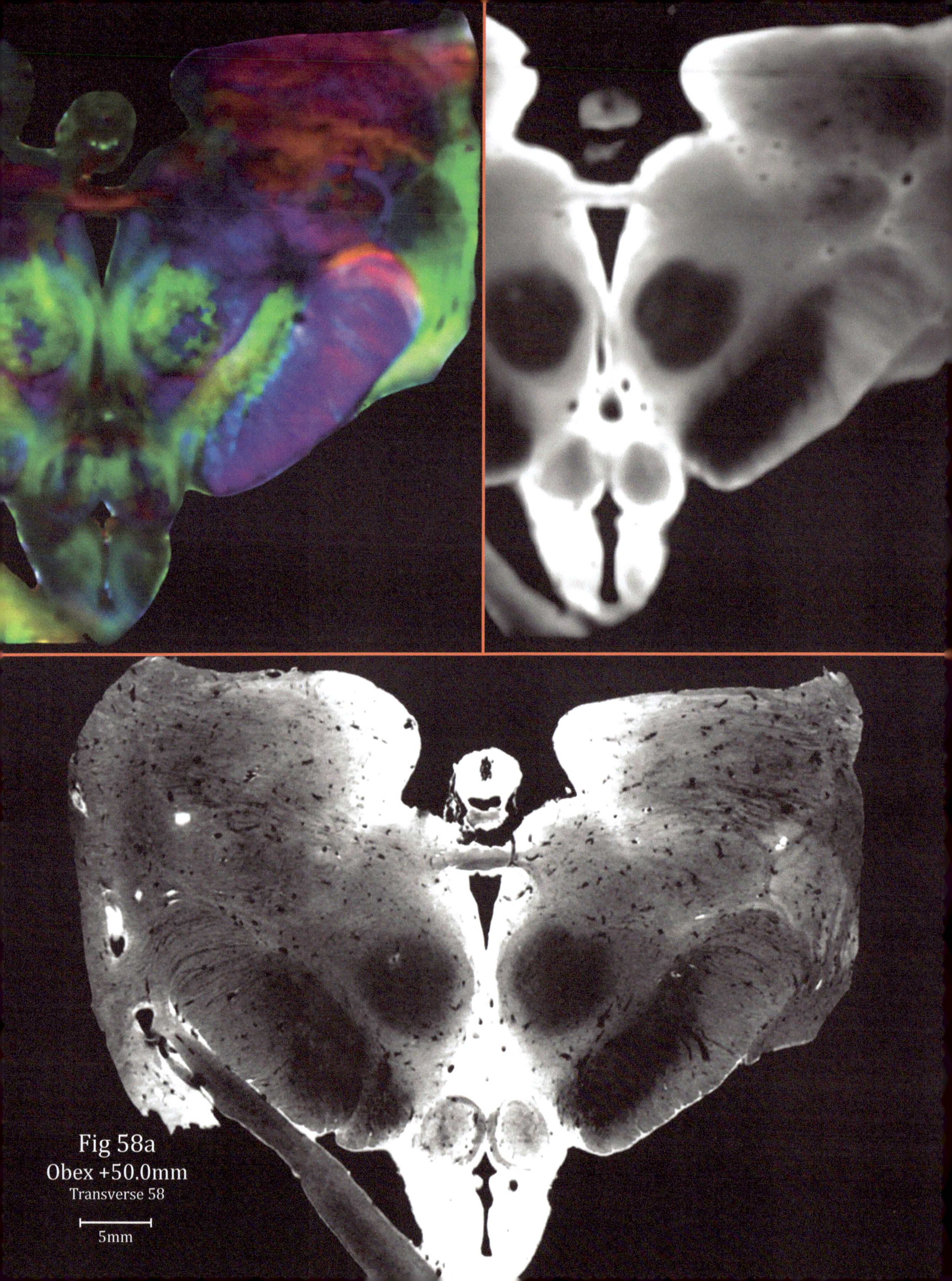

Fig 58a
Obex +50.0mm
Transverse 58

5mm

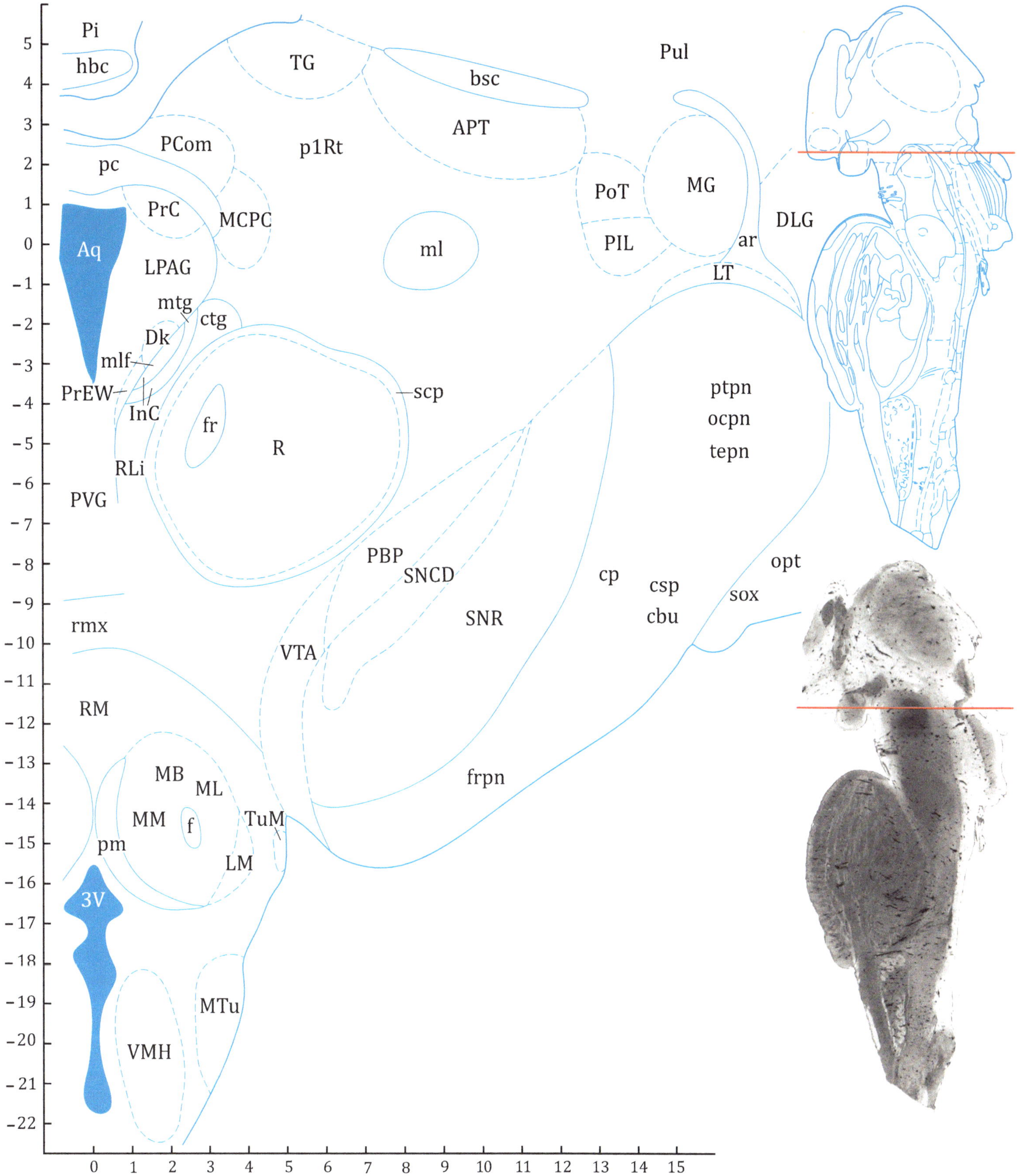

3V 3rd ventricle
APT anterior pretectal nu
Aq aqueduct
ar acoustic radiation
bsc brachium superior colliculus
cbu corticobulbar tract
cp cerebral peduncle
csp corticospinal tract
ctg central tegmental tract
Dk nu of Darkschewitsch
DLG dorsal lat geniculate nu
f fornix
fr fasciculus retroflexus
frpn frontopontine fibers
hbc habenular commissure
InC interstitial nu of Cajal
LM lateral mammillary nu
LPAG lat periaqueductal gray
LT lat terminal nu acc optic tr
MB mammillary body
MCPC magnocell nu post comm
MG med geniculate nu
ML med mammillary nu, lat
ml medial lemniscus
mlf med longitudinal fasciculus
MM med mammillary nu, medial
mtg mammillotegmental tract
MTu medial tuberal nu
ocpn occipitopontine fibers
opt optic tract
p1Rt prosomere 1 reticular formation
PBP parabrachial pigmented nu
pc posterior comm
PCom nu posterior comm
Pi pineal gland
PIL post intralaminar thal nu
pm principal mammillary tr
PoT posterior thal nu, triangular
PrC precommissural nu
PrEW pre-Edinger-Westphal nu
ptpn parietal pontine fibers
Pul pulvinar nuclei
PVG periventricular gray
R red nu
RLi rostral linear nu raphe
RM retromammillary nu
rmx retromammillary decuss
scp superior Cb peduncle
SNCD s nigra, compact, dors tier
SNR s nigra, reticular
sox supraoptic decussation
tepn temporopontine fibers

Fig 58b
Obex +50.0mm
Transverse 58

TG tectal gray
TuM tuberomammillary nucleus
VMH ventmed hypothal nu
VTA vent tegmental area

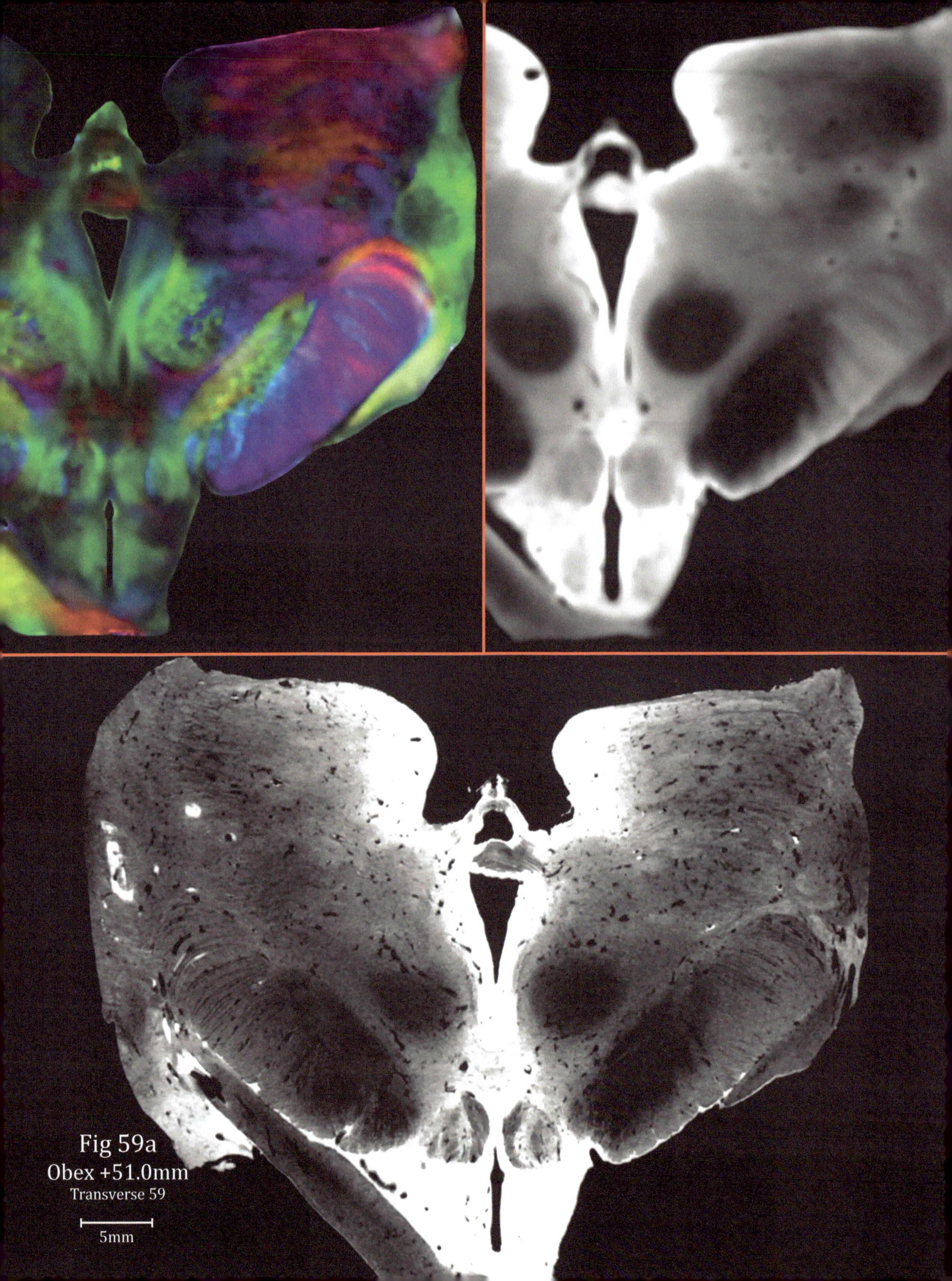

Fig 59a
Obex +51.0mm
Transverse 59

5mm

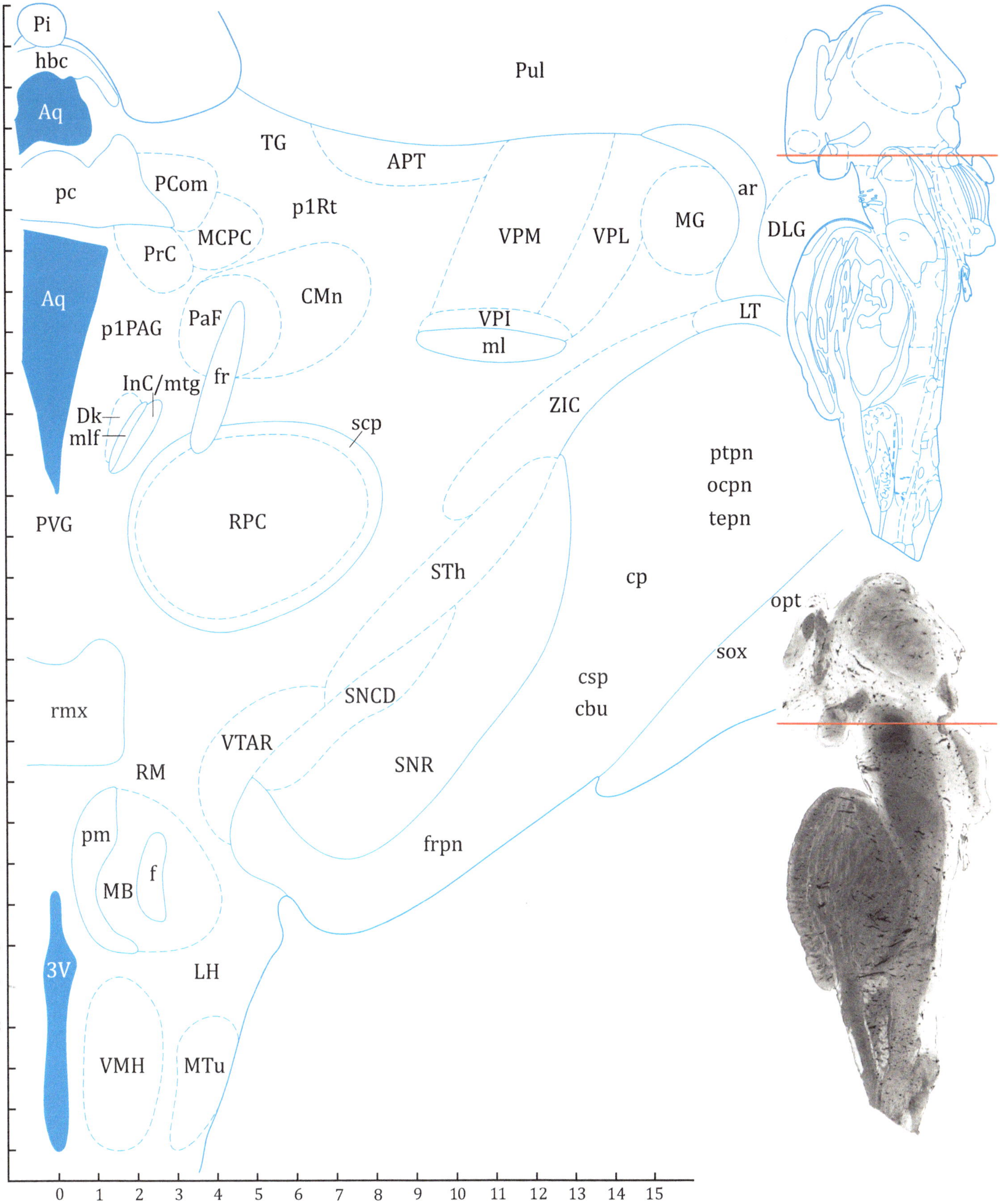

3V 3rd ventricle
APT anterior pretectal nu
Aq aqueduct
ar acoustic radiation
cbu corticobulbar tract
CMn centromedian thalamic nucleus
cp cerebral peduncle
csp corticospinal tract
Dk nu of Darkschewitsch
DLG dorsal lat geniculate nu
f fornix
fr fasciculus retroflexus
frpn frontopontine fibers
hbc habenular commissure
InC/mtg interstitial nu/mam teg tract
LH lateral hypothalamic area
LT lat terminal nu acc optic tr
MB mammillary body
MCPC magnocell nu post comm
MG med geniculate nu
ml medial lemniscus
mlf med longitudinal fasciculus
MTu medial tuberal nu
ocpn occipitopontine fibers
opt optic tract
p1PAG p1 periaqueductal gray
p1Rt prosomere 1 reticular formation
PaF parafascicular thal nu
pc posterior comm
PCom nu posterior comm
Pi pineal gland
pm principal mammillary tr
PrC precommissural nu
ptpn parietal pontine fibers
Pul pulvinar nuclei
PVG periventricular gray
RM retromammillary nu
rmx retromammillary decuss
RPC red nu, parvicell part
scp superior Cb peduncle
SNCD s nigra, compact, dors tier
SNR s nigra, reticular
sox supraoptic decussation
STh subthalamic nu
tepn temporopontine fibers
TG tectal gray
VMH ventmed hypothal nu
VPI ventral posterior inferior nucleus
VPL ventral posterolat thal nu
VPM vent posteromed thal nu
VTAR vent tegmental area, rostr
ZIC zona incerta, caudal

Fig 59b
Obex +51.0mm
Transverse 59

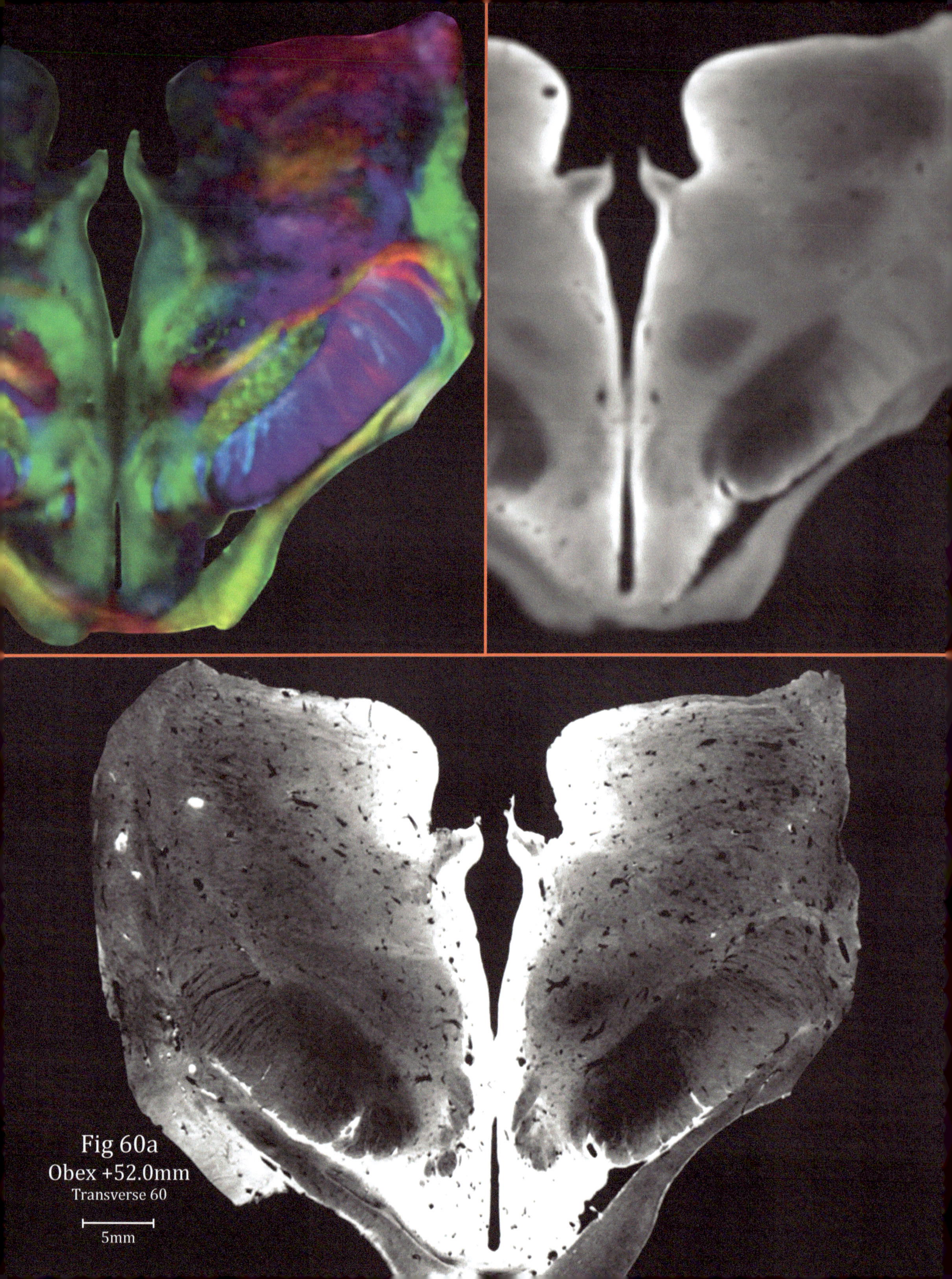

Fig 60a
Obex +52.0mm
Transverse 60

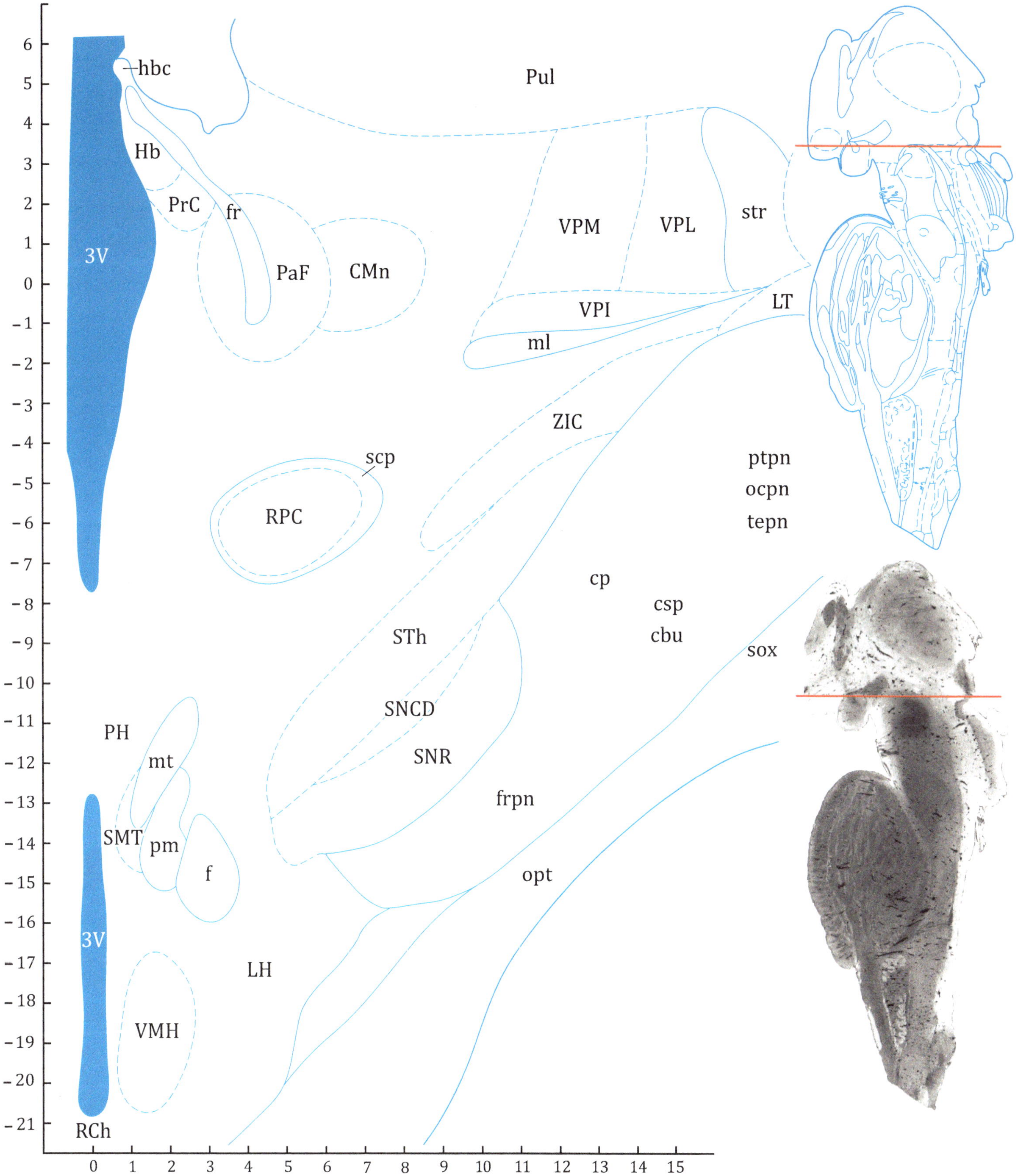

3V 3rd ventricle
cbu corticobulbar tract
CMn centromedian thalamic nucleus
cp cerebral peduncle
csp corticospinal tract
f fornix
fr fasciculus retroflexus
frpn frontopontine fibers
Hb habenular nuclei
hbc habenular commissure
LH lateral hypothalamic area
LT lat terminal nu acc optic tr
ml medial lemniscus
mt mammillothalamic tract
ocpn occipitopontine fibers
opt optic tract
PaF parafascicular thal nu
PH posterior hypothal nu
pm principal mammillary tr
PrC precommissural nu
ptpn parietal pontine fibers
Pul pulvinar nuclei
RCh retrochiasmatic area
RPC red nu, parvicell part
scp superior Cb peduncle
SMT submammillothalamic nu
SNCD s nigra, compact, dors tier
SNR s nigra, reticular
sox supraoptic decussation
STh subthalamic nu
str superior thalamic radiation
tepn temporopontine fibers
VMH ventmed hypothal nu
VPI ventral posterior inferior nucleus
VPL ventral posterolat thal nu
VPM vent posteromed thal nu
ZIC zona incerta, caudal

Fig 60b
Obex +52.0mm
Transverse 60

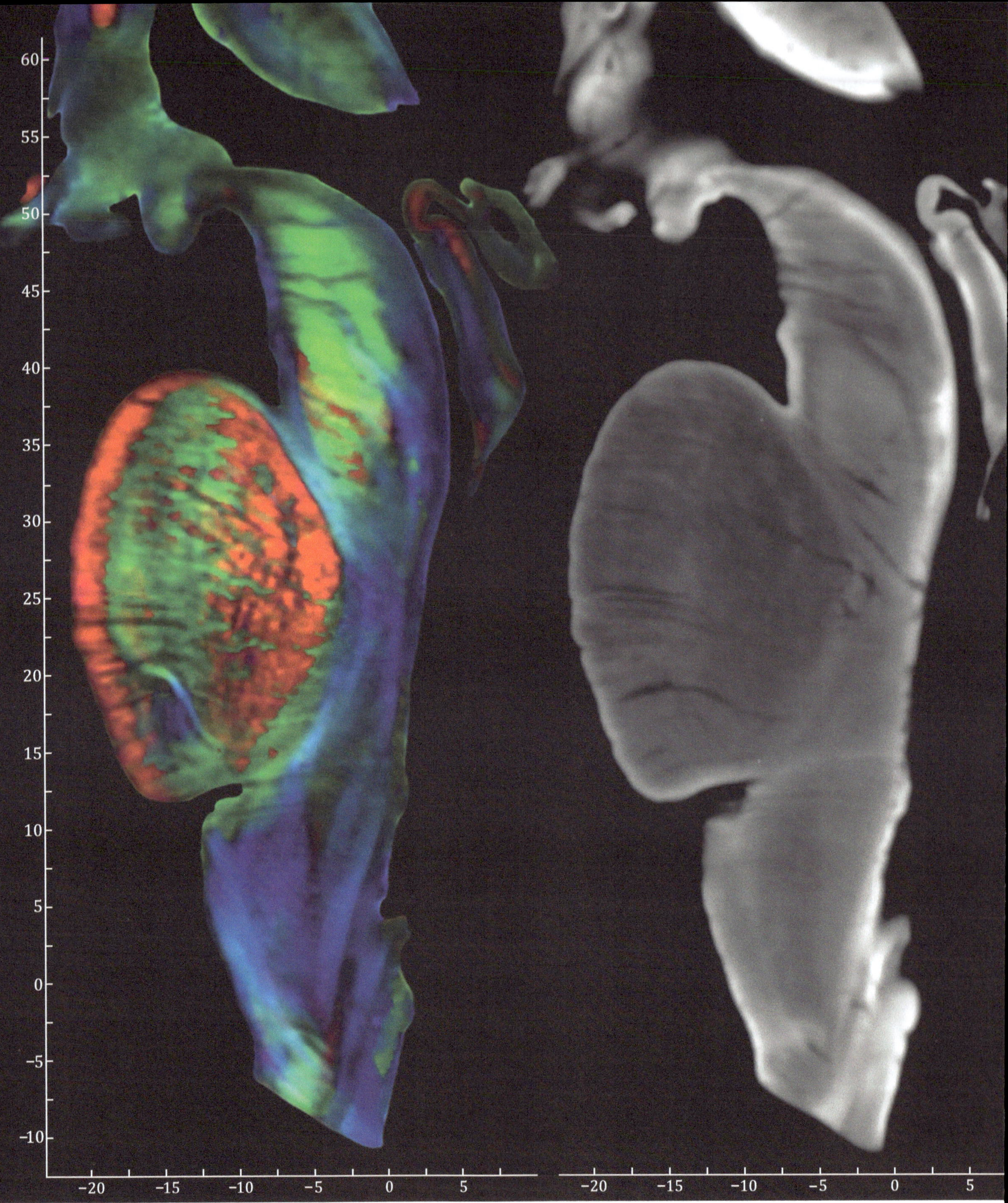

Fig 61a
Obex +0.0mm
Sagittal 01

- **10N** dorsal motor nu of vagus
- **12N** hypoglossal nu
- **3V** 3rd ventricle
- **4V** 4th ventricle
- **AP** area postrema
- **Aq** aqueduct
- **Ar** arcuate nu of the caudal hindbrain
- **Arc** arcuate hypothal nu
- **Au** austral nucleus
- **CeCv** central cervical nu
- **cic** commissure inf colliculus
- **CLi** caudal linear nu raphe
- **cu** cuneate fasciculus
- **DCIC** dors cx inf colliculus
- **DMPAG** dorsomed periaqueductl gray
- **DpWh** deep white superior coll
- **DR** dorsal raphe nu
- **DRC** dorsal raphe nu, caudal
- **Gi** gigantocellular reticular nu
- **GiA** gigantocellular retic, alpha
- **gr** gracile fasciculus
- **Gr** gracile nu
- **hbc** habenular commissure
- **hypothal** hypothalamus
- **I3** interoculomotor nucleus
- **ia** internal arcuate fibers
- **IF** interfascicular nu
- **In** intercalated nu medulla
- **IOM** inf olive, med nu
- **IP** interpedunc nu
- **ipf** interpedunc fossa
- **lfp** longitudinal fasciculus pons
- **LPAG** lat periaqueductal gray
- **MB** mammillary body
- **ml** medial lemniscus
- **mlf** med longitudinal fasciculus
- **MnR** median raphe nu
- **Mx** matrix region medulla
- **Nt** noto cuneate nucleus
- **ocx** olivocerebellar bundle decussation
- **p1PVG** p1 periventricular gray
- **pc** posterior comm
- **Pi** pineal gland
- **PiRe** pineal recess
- **PlGl** pleioglia periaqueductal gray
- **pm** principal mammillary tr

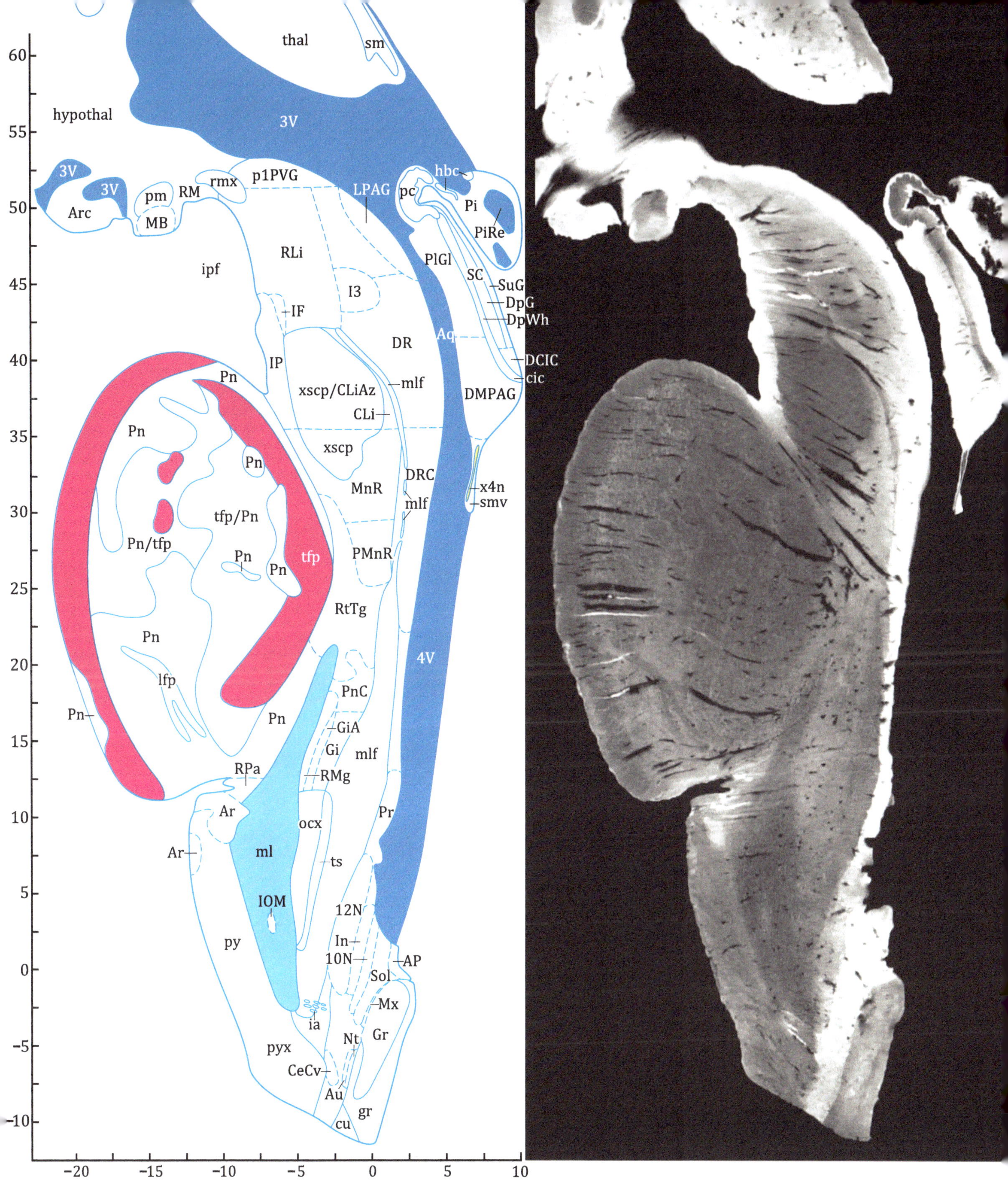

PMnR paramedian raphe nu
Pn pontine nuclei
Pn/tfp pontine nu/transv fibres pons
PnC pontine reticular nu, caudal
Pr prepositus nu
py pyramidal tr
pyx pyramidal decussation
RLi rostral linear nu raphe
RM retromammillary nu
RMg raphe magnus nu
rmx retromammillary decuss

RPa raphe pallidus nu
RtTg reticulotegmental nu pons
SC superior colliculus
sm stria medullaris thal
smv superior medullary velum
Sol nu of solitary tract
SuG superficial gray sup coll
tfp transverse fibers pons
tfp/Pn transv fibers pons/pontine nu
thal thalamus
ts tectospinal tr

x4n decussation of the trochlear nerve
xscp decussation sup cereb ped
xscp/CLiAz dec scp/caud lin nu, azygos

Fig 61b
Obex +0.0mm
Sagittal 01

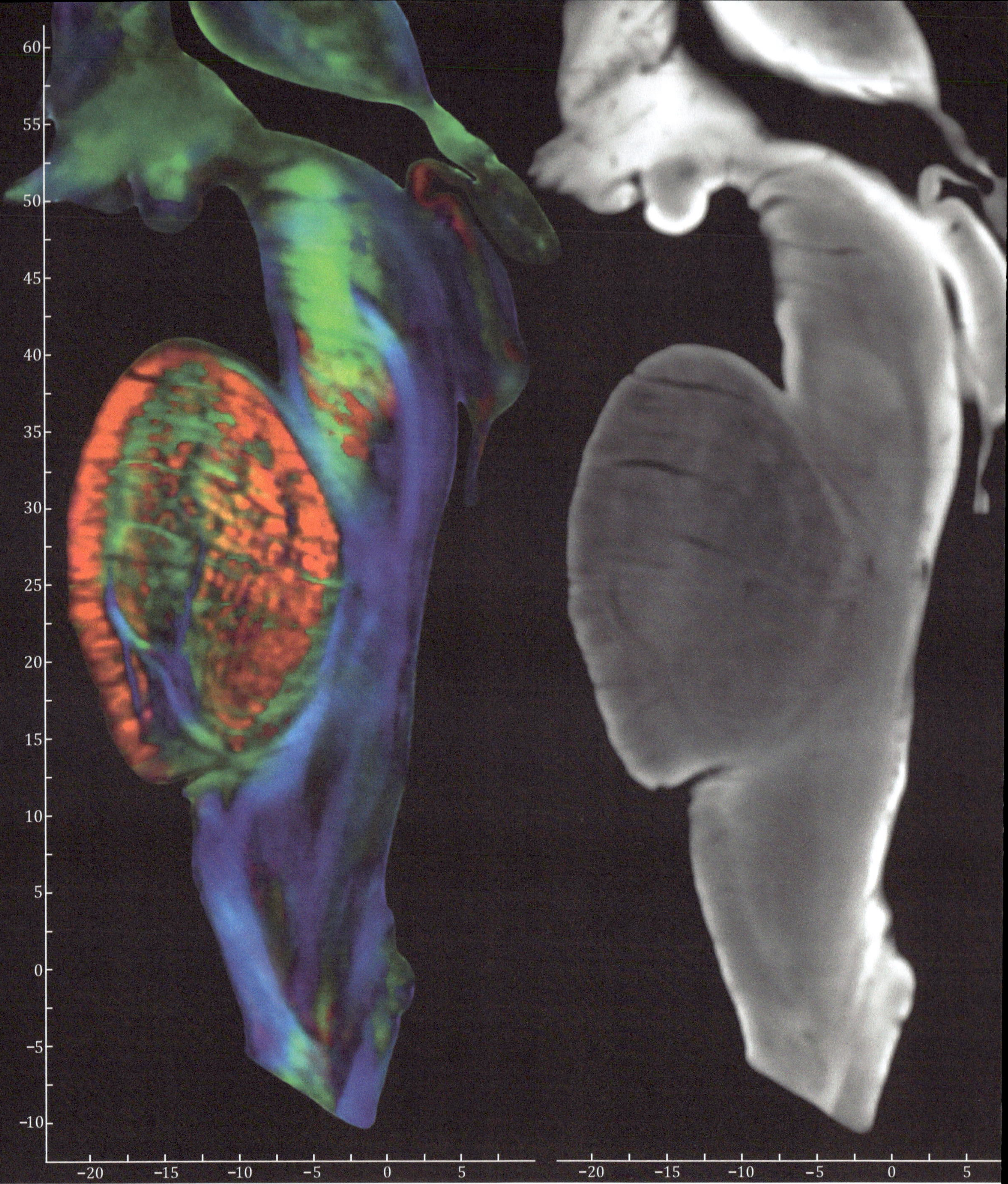

Fig 62a
Obex +0.5mm
Sagittal 02

10N dorsal motor nu of vagus
12N hypoglossal nu
3n oculomotor nerve
3N oculomotor nu
3V 3rd ventricle
4n trochlear nerve
4V 4th ventricle
ami amiculum of the olive
AP area postrema
Aq aqueduct
Ar arcuate nu of the caudal hindbrain
CeCv central cervical nu
CG central gray
cic commissure inf colliculus
CLi caudal linear nu raphe
CLiAz caudal linear nu of raphe, azygos
Cu cuneate nu
cu cuneate fasciculus
DCIC dors cx inf colliculus
DLPAG dorsolat periaqueductal gray
DMPAG dorsomed periaqueductl gray
DpG deep gray superior coll
DPGi dors paragigantocellular nu
DpWh deep white superior coll
DR dorsal raphe nu
DRC dorsal raphe nu, caudal
DTg dorsal tegmental nu
EW Edinger-Westphal nu
fr fasciculus retroflexus
g7 genu facial nerve
Gi gigantocellular reticular nu
GiA gigantocellular retic, alpha
gr gracile fasciculus
Gr gracile nu
Hb habenular nuclei
hbc habenular commissure
hypothal hypothalamus
I3 interoculomotor nucleus
ia internal arcuate fibers
In intercalated nu medulla
InG intermed gray layer SC
InWh intermediate white layer SC
IOBe inf olive, beta subnu
IODM inf olive, dorsomed cell group
IOK inf olive, cap of Kooy med nu
IOM inf olive, med nu

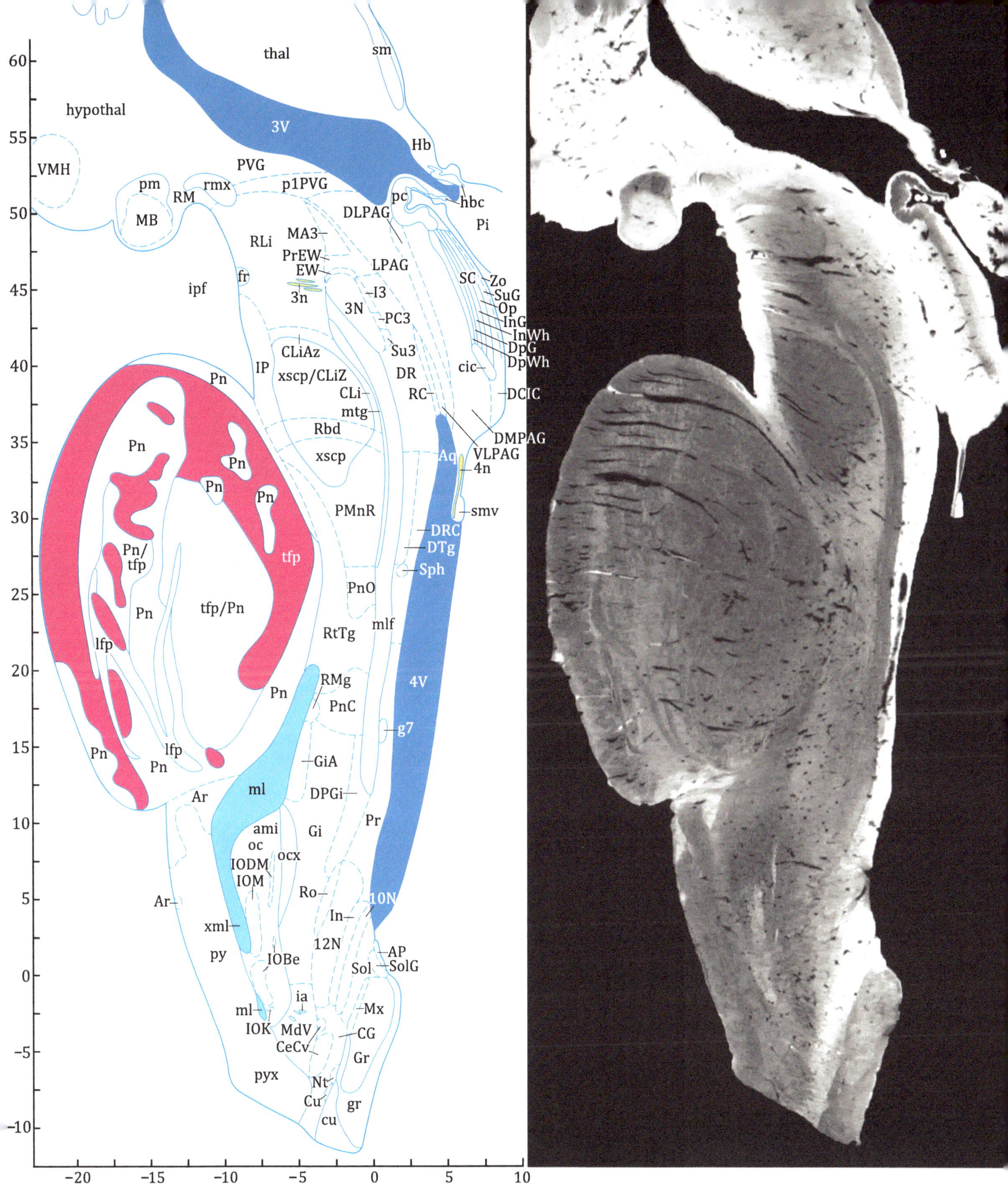

IP interpedunc nu
ipf interpedunc fossa
lfp longitudinal fasciculus pons
LPAG lat periaqueductal gray
MA3 med access oculomotor nu
MB mammillary body
MdV medullary reticular nu, vent
ml medial lemniscus
mlf med longitudinal fasciculus
mtg mammillotegmental tract
Mx matrix region medulla
Nt noto cuneate nucleus
oc olivocerebellar tract
ocx olivocerebellar bundle decussation
Op optic nerve layer sup coll
p1PVG p1 periventricular gray
pc posterior comm
PC3 parvicellular oculomotor nucleus
Pi pineal gland
pm principal mammillary tr
PMnR paramedian raphe nu
Pn pontine nuclei
Pn/tfp pontine nu/transv fibres pons
PnC pontine reticular nu, caudal
PnO pontine reticular nu, oral
Pr prepositus nu
PrEW pre-Edinger-Westphal nu
PVG periventricular gray
py pyramidal tr
pyx pyramidal decussation
Rbd rhabdoid nu
RC raphe cap
RLi rostral linear nu raphe
RM retromammillary nu
RMg raphe magnus nu
rmx retromammillary decuss
Ro nu of Roller
RtTg reticulotegmental nu pons
SC superior colliculus
sm stria medullaris thal
smv superior medullary velum
Sol nu of solitary tract
SolG nu sol tract, gelatinous
Sph sphenoid nu
Su3 supraoculomotor PAG
See List of Structures

Fig 62b
Obex +0.5mm
Sagittal 02

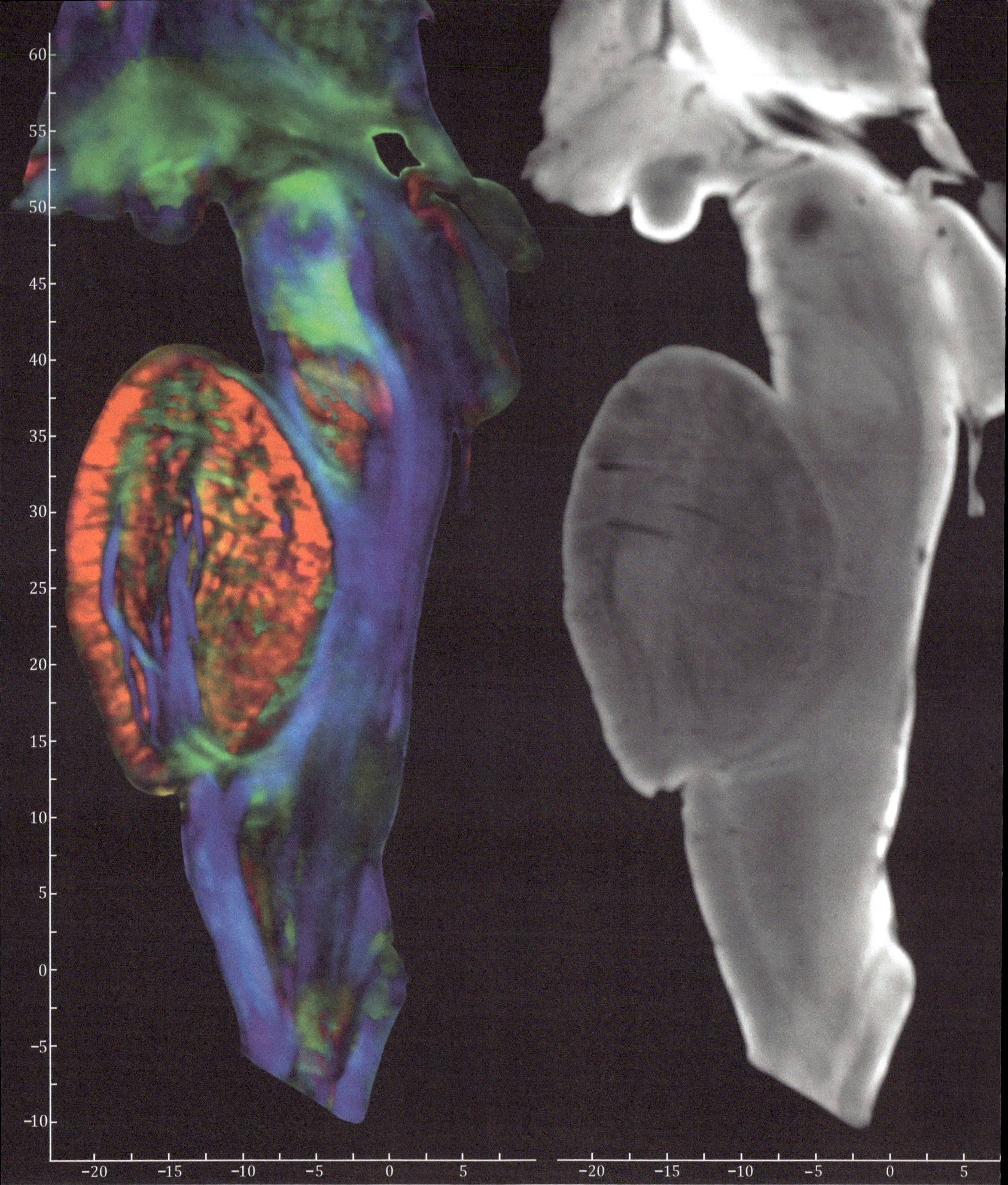

Fig 63a
Obex +1.0mm
Sagittal 03

10N dorsal motor nu of vagus
12N hypoglossal nu
3n oculomotor nerve
3N oculomotor nu
3V 3rd ventricle
4N trochlear nu
4n trochlear nerve
4V 4th ventricle
9Sp Rexed's lamina
ami amiculum of the olive
Aq aqueduct
Ar arcuate nu of the caudal hindbrain
CeCv central cervical nu
CGPn central gray pons
cic commissure inf colliculus
CLi caudal linear nu raphe
CLiAz caudal linear nu of raphe, azygos
Cu cuneate nu
cu cuneate fasciculus
DCIC dors cx inf colliculus
Dk nu of Darkschewitsch
DLPAG dorsolat periaqueductal gray
DMPAG dorsomed periaqueductl gray
DpG deep gray superior coll
DPGi dors paragigantocellular nu
DpWh deep white superior coll
DR dorsal raphe nu
DTg dorsal tegmental nu
EL endolemniscal nucleus
f fornix
fr **fasciculus retroflexus**
g7 genu facial nerve
Gi gigantocellular reticular nu
GiA gigantocellular retic, alpha
gr gracile fasciculus
Gr gracile nu
Hb habenular nuclei
hbc habenular commissure
hio hilus of the inferior olive
hypothal hypothalamus
ia **internal arcuate fibers**
In intercalated nu medulla
InG intermed gray layer SC
InWh intermediate white layer SC
IOBe inf olive, beta subnu
IOC inf olive, subnu C of med nu

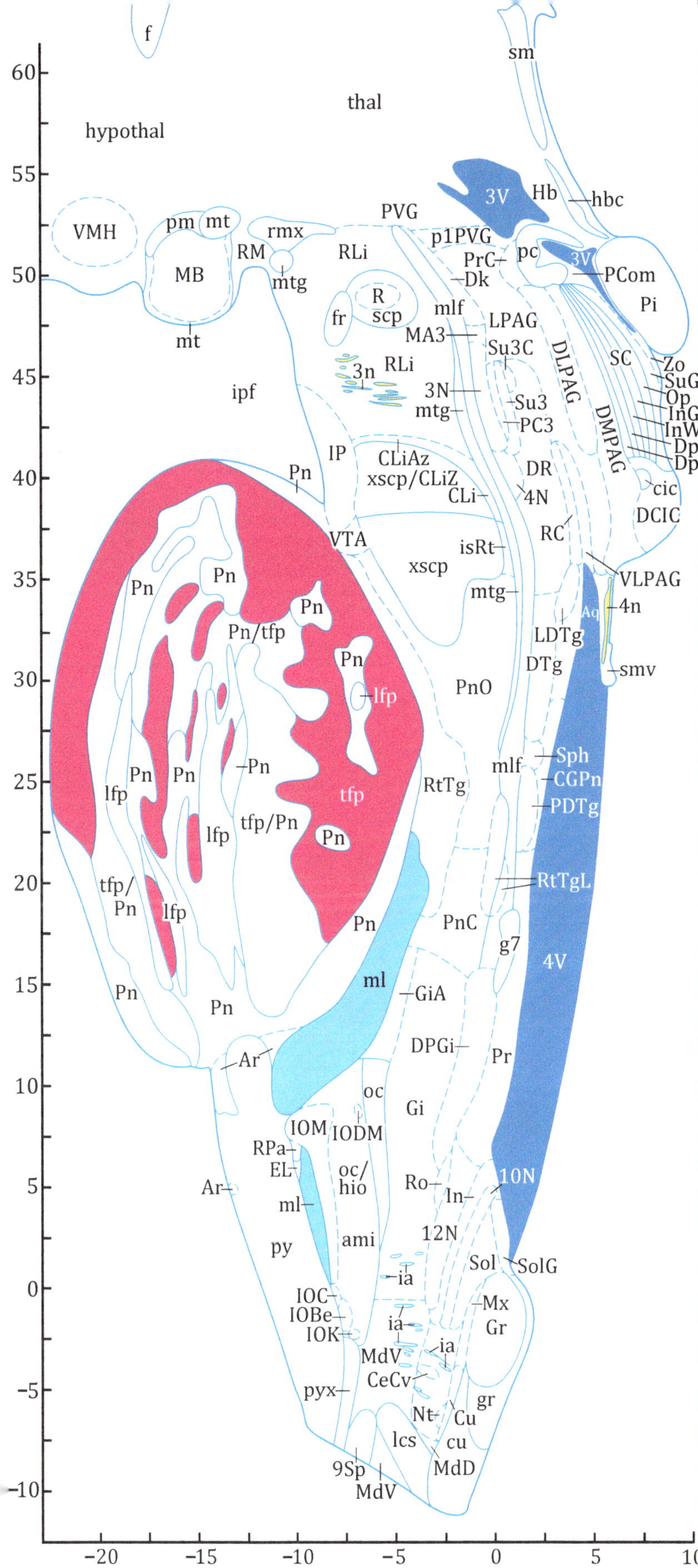

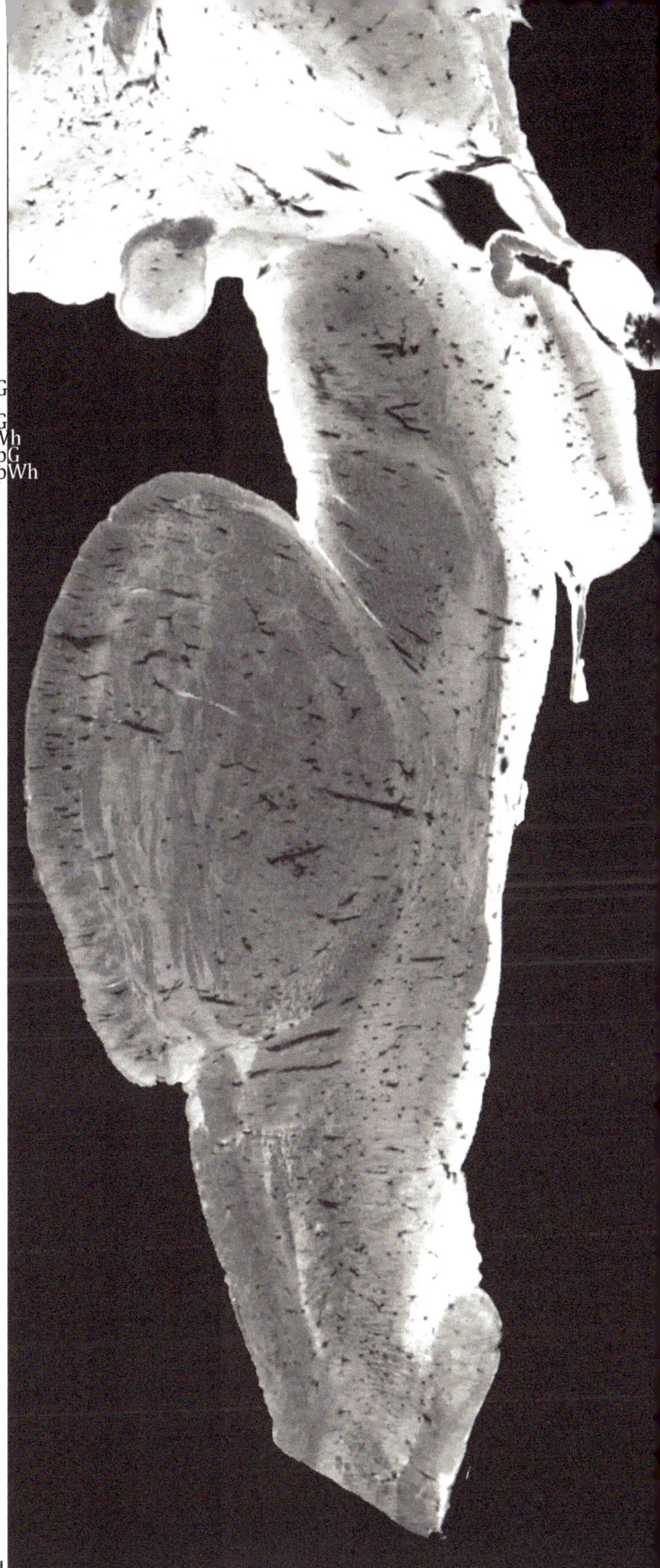

IODM inf olive, dorsomed cell group
IOK inf olive, cap of Kooy med nu
IOM inf olive, med nu
IP interpedunc nu
ipf interpedunc fossa
isRt isthmic reticular formation
lcs lateral corticospinal tract
LDTg laterodor tegmental nu
lfp longitudinal fasciculus pons
LPAG lat periaqueductal gray
MA3 med access oculomotor nu

MB mammillary body
MdD medullary reticular nu, dors
MdV medullary reticular nu, vent
ml medial lemniscus
mlf med longitudinal fasciculus
mt mammillothalamic tract
mtg mammillotegmental tract
Mx matrix region medulla
Nt noto cuneate nucleus
oc olivocerebellar tract
oc/hio olivocerebel tr/hilus inf olive

Op optic nerve layer sup coll
p1PVG p1 periventricular gray
pc posterior comm
PC3 parvicellular oculomotor nucleus
PCom nu posterior comm
PDTg posterodors tegmental nu
Pi pineal gland
pm principal mammillary tr
Pn pontine nuclei
Pn/tfp pontine nu/transv fibres pons
PnC pontine reticular nu, caudal

PnO pontine reticular nu, oral
Pr prepositus nu
PrC precommissural nu
PVG periventricular gray
py pyramidal tr
pyx pyramidal decussation
R red nu
RC raphe cap
RLi rostral linear nu raphe
RM retromammillary nu
rmx retromammillary decuss

Fig 63b
Obex +1.0mm
Sagittal 03

Ro nu of Roller
See List of Structures

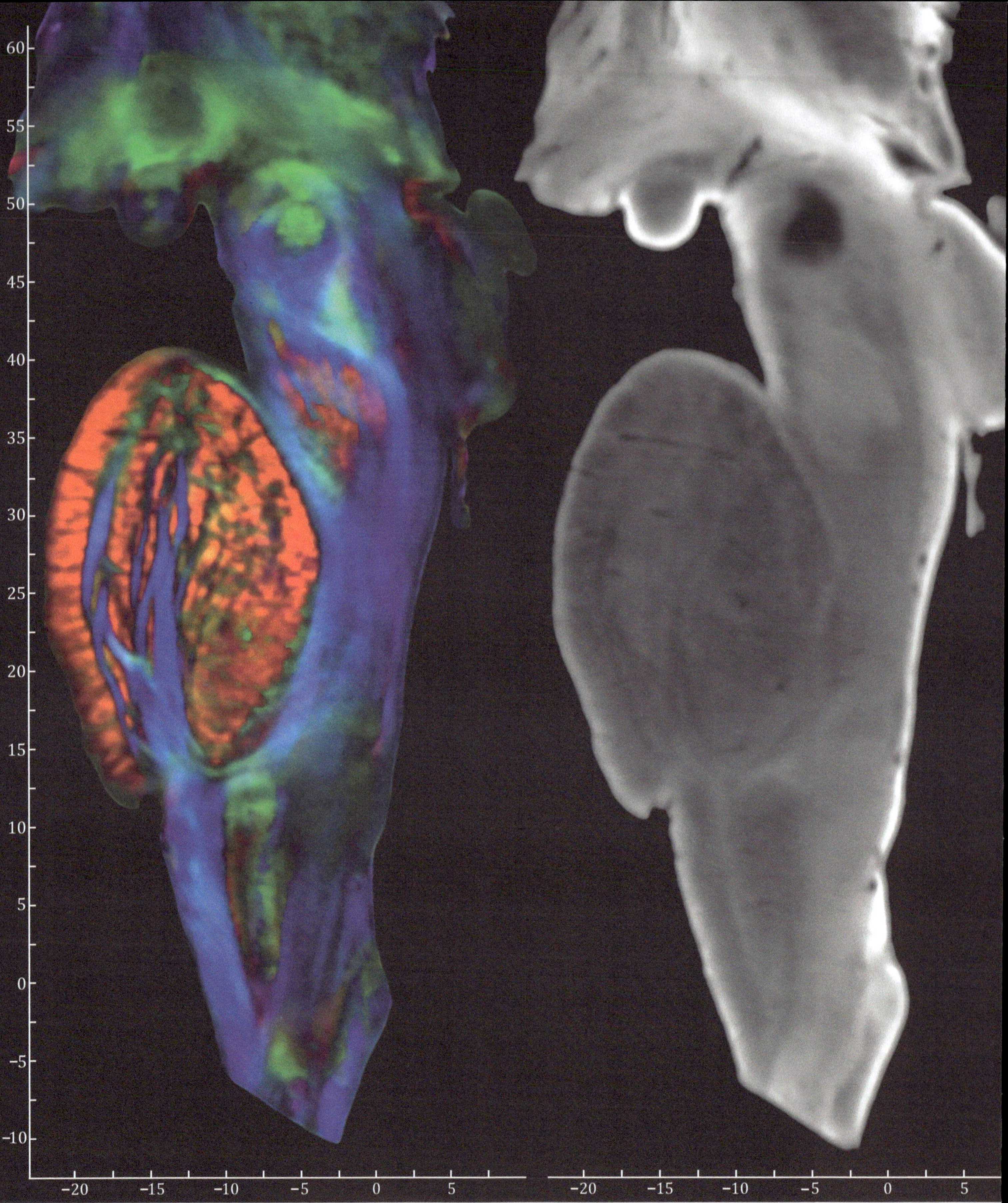

Fig 64a
Obex +1.5mm
Sagittal 04

10N dorsal motor nu of vagus
12n root of hypoglossal nerve
3n oculomotor nerve
3V 3rd ventricle
4N trochlear nu
4V 4th ventricle
6n root of abducens nerve
7n facial nerve
9Sp Rexed's lamina
ami amiculum of the olive
Ar arcuate nu of the caudal hindbrain
B9 B9 serotonin cells
CGPn central gray pons
cic commissure inf colliculus
Ct conterminal nu
ctg central tegmental tract
Cu cuneate nu
cu cuneate fasciculus
DCIC dors cx inf colliculus
Dk nu of Darkschewitsch
DLPAG dorsolat periaqueductal gray
DMTg dorsomed tegmental area
DpG deep gray superior coll
DPGi dors paragigantocellular nu
DpWh deep white superior coll
DR dorsal raphe nu
DTg dorsal tegmental nu
EL endolemniscal nucleus
f fornix
fr fasciculus retroflexus
g7 genu facial nerve
Gi gigantocellular reticular nu
GiA gigantocellular retic, alpha
GiV gigantocellular retic, vent
gr gracile fasciculus
Gr gracile nu
Hb habenular nuclei
hio hilus of the inferior olive
hypothal hypothalamus
ia internal arcuate fibers
In intercalated nu medulla
InC/mtg interstitial nu/mam teg tract
InG intermed gray layer SC
InWh intermediate white layer SC
IOBe inf olive, beta subnu
IOD inf olive, dorsal nu

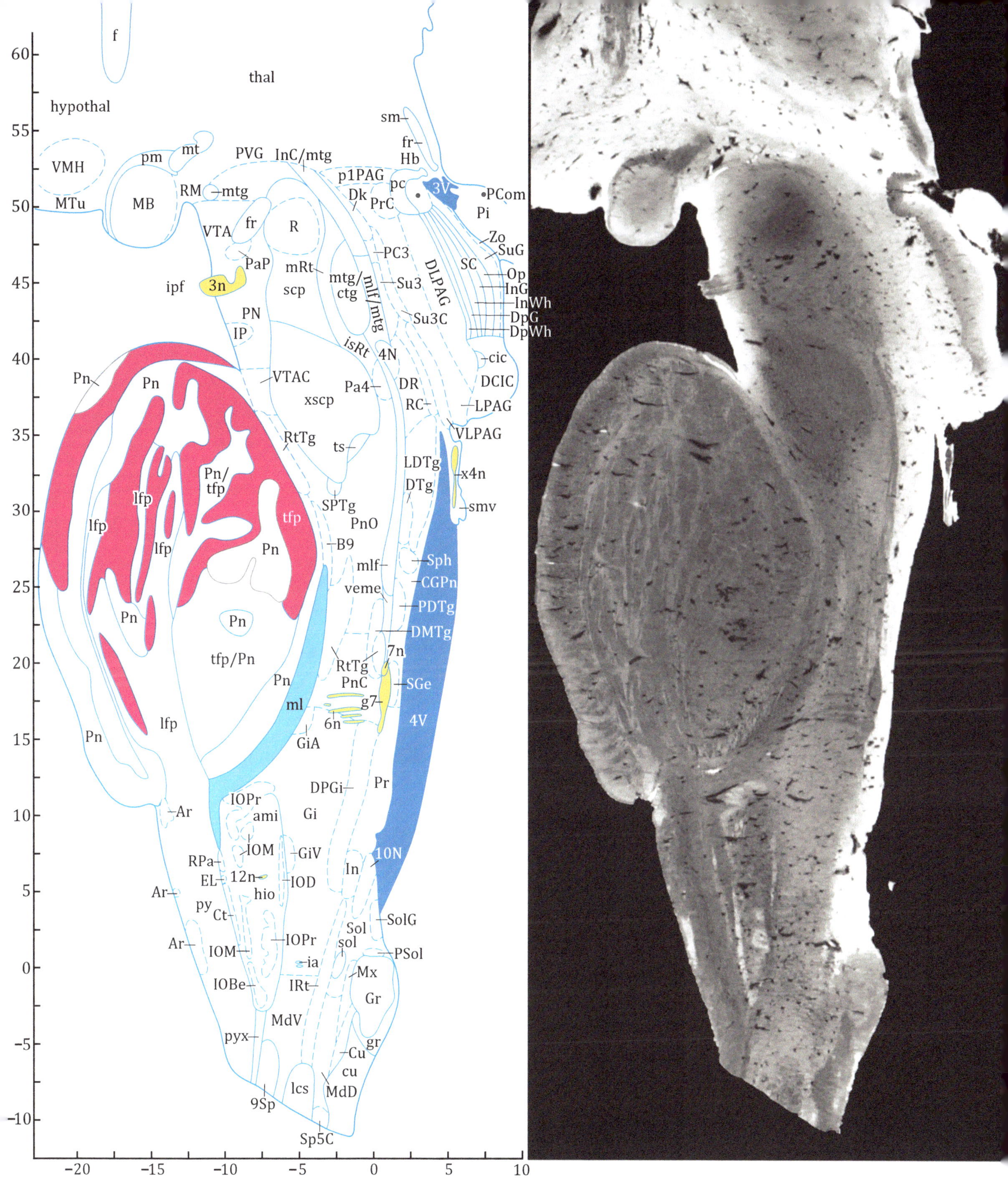

IOM inf olive, med nu
IOPr inf olive, principal nu
IP interpeduncular nu
ipf interpeduncular fossa
IRt intermed reticular nu
isRt isthmic reticular formation
lcs lateral corticospinal tract
LDTg laterodor tegmental nu
lfp longitudinal fasciculus pons
LPAG lat periaqueductal gray
MB mammillary body

MdD medullary reticular nu, dors
MdV medullary reticular nu, vent
ml medial lemniscus
mlf med longitudinal fasciculus
mlf/mtg medial long fas/mam teg tr
mRt mesencephalic retic form
mt mammillothalamic tract
mtg mammillotegmental tract
mtg/ctg mam teg tr/central teg tr
MTu medial tuberal nu
Mx matrix region medulla

Op optic nerve layer sup coll
p1PAG p1 periaqueductal gray
Pa4 paratrochlear nu
PaP parapeduncular nucleus
pc posterior comm
PC3 parvicellular oculomotor nucleus
PCom nu posterior comm
PDTg posterodors tegmental nu
Pi pineal gland
pm principal mammillary tr
PN paranigral nu of the VTA

Pn pontine nuclei
Pn/tfp pontine nu/transv fibres pons
PnC pontine reticular nu, caudal
PnO pontine reticular nu, oral
Pr prepositus nu
PrC precommissural nu
PSol parasolitary nu
PVG periventricular gray
py pyramidal tr
pyx pyramidal decussation
R red nu

RC raphe cap
See List of Structure

Fig 64b
Obex +1.5mm
Sagittal 04

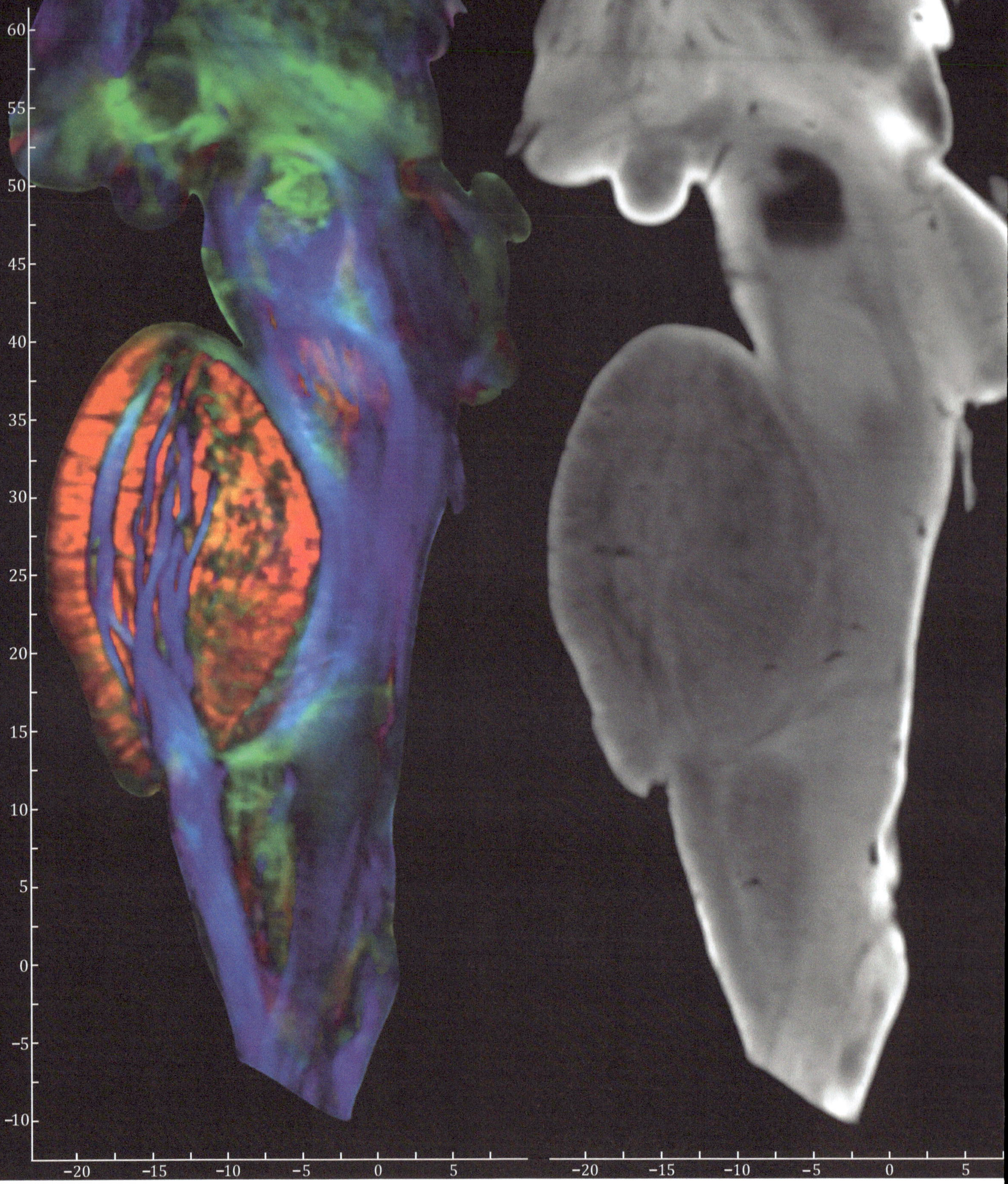

Fig 65a
Obex +2.0mm
Sagittal 05

10N dorsal motor nu of vagus
3n oculomotor nerve
4n trochlear nerve
4V 4th ventricle
6N abducens nu
6n root of abducens nerve
ami amiculum of the olive
Ar arcuate nu of the caudal hindbrain
B9 B9 serotonin cells
CGB central gray, beta part
CGPn central gray pons
cic commissure inf colliculus
Ct conterminal nu
ctg central tegmental tract
Cu cuneate nu
cu cuneate fasciculus
DCIC dors cx inf colliculus
Dk nu of Darkschewitsch
DLPAG dorsolat periaqueductal gray
DMTg dorsomed tegmental area
DpG deep gray superior coll
DPGi dors paragigantocellular nu
DpWh deep white superior coll
DR dorsal raphe nu
ECIC external cx inferior coll
f fornix
fr fasciculus retroflexus
g7 genu facial nerve
Gi gigantocellular reticular nu
GiV gigantocellular retic, vent
gr gracile fasciculus
Gr gracile nu
Hb habenular nuclei
hio hilus of the inferior olive
hypothal hypothalamus
ia internal arcuate fibers
ID intradecussational nu scp
In intercalated nu medulla
InC/mtg interstitial nu/mam teg tract
InG intermed gray layer SC
InWh intermediate white layer SC
IOA inf olive, subnu A medial nu
IOB inf olive, subnu B med nu
IOD inf olive, dorsal nu
IOM inf olive, med nu
IOPr inf olive, principal nu

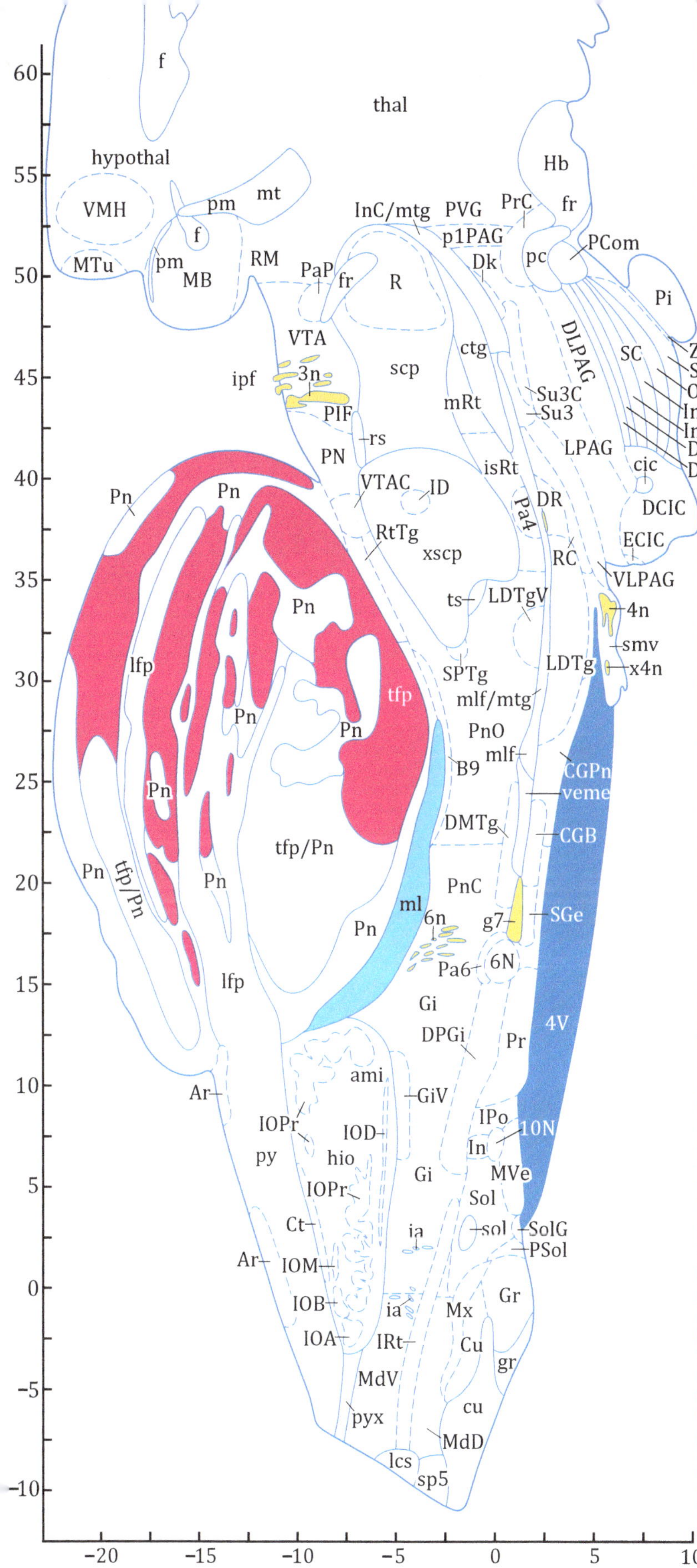

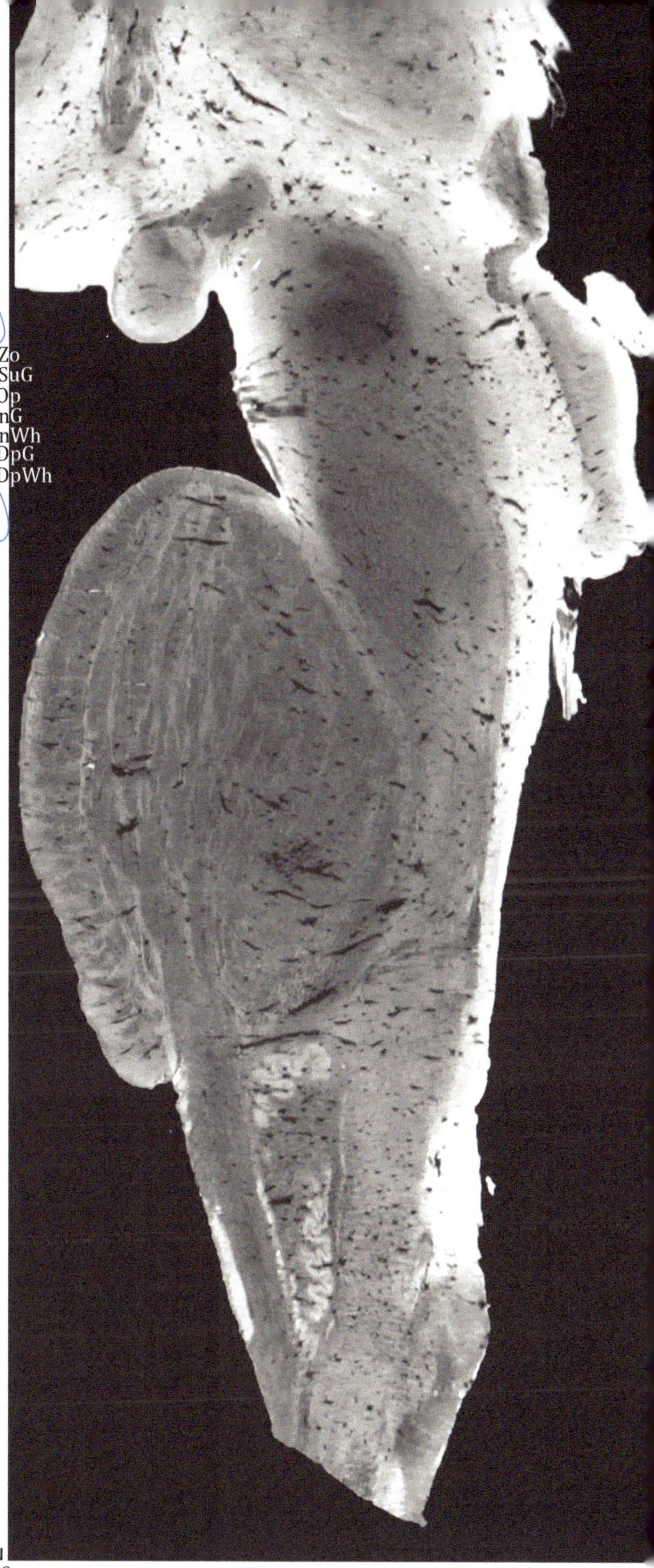

ipf interpedunc fossa
IPo interpositus
IRt intermed reticular nu
isRt isthmic reticular formation
lcs lateral corticospinal tract
LDTg laterodor tegmental nu
LDTgV laterodor tegmental nu, vent
lfp longitudinal fasciculus pons
LPAG lat periaqueductal gray
MB mammillary body
MdD medullary reticular nu, dors

MdV medullary reticular nu, vent
ml medial lemniscus
mlf med longitudinal fasciculus
mlf/mtg medial long fas/mam teg tr
mRt mesencephalic retic form
mt mammillothalamic tract
MTu medial tuberal nu
MVe medial vestibular nu
Mx matrix region medulla
Op optic nerve layer sup coll
p1PAG p1 periaqueductal gray

Pa4 paratrochlear nu
Pa6 paraabducens nu
PaP parapeduncular nucleus
pc posterior comm
PCom nu posterior comm
Pi pineal gland
PIF parainterfascicular nu VTA
pm principal mammillary tr
Pn pontine nuclei
PN paranigral nu of the VTA
PnC pontine reticular nu, caudal

PnO pontine reticular nu, oral
Pr prepositus nu
PrC precommissural nu
PSol parasolitary nu
PVG periventricular gray
py pyramidal tr
pyx pyramidal decussation
R red nu
RC raphe cap
RM retromammillary nu
rs rubrospinal tract

RtTg reticulotegmental nu pons
See List of Structures

Fig 65b
Obex +2.0mm
Sagittal 05

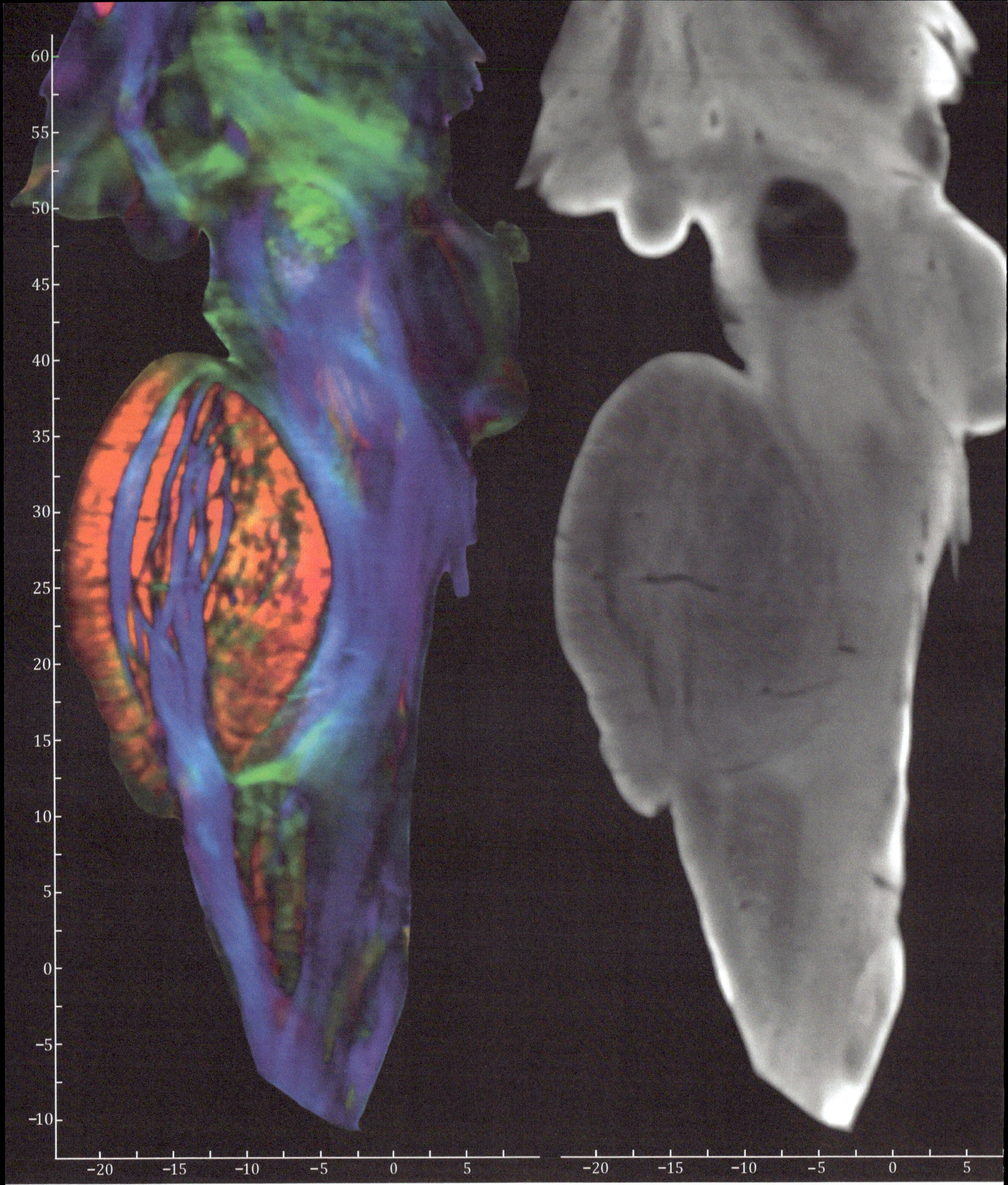

Fig 66a
Obex +2.5mm
Sagittal 06

10N dorsal motor nu of vagus
3n oculomotor nerve
4n trochlear nerve
4V 4th ventricle
6n root of abducens nerve
6N abducens nu
7n facial nerve
ami amiculum of the olive
Aq aqueduct
Ar arcuate nu of the caudal hindbrain
B9 B9 serotonin cells
CGB central gray, beta part
CGPn central gray pons
cp cerebral peduncle
Ct conterminal nu
ctg central tegmental tract
Cu cuneate nu
cu cuneate fasciculus
DCIC dors cx inf colliculus
DLPAG dorsolat periaqueductal gray
DMTg dorsomed tegmental area
DpG deep gray superior coll
DPGi dors paragigantocellular nu
DpWh deep white superior coll
DR dorsal raphe nu
ECIC external cx inferior coll
f fornix
fr fasciculus retroflexus
g7 genu facial nerve
Gi gigantocellular reticular nu
GiV gigantocellular retic, vent
gr gracile fasciculus
Gr gracile nu
Hb habenular nuclei
hio hilus of the inferior olive
hypothal hypothalamus
InC interstitial nu of Cajal
InG intermed gray layer SC
InWh intermediate white layer SC
IOA inf olive, subnu A medial nu
IOD inf olive, dorsal nu
IOPr inf olive, principal nu
IPo interpositus
IRt intermed reticular nu
isRt isthmic reticular formation
lcs lateral corticospinal tract

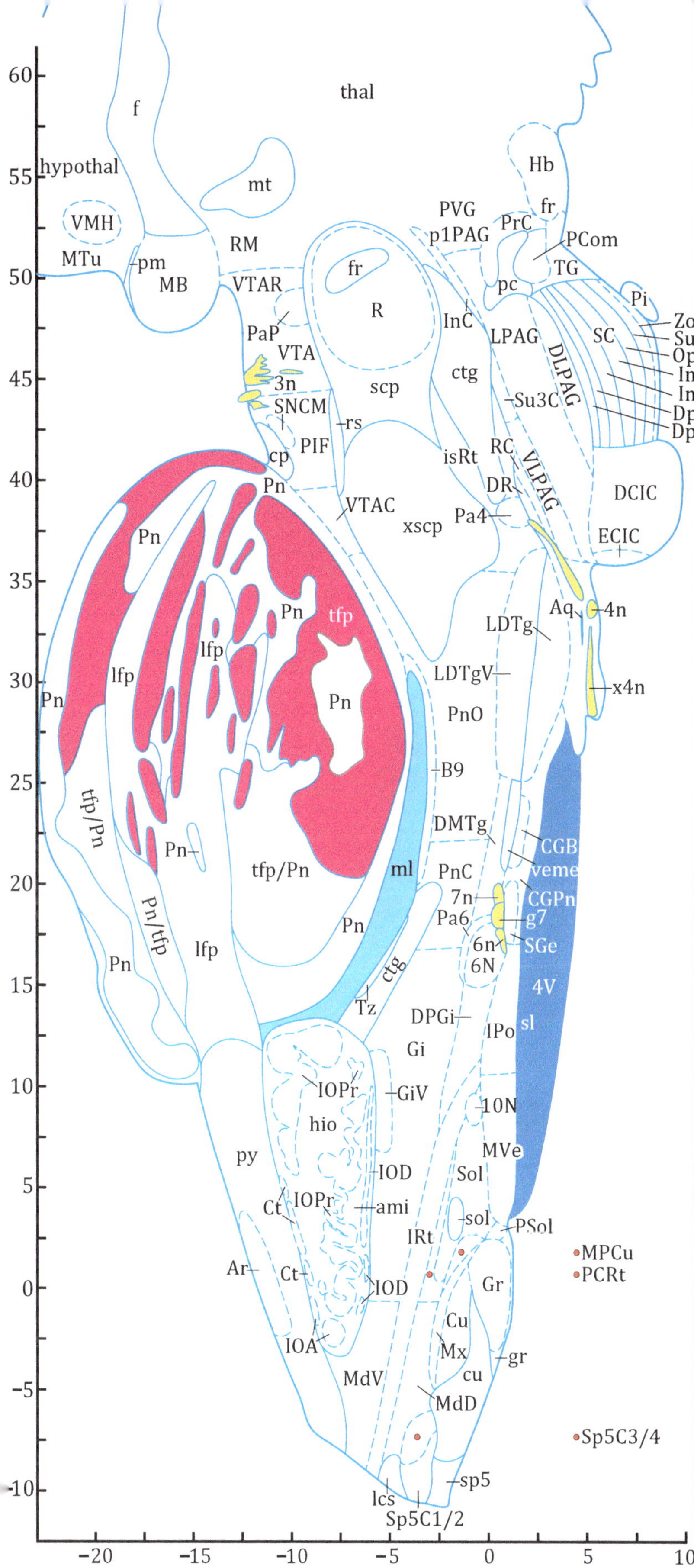

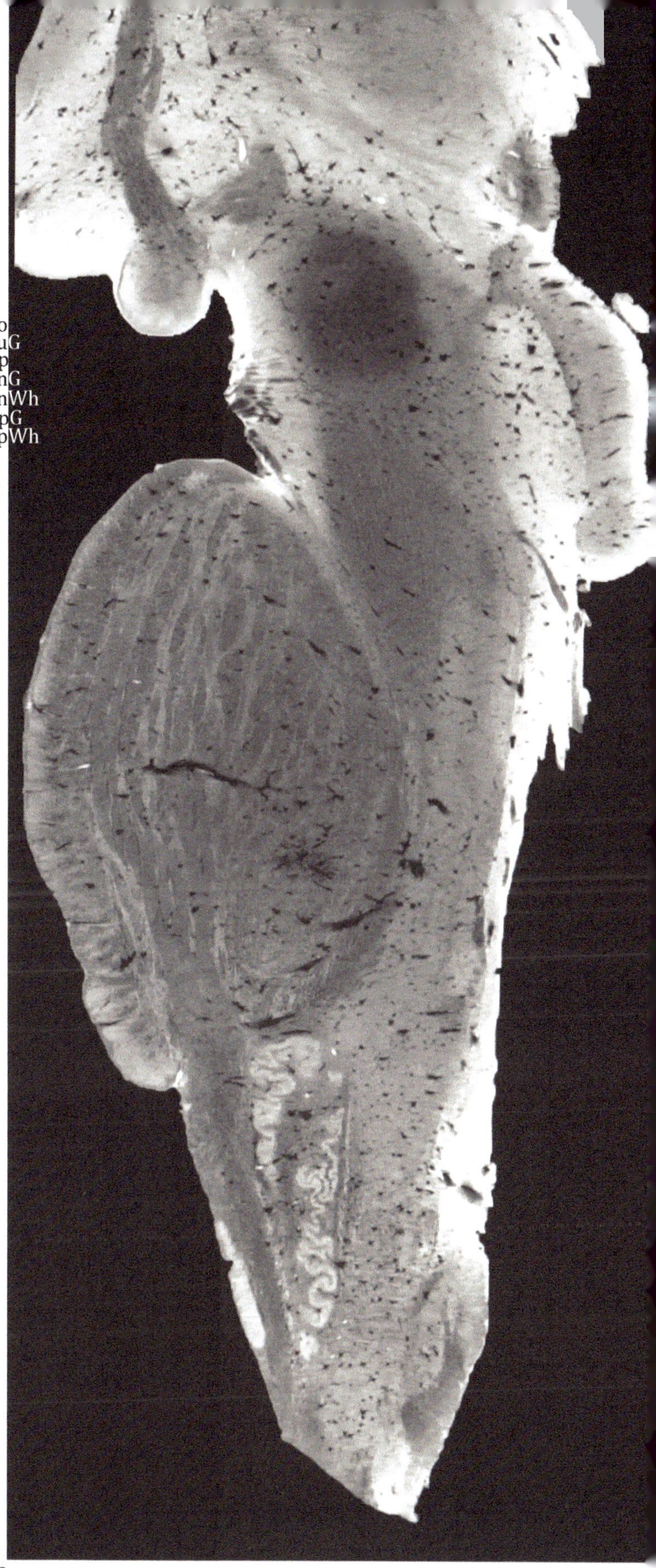

.DTg laterodor tegmental nu
.DTgV laterodor tegmental nu, vent
fp longitudinal fasciculus pons
.PAG lat periaqueductal gray
MB mammillary body
MdD medullary reticular nu, dors
MdV medullary reticular nu, vent
nl medial lemniscus
MPCu medial pericuneate nucleus
nt mammillothalamic tract
MTu medial tuberal nu

MVe medial vestibular nu
Mx matrix region medulla
Op optic nerve layer sup coll
p1PAG p1 periaqueductal gray
Pa4 paratrochlear nu
Pa6 paraabducens nu
PaP parapeduncular nucleus
pc posterior comm
PCom nu posterior comm
PCRt parvicell reticular nu
Pi pineal gland

PIF parainterfascicular nu VTA
pm principal mammillary tr
Pn pontine nuclei
Pn/tfp pontine nu/transv fibres pons
PnC pontine reticular nu, caudal
PnO pontine reticular nu, oral
PrC precommissural nu
PSol parasolitary nu
PVG periventricular gray
py pyramidal tr
R red nu

RC raphe cap
RM retromammillary nu
rs rubrospinal tract
SC superior colliculus
scp superior Cb peduncle
SGe supragenual nu
sl sulcus limitans
SNCM s nigra, compact, med tier
sol solitary tract
Sol nu of solitary tract
sp5 spinal trigem tract

Fig 66b
Obex +2.5mm
Sagittal 06

Sp5C1/2 spinal 5 nu, caud, lam 1 & 2

See List of Structures

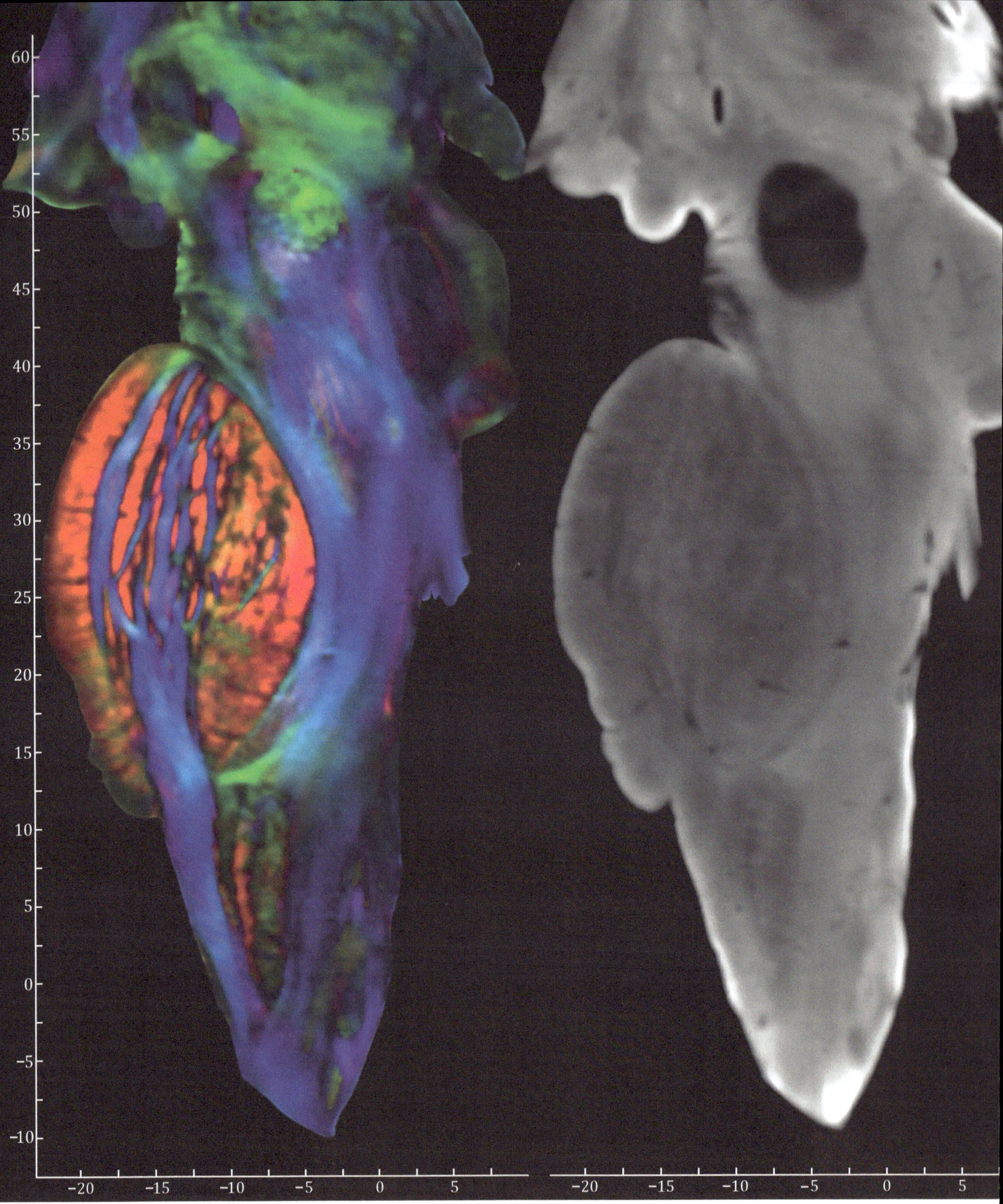

Fig 67a
Obex +3.0mm
Sagittal 07

3n oculomotor nerve
4n trochlear nerve
4V 4th ventricle
6N abducens nu
7n facial nerve
AmbL ambiguus nu, loose part
ami amiculum of the olive
Ar arcuate nu of the caudal hindbrain
B9 B9 serotonin cells
CGPn central gray pons
CIC central nu inf colliculus
CnF cuneiform nu
cp cerebral peduncle
Crb cribiform nucleus
ctg central tegmental tract
Cu cuneate nu
cu cuneate fasciculus
DCIC dors cx inf colliculus
DMTg dorsomed tegmental area
DpG deep gray superior coll
DPGi dors paragigantocellular nu
DpWh deep white superior coll
EC epicoeruleus nucleus
ECIC external cx inferior coll
f fornix
fr fasciculus retroflexus
Gi gigantocellular reticular nu
GiV gigantocellular retic, vent
gr gracile fasciculus
Gr gracile nu
Hb habenular nuclei
hio hilus of the inferior olive
hypothal hypothalamus
InG intermed gray layer SC
InWh intermediate white layer SC
IOA inf olive, subnu A medial nu
IOD inf olive, dorsal nu
IOPr inf olive, principal nu
IPo interpositus
IRt intermed reticular nu
isRt isthmic reticular formation
LC locus coeruleus
lcs lateral corticospinal tract
LDTg laterodor tegmental nu
LDTgV laterodor tegmental nu, vent
lfp longitudinal fasciculus pons

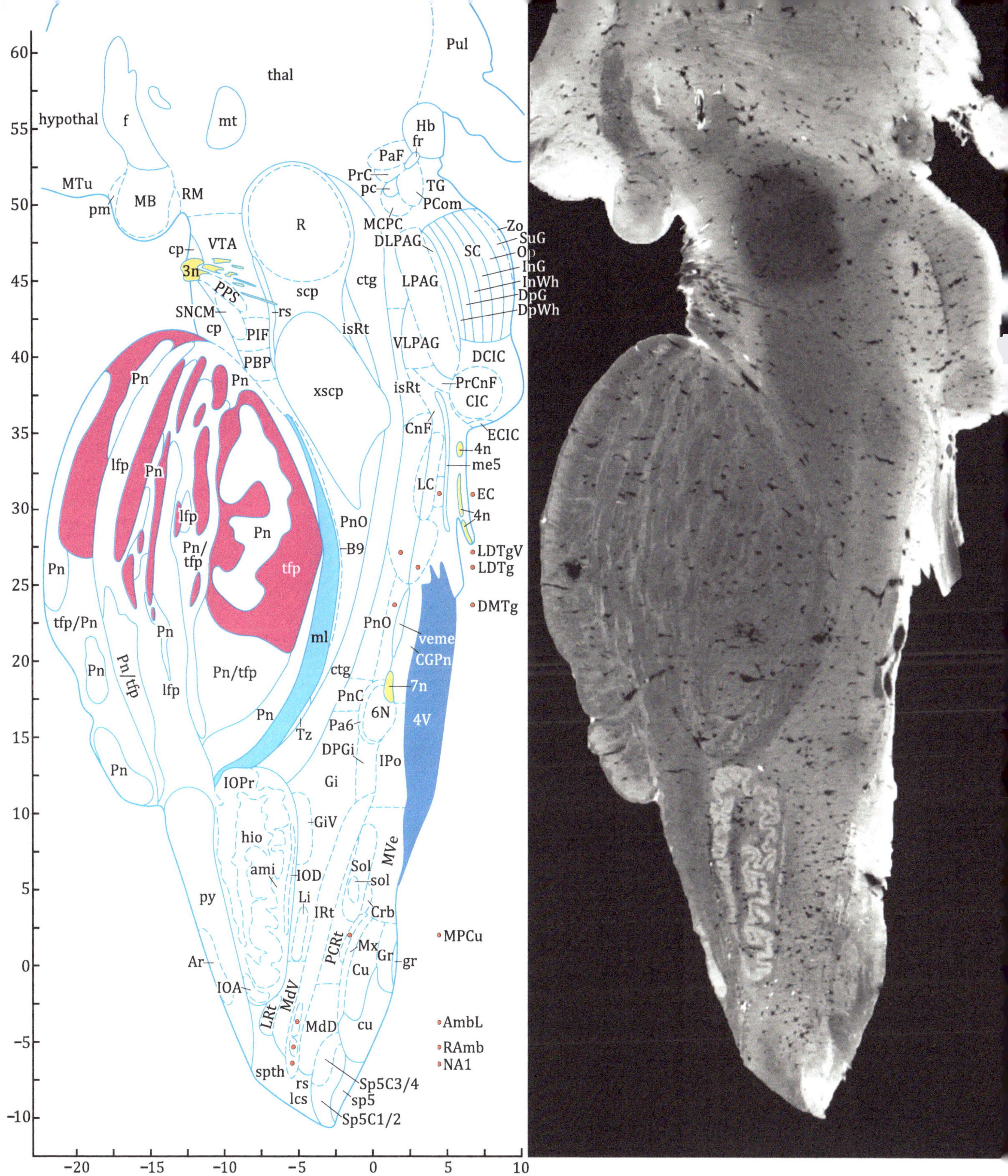

Li linear nu of the medulla
LPAG lat periaqueductal gray
LRt lateral reticular nu
MB mammillary body
MCPC magnocell nu post comm
MdD medullary reticular nu, dors
MdV medullary reticular nu, vent
me5 mesencephalic trigem tract
ml medial lemniscus
MPCu medial pericuneate nucleus
mt mammillothalamic tract
MTu medial tuberal nu
MVe medial vestibular nu
Mx matrix region medulla
NA1 A1 noradrenaline cells
Op optic nerve layer sup coll
Pa6 paraabducens nu
PaF parafascicular thal nu
PBP parabrachial pigmented nu
pc posterior comm
PCom nu posterior comm
PCRt parvicell reticular nu
PIF parainterfascicular nu VTA
pm principal mammillary tr
Pn pontine nuclei
Pn/tfp pontine nu/transv fibres pons
PnC pontine reticular nu, caudal
PnO pontine reticular nu, oral
PPS posterior perforated substance
PrC precommissural nu
PrCnF precuneiform area
Pul pulvinar nuclei
py pyramidal tr
R red nu
RAmb retroambiguus nu
RM retromammillary nu
rs rubrospinal tract
SC superior colliculus
scp superior Cb peduncle
SNCM s nigra, compact, med tier
sol solitary tract
Sol nu of solitary tract
sp5 spinal trigem tract
Sp5C1/2 spinal 5 nu, caud, lam 1 & 2
Sp5C3/4 spinal 5 nu, caud, lam 3 & 4

See List of Structures

Fig 67b
Obex +3.0mm
Sagittal 07

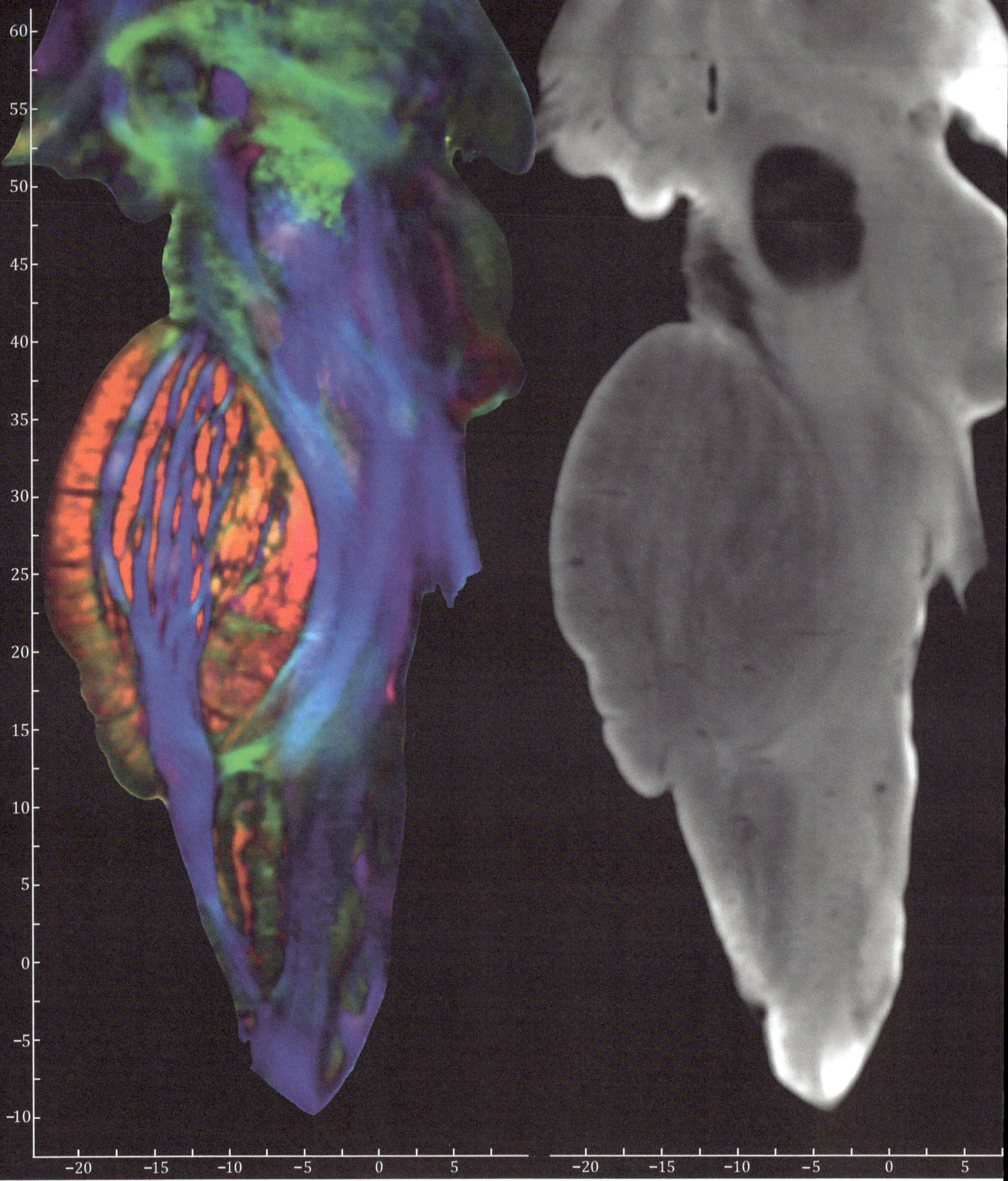

Fig 68a
Obex +3.5mm
Sagittal 08

3n oculomotor nerve
4n trochlear nerve
4V 4th ventricle
6n root of abducens nerve
6N abducens nu
7n facial nerve
AmbL ambiguus nu, loose part
ami amiculum of the olive
Ar arcuate nu of the caudal hindbrain
B9 B9 serotonin cells
CGPn central gray pons
CIC central nu inf colliculus
CnF cuneiform nu
cp cerebral peduncle
Crb cribiform nucleus
ctg central tegmental tract
Cu cuneate nu
cu cuneate fasciculus
DCIC dors cx inf colliculus
DMTg dorsomed tegmental area
DpG deep gray superior coll
DpWh deep white superior coll
ECIC external cx inferior coll
f fornix
fr fasciculus retroflexus
Gi gigantocellular reticular nu
Gr gracile nu
hio hilus of the inferior olive
hypothal hypothalamus
InG intermed gray layer SC
InWh intermediate white layer SC
IOD inf olive, dorsal nu
IOPr inf olive, principal nu
IPo interpositus
IRt intermed reticular nu
isRt isthmic reticular formation
LC locus coeruleus
lfp longitudinal fasciculus pons
LH lateral hypothalamic area
Li linear nu of the medulla
LPAG lat periaqueductal gray
LPB lateral parabrachial nu
LRt lateral reticular nu
MB mammillary body
mcp middle cerebellar peduncle
MCPC magnocell nu post comm

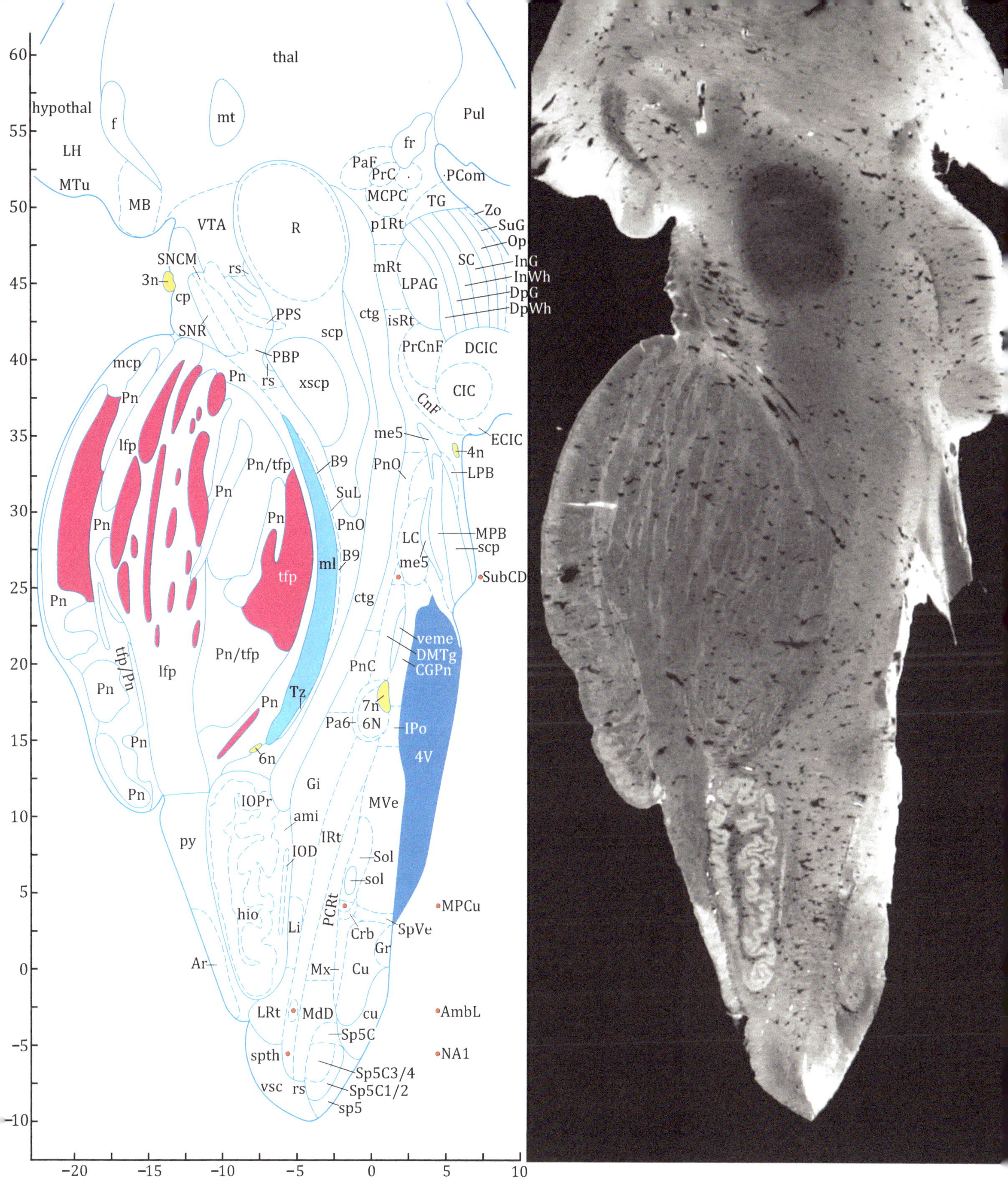

MdD medullary reticular nu, dors
me5 mesencephalic trigem tract
ml medial lemniscus
MPCu medial pericuneate nucleus
mRt mesencephalic retic form
mt mammillothalamic tract
MTu medial tuberal nu
MVe medial vestibular nu
Mx matrix region medulla
NA1 A1 noradrenaline cells
Op optic nerve layer sup coll

p1Rt prosomere 1 reticular formation
Pa6 paraabducens nu
PaF parafascicular thal nu
PBP parabrachial pigmented nu
PCom nu posterior comm
PCRt parvicell reticular nu
Pn pontine nuclei
Pn/tfp pontine nu/transv fibres pons
PnC pontine reticular nu, caudal
PnO pontine reticular nu, oral
PPS posterior perforated substance

PrC precommissural nu
PrCnF precuneiform area
Pul pulvinar nuclei
py pyramidal tr
R red nu
rs rubrospinal tract
SC superior colliculus
scp superior Cb peduncle
SNCM s nigra, compact, med tier
SNR s nigra, reticular
Sol nu of solitary tract

sol solitary tract
sp5 spinal trigem tract
Sp5C spinal trigem nu, caudal
Sp5C1/2 spinal 5 nu, caud, lam 1 & 2
Sp5C3/4 spinal 5 nu, caud, lam 3 & 4
spth spinothalamic tract
SpVe spinal vestibular nu
SubCD subcoeruleus nu, dorsal
SuG superficial gray sup coll
SuL supralemniscal nucleus
tfp transverse fibers pons

Fig 68b
Obex +3.5mm
Sagittal 08

tfp/Pn transv fibers pons/pontine nu
See List of Structures

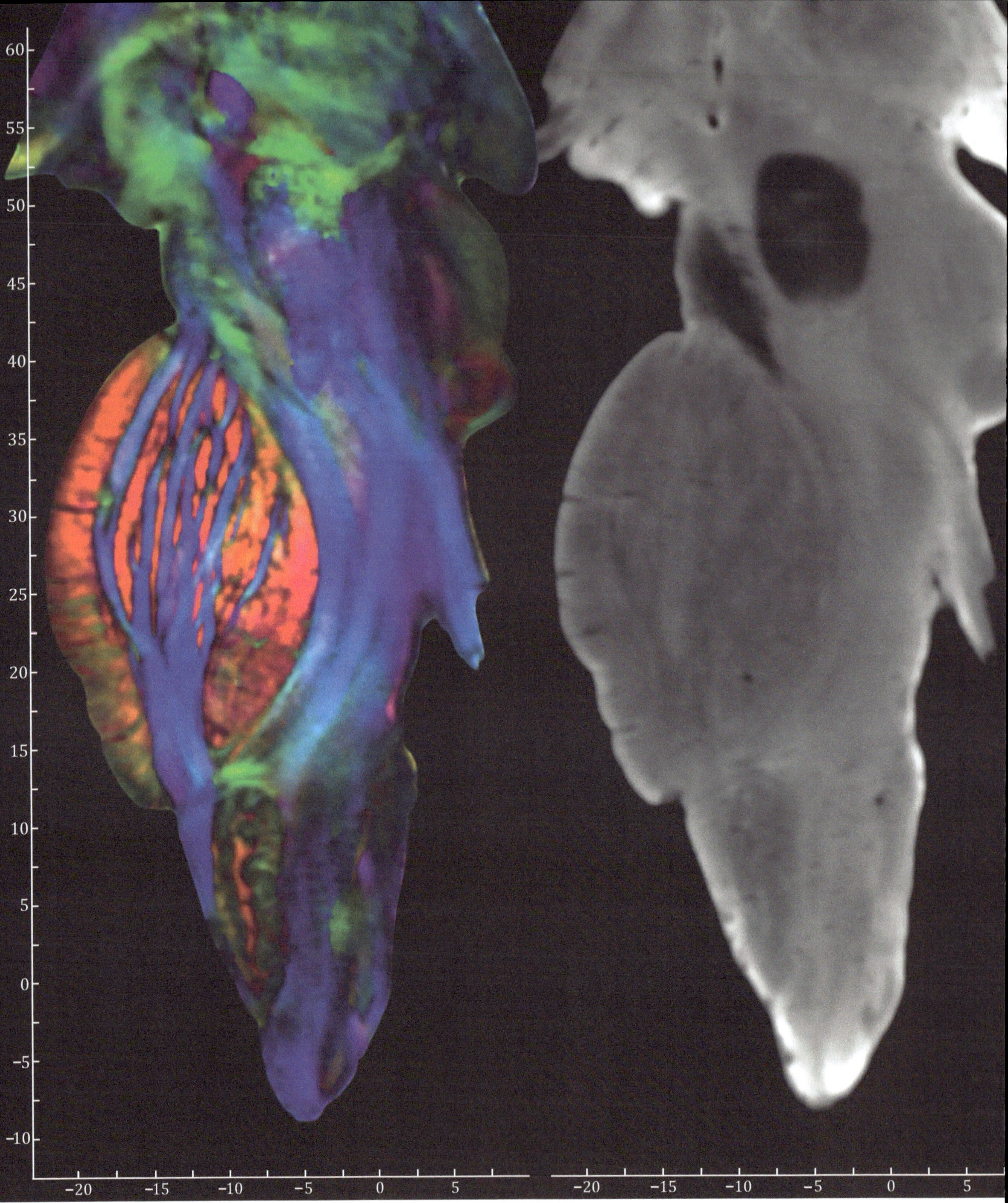

Fig 69a
Obex +4.0mm
Sagittal 09

3n oculomotor nerve
4n trochlear nerve
4V 4th ventricle
5Sol trigeminal-solitary trans
6n root of abducens nerve
6N abducens nu
7n facial nerve
AmbL ambiguus nu, loose part
AmbSC ambiguus nu, subcompact
ami amiculum of the olive
Ar arcuate nu of the caudal hindbrain
B9 B9 serotonin cells
CGPn central gray pons
CIC central nu inf colliculus
CnF cuneiform nu
cp cerebral peduncle
ctg central tegmental tract
Cu cuneate nu
cu cuneate fasciculus
DCIC dors cx inf colliculus
DMTg dorsomed tegmental area
DpG deep gray superior coll
DpWh deep white superior coll
dsc dorsal spinocereb tract
ECIC external cx inferior coll
fr fasciculus retroflexus
Gi gigantocellular reticular nu
hio hilus of the inferior olive
hypothal hypothalamus
InG intermed gray layer SC
InWh intermediate white layer SC
IOD inf olive, dorsal nu
IOPr inf olive, principal nu
IRt intermed reticular nu
IS inferior salivatory nu
isRt isthmic reticular formation
LC locus coeruleus
lfp longitudinal fasciculus pons
LH lateral hypothalamic area
Li linear nu of the medulla
LPAG lat periaqueductal gray
LPB lateral parabrachial nu
LPGi lat paragigantocellular nu
LRt lateral reticular nu
MB mammillary body
mcp middle cerebellar peduncle

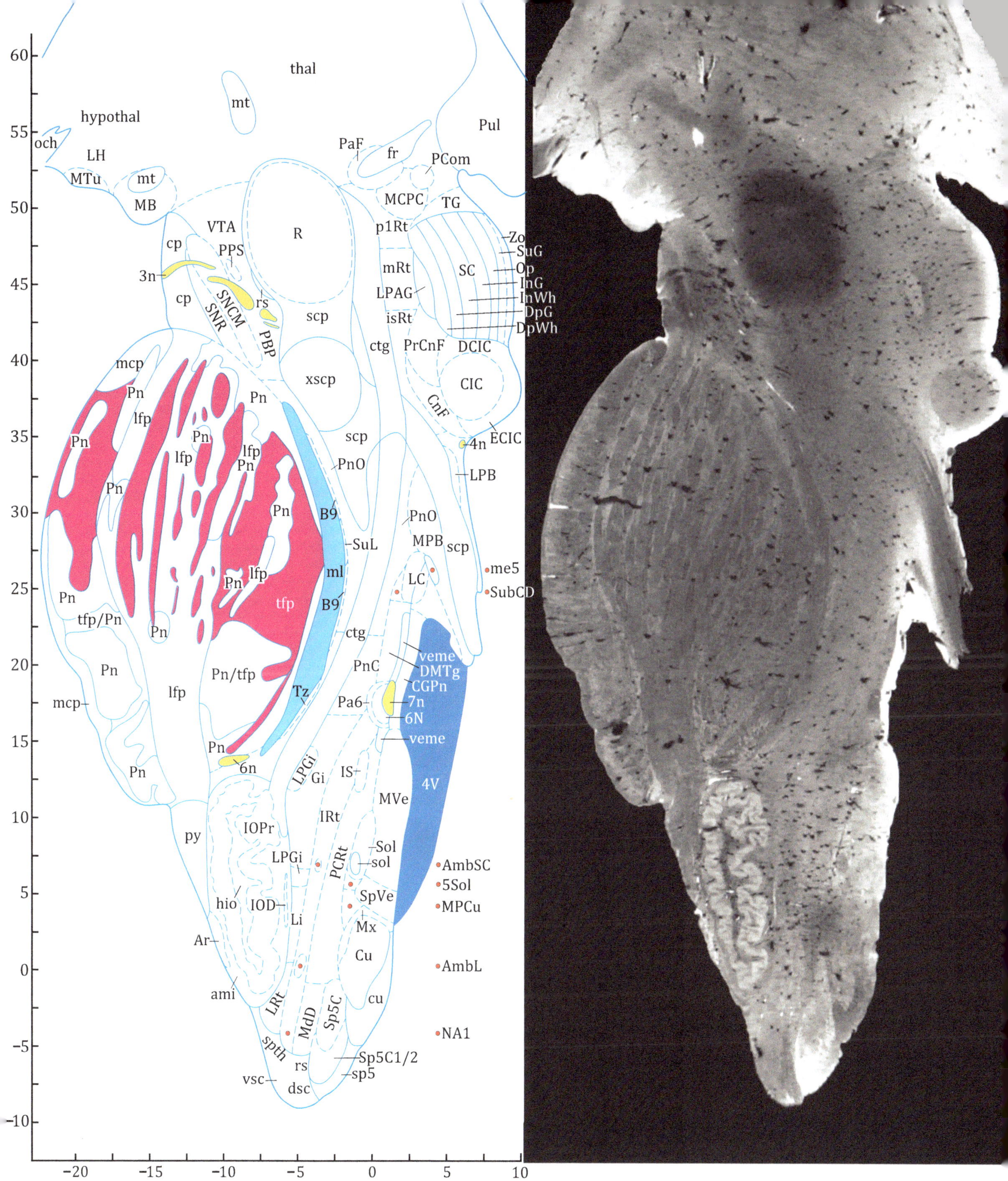

MCPC magnocell nu post comm
MdD medullary reticular nu, dors
me5 mesencephalic trigem tract
ml medial lemniscus
MPB medial parabrachial nu
MPCu medial pericuneate nucleus
mRt mesencephalic retic form
mt mammillothalamic tract
MTu medial tuberal nu
MVe medial vestibular nu
Mx matrix region medulla

NA1 A1 noradrenaline cells
och optic chiasm
Op optic nerve layer sup coll
p1Rt prosomere 1 reticular formation
Pa6 paraabducens nu
PaF parafascicular thal nu
PBP parabrachial pigmented nu
PCom nu posterior comm
PCRt parvicell reticular nu
Pn pontine nuclei
Pn/tfp pontine nu/transv fibres pons

PnC pontine reticular nu, caudal
PnO pontine reticular nu, oral
PPS posterior perforated substance
PrCnF precuneiform area
Pul pulvinar nuclei
py pyramidal tr
R red nu
rs rubrospinal tract
SC superior colliculus
scp superior Cb peduncle
SNCM s nigra, compact, med tier

SNR s nigra, reticular
sol solitary tract
Sol nu of solitary tract
sp5 spinal trigem tract
Sp5C spinal trigem nu, caudal
Sp5C1/2 spinal 5 nu, caud, lam 1 & 2
spth spinothalamic tract
SpVe spinal vestibular nu
SubCD subcoeruleus nu, dorsal
SuG superficial gray sup coll
SuL supralemniscal nucleus

tfp transverse fibers pons
See List of Structures

Fig 69b
Obex +4.0mm
Sagittal 09

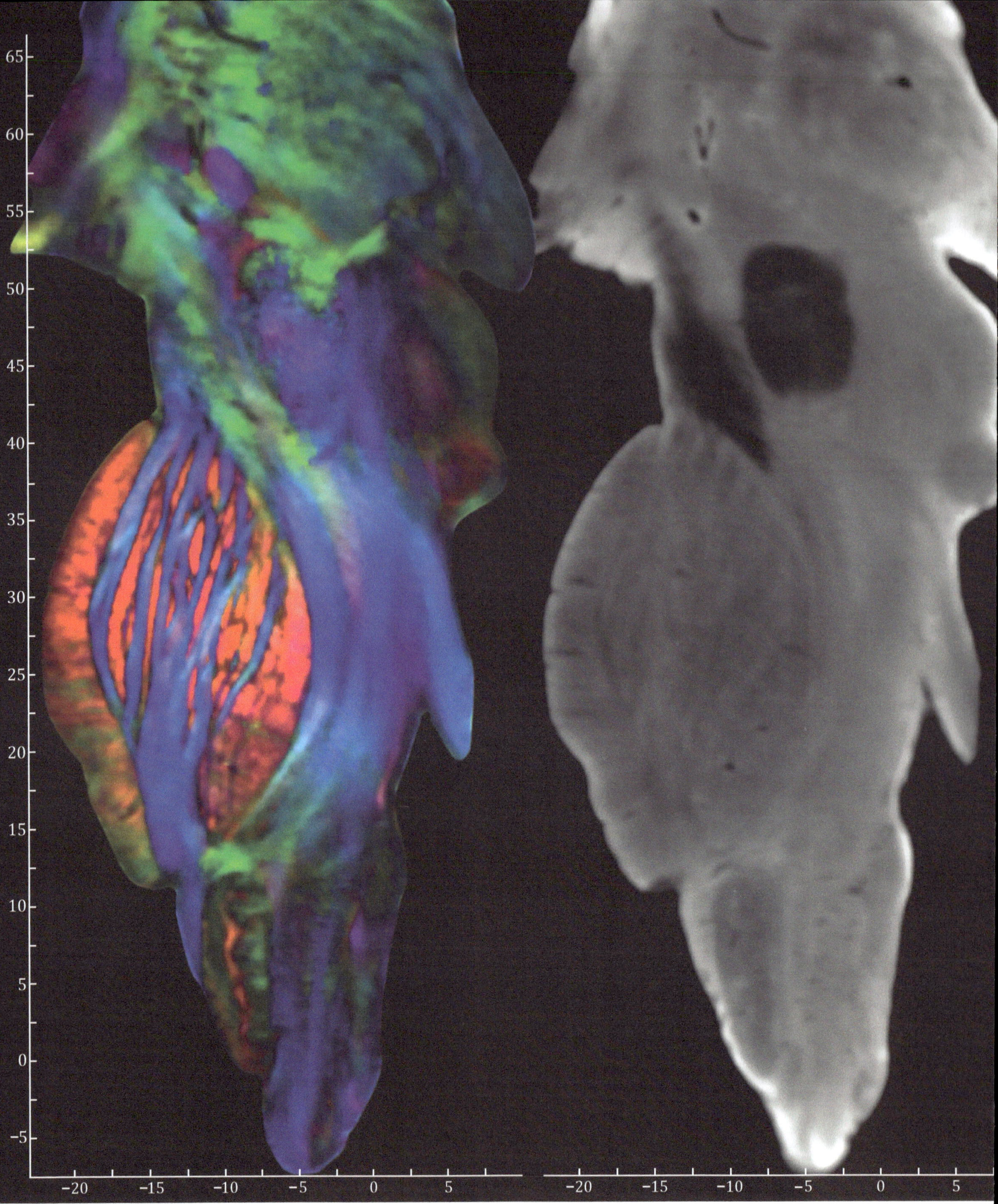

Fig 70a
Obex +4.5mm
Sagittal 10

4V 4th ventricle
5Sol trigeminal-solitary trans
6n root of abducens nerve
7N facial nu
7n facial nerve
ac anterior commissure
AmbL ambiguus nu, loose part
AmbSC ambiguus nu, subcompact
ami amiculum of the olive
Ar arcuate nu of the caudal hindbrain
asc7 ascending fibers facial n
B9 B9 serotonin cells
CGPn central gray pons
CIC central nu inf colliculus
CnF cuneiform nu
cp cerebral peduncle
ctg central tegmental tract
Cu cuneate nu
cu cuneate fasciculus
DCIC dors cx inf colliculus
DpG deep gray superior coll
DPO dorsal periolivary region
DpWh deep white superior coll
dsc dorsal spinocereb tract
ECIC external cx inferior coll
ECu external cuneate nu
fr fasciculus retroflexus
Gam gamma pontine nucleus
Gi gigantocellular reticular nu
hio hilus of the inferior olive
hypothal hypothalamus
InG intermed gray layer SC
InWh intermediate white layer SC
IOPr inf olive, principal nu
IRt intermed reticular nu
IS inferior salivatory nu
isRt isthmic reticular formation
LC locus coeruleus
lfp longitudinal fasciculus pons
lfp/Pn long fasciculus pons/pontine nu
LH lateral hypothalamic area
Li linear nu of the medulla
LPB lateral parabrachial nu
LPGi lat paragigantocellular nu
LRt lateral reticular nu
mcp middle cerebellar peduncle

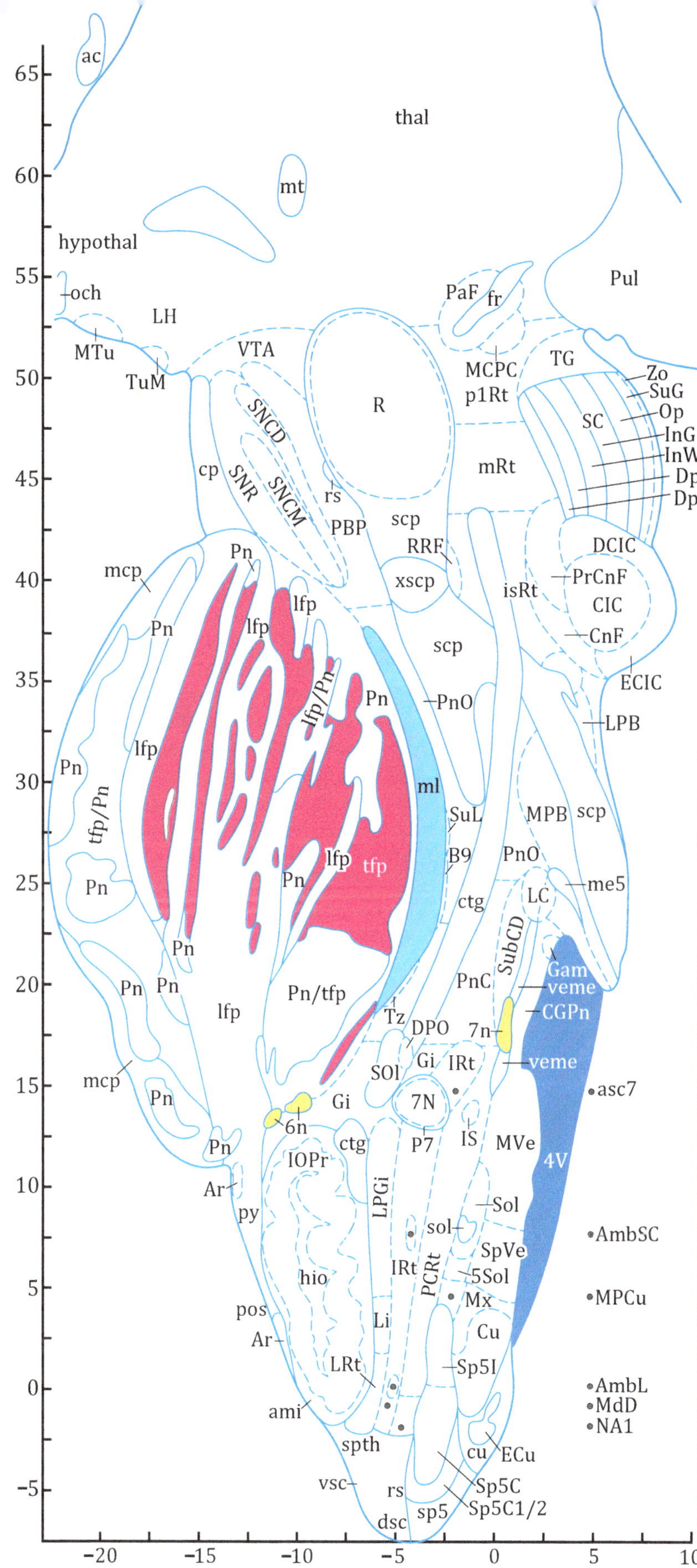

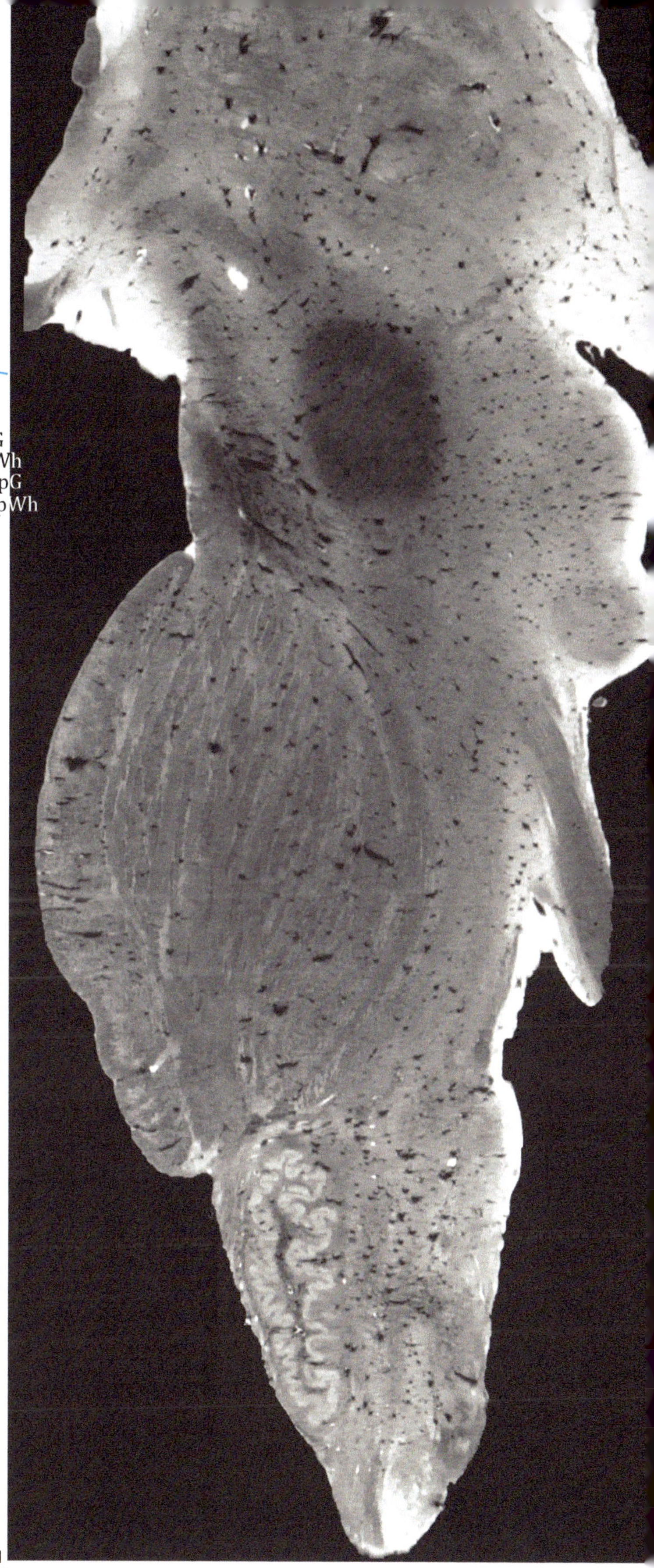

MCPC magnocell nu post comm
MdD medullary reticular nu, dors
me5 mesencephalic trigem tract
ml medial lemniscus
MPB medial parabrachial nu
MPCu medial pericuneate nucleus
mRt mesencephalic retic form
mt mammillothalamic tract
MTu medial tuberal nu
MVe medial vestibular nu
Mx matrix region medulla
NA1 A1 noradrenaline cells
och optic chiasm
Op optic nerve layer sup coll
p1Rt prosomere 1 reticular formation
P7 perifacial zone
PaF parafascicular thal nu
PBP parabrachial pigmented nu
PCRt parvicell reticular nu
Pn pontine nuclei
Pn/tfp **pontine nu/transv fibres pons**
PnC pontine reticular nu, caudal
PnO pontine reticular nu, oral
pos preolivary sulcus
PrCnF precuneiform area
Pul pulvinar nuclei
py pyramidal tr
R red nu
RRF **retrorubral field**
rs rubrospinal tract
SC superior colliculus
scp superior Cb peduncle
SNCD s nigra, compact, dors tier
SNCM s nigra, compact, med tier
SNR s nigra, reticular
SOl superior olive, lateral
sol solitary tract
Sol nu of solitary tract
sp5 spinal trigem tract
Sp5C spinal trigem nu, caudal
Sp5C1/2 spinal 5 nu, caud, lam 1 & 2
Sp5I spinal trigem nu, interpolar
spth spinothalamic tract
SpVe spinal vestibular nu

Fig 70b
Obex +4.5mm
Sagittal 10

SubCD subcoeruleus nu, dorsal

See List of Structures

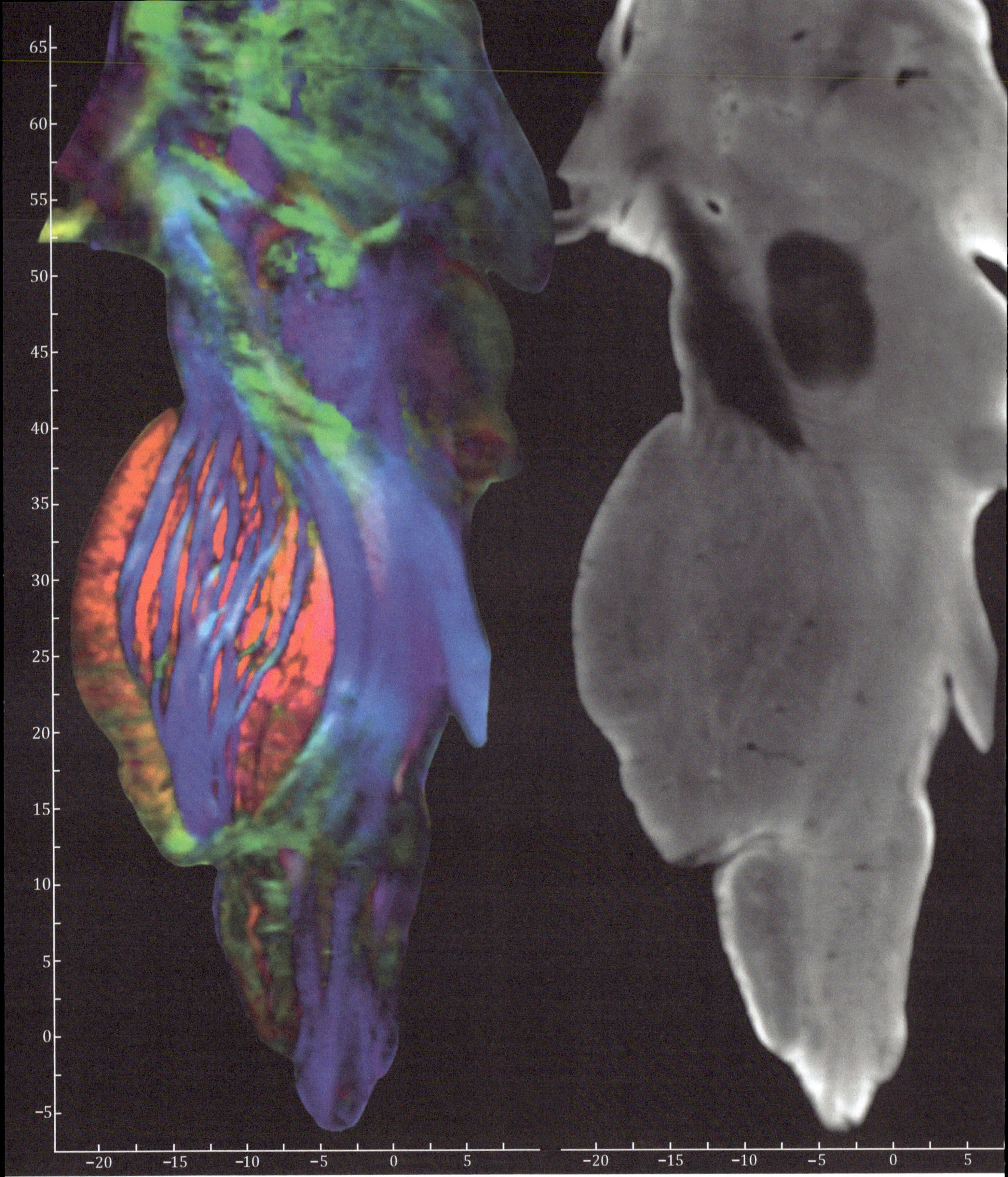

Fig 71a
Obex +5.0mm
Sagittal 11

5Sol trigeminal-solitary trans
6n root of abducens nerve
7N facial nu
7n facial nerve
ac anterior commissure
Ad1/NA1 Ad1 adren NA1 noradr
AmbSC ambiguus nu, subcompact
ami amiculum of the olive
Arc arcuate hypothal nu
B9 B9 serotonin cells
CGPn central gray pons
CIC central nu inf colliculus
CMn centromedian thalamic nucleus
CnF cuneiform nu
cp cerebral peduncle
Ct conterminal nu
ctg central tegmental tract
Cu cuneate nu
cu cuneate fasciculus
DpG deep gray superior coll
DPO dorsal periolivary region
DpWh deep white superior coll
dsc dorsal spinocereb tract
ECIC external cx inferior coll
ECu external cuneate nu
EVe nu origin efferents ve n
fr fasciculus retroflexus
Gam gamma pontine nucleus
hio hilus of the inferior olive
hypothal hypothalamus
ic internal capsule
InG intermed gray layer SC
InWh intermediate white layer SC
IOPr inf olive, principal nu
IRt intermed reticular nu
IS inferior salivatory nu
isRt isthmic reticular formation
JxO juxtaolivary nu
LC locus coeruleus
lfp longitudinal fasciculus pons
lfp/Pn long fasciculus pons/pontine nu
LH lateral hypothalamic area
Li linear nu of the medulla
ll lateral lemniscus
LPB lateral parabrachial nu
LPGi lat paragigantocellular nu

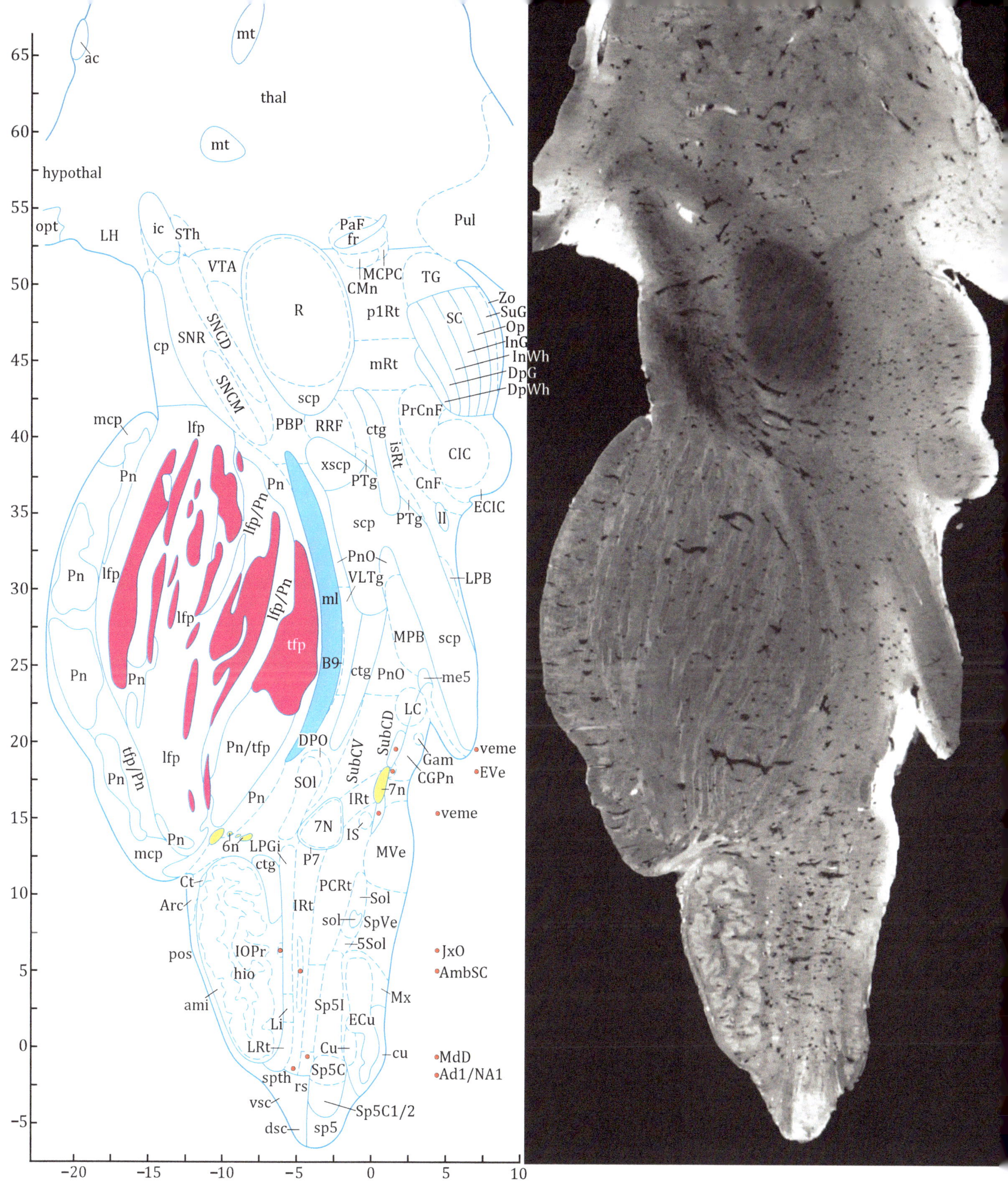

LRt lateral reticular nu
mcp middle cerebellar peduncle
MCPC magnocell nu post comm
MdD medullary reticular nu, dors
me5 mesencephalic trigem tract
ml medial lemniscus
MPB medial parabrachial nu
mRt mesencephalic retic form
mt mammillothalamic tract
MVe medial vestibular nu
Mx matrix region medulla

Op optic nerve layer sup coll
opt optic tract
p1Rt prosomere 1 reticular formation
P7 perifacial zone
PaF parafascicular thal nu
PBP parabrachial pigmented nu
PCRt parvicell reticular nu
Pn pontine nuclei
Pn/tfp pontine nu/transv fibres pons
PnO pontine reticular nu, oral
pos preolivary sulcus

PrCnF precuneiform area
PTg pedunculopontine tegmental
Pul pulvinar nuclei
R red nu
RRF retrorubral field
rs rubrospinal tract
SC superior colliculus
scp superior Cb peduncle
SNCD s nigra, compact, dors tier
SNCM s nigra, compact, med tier
SNR s nigra, reticular

SOl superior olive, lateral
sol solitary tract
Sol nu of solitary tract
sp5 spinal trigem tract
Sp5C spinal trigem nu, caudal
Sp5C1/2 spinal 5 nu, caud, lam 1 & 2
Sp5I spinal trigem nu, interpolar
spth spinothalamic tract
SpVe spinal vestibular nu
STh subthalamic nu
SubCD subcoeruleus nu, dorsal

Fig 71b
Obex +5.0mm
Sagittal 11

SubCV subcoeruleus nu, ventral

See List of Structures

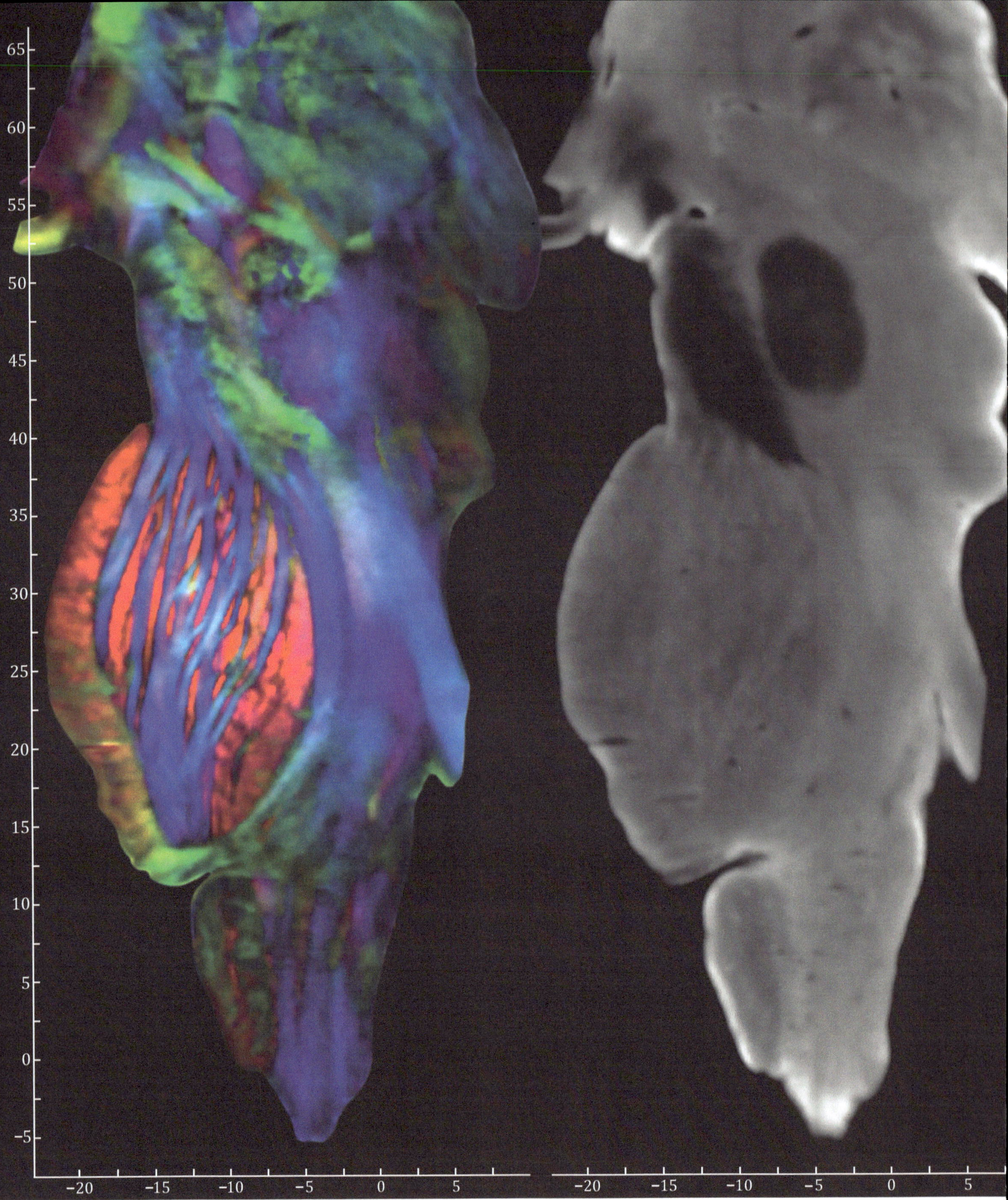

Fig 72a
Obex +5.5mm
Sagittal 12

10n vagus nerve
5Sol trigeminal-solitary trans
7n facial nerve
7N facial nu
Ad1/NA1 Ad1 adren NA1 noradr
AmbSC ambiguus nu, subcompact
ami amiculum of the olive
CGPn central gray pons
CIC central nu inf colliculus
CMn centromedian thalamic nucleus
CnF cuneiform nu
cp cerebral peduncle
ctg central tegmental tract
cu cuneate fasciculus
DpG deep gray superior coll
DPO dorsal periolivary region
dsc dorsal spinocereb tract
ECIC external cx inferior coll
ECu external cuneate nu
Gam gamma pontine nucleus
hio hilus of the inferior olive
hypothal hypothalamus
icp inf cerebellar peduncle
InG intermed gray layer SC
InWh intermediate white layer SC
IOPr inf olive, principal nu
IRt intermed reticular nu
isRt isthmic reticular formation
JxO juxtaolivary nu
LC locus coeruleus
lfp longitudinal fasciculus pons
LH lateral hypothalamic area
Li linear nu of the medulla
ll lateral lemniscus
LPB lateral parabrachial nu
LPCu lateral pericuneate nucleus
LPGi lat paragigantocellular nu
LRt lateral reticular nu
LRtS5 lat reticular nu, subtrigeminal
mcp middle cerebellar peduncle
me5 mesencephalic trigem tract
ml medial lemniscus
MPB medial parabrachial nu
mRt mesencephalic retic form
mt mammillothalamic tract
MVe medial vestibular nu

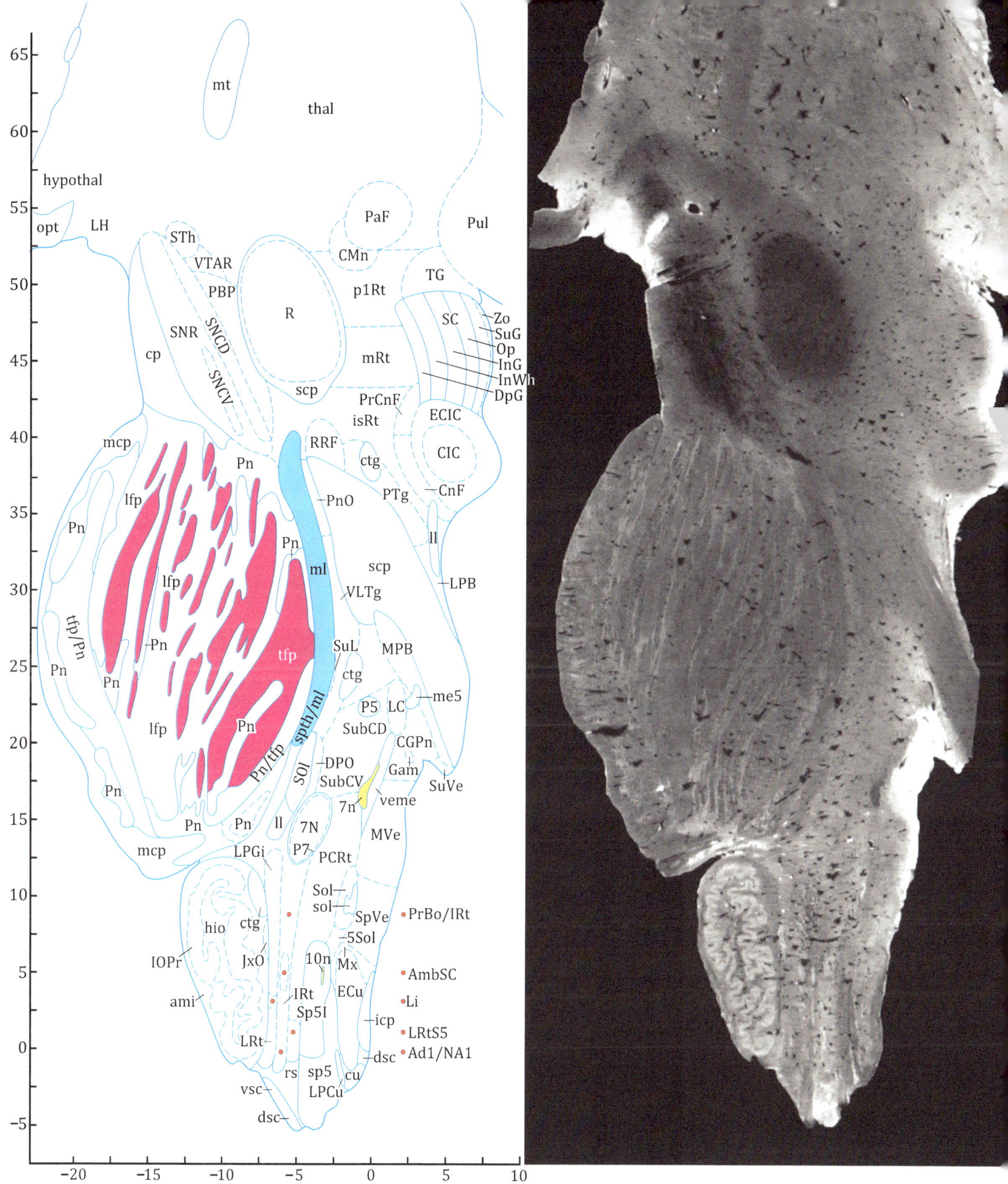

Mx matrix region medulla
Op optic nerve layer sup coll
opt optic tract
p1Rt prosomere 1 reticular formation
P5 peritrigeminal zone
P7 perifacial zone
PaF parafascicular thal nu
PBP parabrachial pigmented nu
PCRt parvicell reticular nu
Pn pontine nuclei
Pn/tfp pontine nu/transv fibres pons
PnO pontine reticular nu, oral
PrBo/IRt pre-Bötzinger/int reticular nu
PrCnF precuneiform area
PTg pedunculopontine tegmental
Pul pulvinar nuclei
R red nu
RRF retrorubral field
rs rubrospinal tract
SC superior colliculus
scp superior Cb peduncle
SNCD s nigra, compact, dors tier
SNCV s nigra, compact, vent tier
SNR s nigra, reticular
SOl superior olive, lateral
sol solitary tract
Sol nu of solitary tract
sp5 spinal trigem tract
Sp5I spinal trigem nu, interpolar
spth/ml spinothal tract/med lemniscus
SpVe spinal vestibular nu
STh subthalamic nu
SubCD subcoeruleus nu, dorsal
SubCV subcoeruleus nu, ventral
SuG superficial gray sup coll
SuL supralemniscal nucleus
SuVe superior vestibular nu
tfp transverse fibers pons
tfp/Pn transv fibers pons/pontine nu
TG tectal gray
thal thalamus
veme vestibulomesencephalic tr
VLTg ventrolateral tegmental area
vsc ventral spinocerebellar tr
VTAR vent tegmental area, rostr
Zo zona layer sup coll

Fig 72b
Obex +5.5mm
Sagittal 12

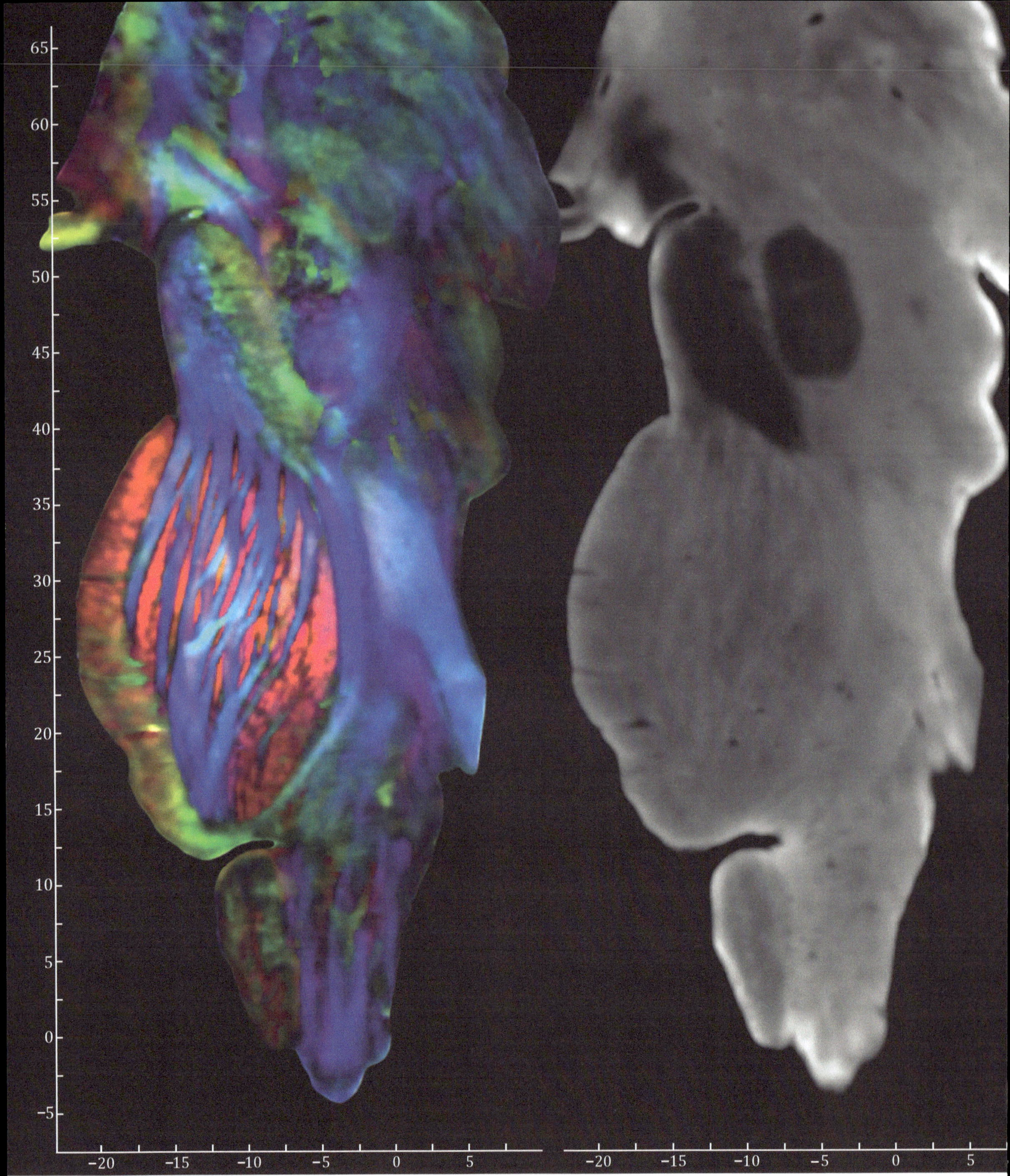

Fig 73a
Obex +6.0mm
Sagittal 13

- 10n vagus nerve
- 5N motor trigeminal nu
- 5Sol trigeminal-solitary trans
- 7n facial nerve
- 7N facial nu
- Ad1 Ad1 adrenalin cells
- ami amiculum of the olive
- CGPn central gray pons
- CIC central nu inf colliculus
- CnF cuneiform nu
- cp cerebral peduncle
- ctg central tegmental tract
- DpG deep gray superior coll
- DPO dorsal periolivary region
- dsc dorsal spinocereb tract
- ECIC external cx inferior coll
- ECu external cuneate nu
- Gam gamma pontine nucleus
- hio hilus of the inferior olive
- hypothal hypothalamus
- icp inf cerebellar peduncle
- InG intermed gray layer SC
- InWh intermediate white layer SC
- IOPr inf olive, principal nu
- IRt intermed reticular nu
- isRt isthmic reticular formation
- lfp longitudinal fasciculus pons
- LH lateral hypothalamic area
- ll lateral lemniscus
- LPB lateral parabrachial nu
- LPCu lateral pericuneate nucleus
- LPGi lat paragigantocellular nu
- LRt lateral reticular nu
- LRtS5 lat reticular nu, subtrigeminal
- mcp middle cerebellar peduncle
- me5 mesencephalic trigem tract
- ml medial lemniscus
- MPB medial parabrachial nu
- mRt mesencephalic retic form
- mt mammillothalamic tract
- MVe medial vestibular nu
- Mx matrix region medulla
- Op optic nerve layer sup coll
- opt optic tract
- p1Rt prosomere 1 reticular formation
- P5 peritrigeminal zone

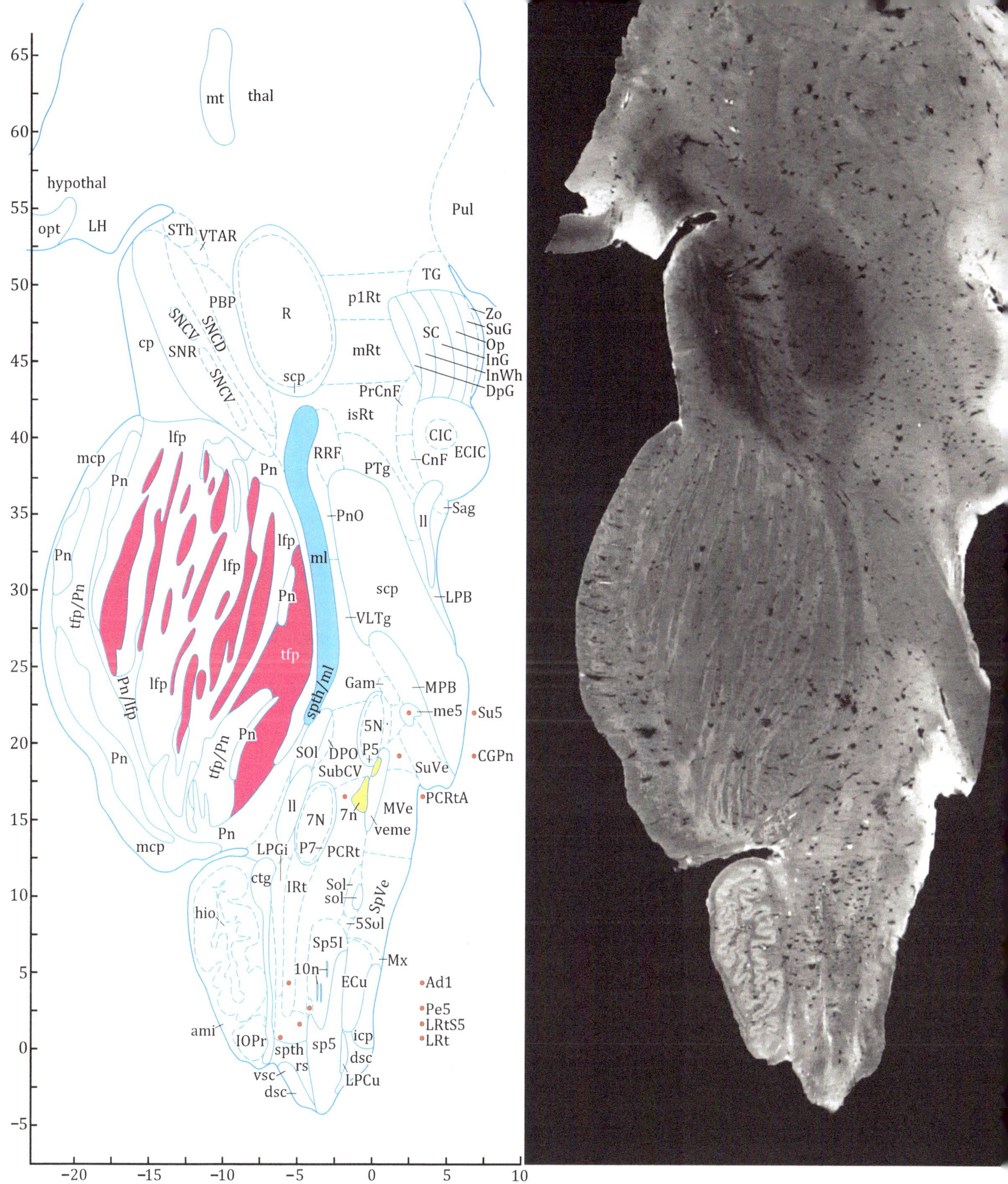

P7 perifacial zone
PBP parabrachial pigmented nu
PCRt parvicell reticular nu
PCRtA parvicell reticular nu, alpha
Pe5 peritrigeminal nucleus
Pn pontine nuclei
Pn/lfp pontine nu/long fasciculus pons
PnO pontine reticular nu, oral
PrCnF precuneiform area
PTg pedunculopontine tegmental
Pul pulvinar nuclei
R red nu
RRF retrorubral field
rs rubrospinal tract
Sag sagulum nu
SC superior colliculus
scp superior Cb peduncle
SNCD s nigra, compact, dors tier
SNCV s nigra, compact, vent tier
SNR s nigra, reticular
SOl superior olive, lateral
sol solitary tract
Sol nu of solitary tract
sp5 spinal trigem tract
Sp5I spinal trigem nu, interpolar
spth spinothalamic tract
spth/ml spinothal tract/med lemniscus
SpVe spinal vestibular nu
STh subthalamic nu
Su5 supratrigeminal nu
SubCV subcoeruleus nu, ventral
SuG superficial gray sup coll
SuVe superior vestibular nu
tfp transverse fibers pons
tfp/Pn transv fibers pons/pontine nu
TG tectal gray
thal thalamus
veme vestibulomesencephalic tr
VLTg ventrolateral tegmental area
vsc ventral spinocerebellar tr
VTAR vent tegmental area, rostr
Zo zona layer sup coll

Fig 73b
Obex +6.0mm
Sagittal 13

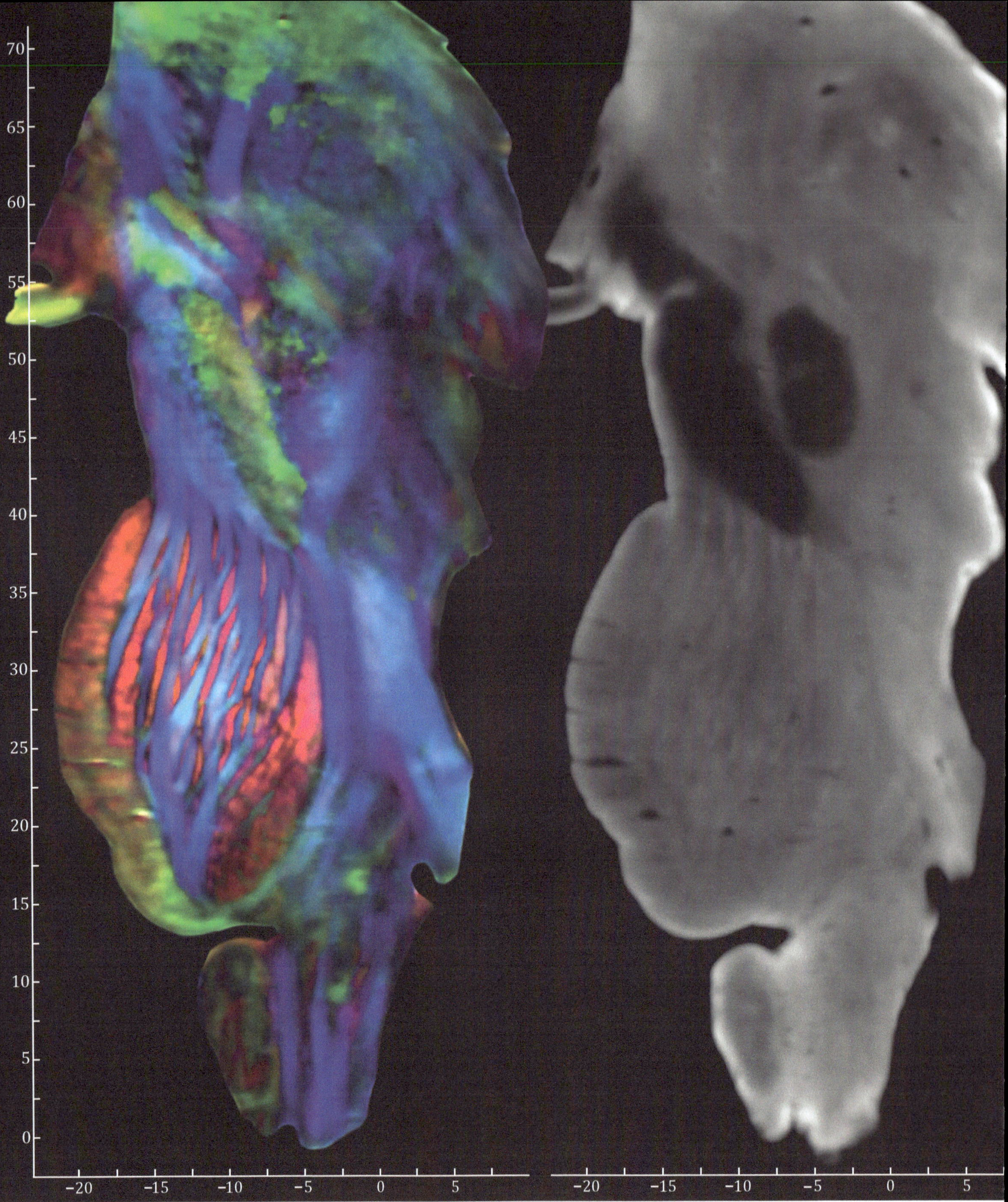

Fig 74a
Obex +6.5mm
Sagittal 14

10n vagus nerve
5N motor trigeminal nu
5Sol trigeminal-solitary trans
7n facial nerve
7N facial nu
Ad1 Ad1 adrenalin cells
ami amiculum of the olive
bic brachium inf colliculus
Cb cerebellum
CIC central nu inf colliculus
CnF cuneiform nu
cp cerebral peduncle
ctg central tegmental tract
dsc dorsal spinocereb tract
ECIC external cx inferior coll
ECu external cuneate nu
Gam gamma pontine nucleus
hio hilus of the inferior olive
hypothal hypothalamus
icp inf cerebellar peduncle
InG intermed gray layer SC
InWh intermediate white layer SC
IOPr inf olive, principal nu
IRt intermed reticular nu
isRt isthmic reticular formation
lfp longitudinal fasciculus pons
LH lateral hypothalamic area
ll lateral lemniscus
LPB lateral parabrachial nu
LPCu lateral pericuneate nucleus
LPGi lat paragigantocellular nu
LRt lateral reticular nu
LRtS5 lat reticular nu, subtrigeminal
mcp middle cerebellar peduncle
me5 mesencephalic trigem tract
ml medial lemniscus
MPB medial parabrachial nu
mRt mesencephalic retic form
mt mammillothalamic tract
MVe medial vestibular nu
oc olivocerebellar tract
Op optic nerve layer sup coll
opt optic tract
p1Rt prosomere 1 reticular formation
P5 peritrigeminal zone
P7 perifacial zone

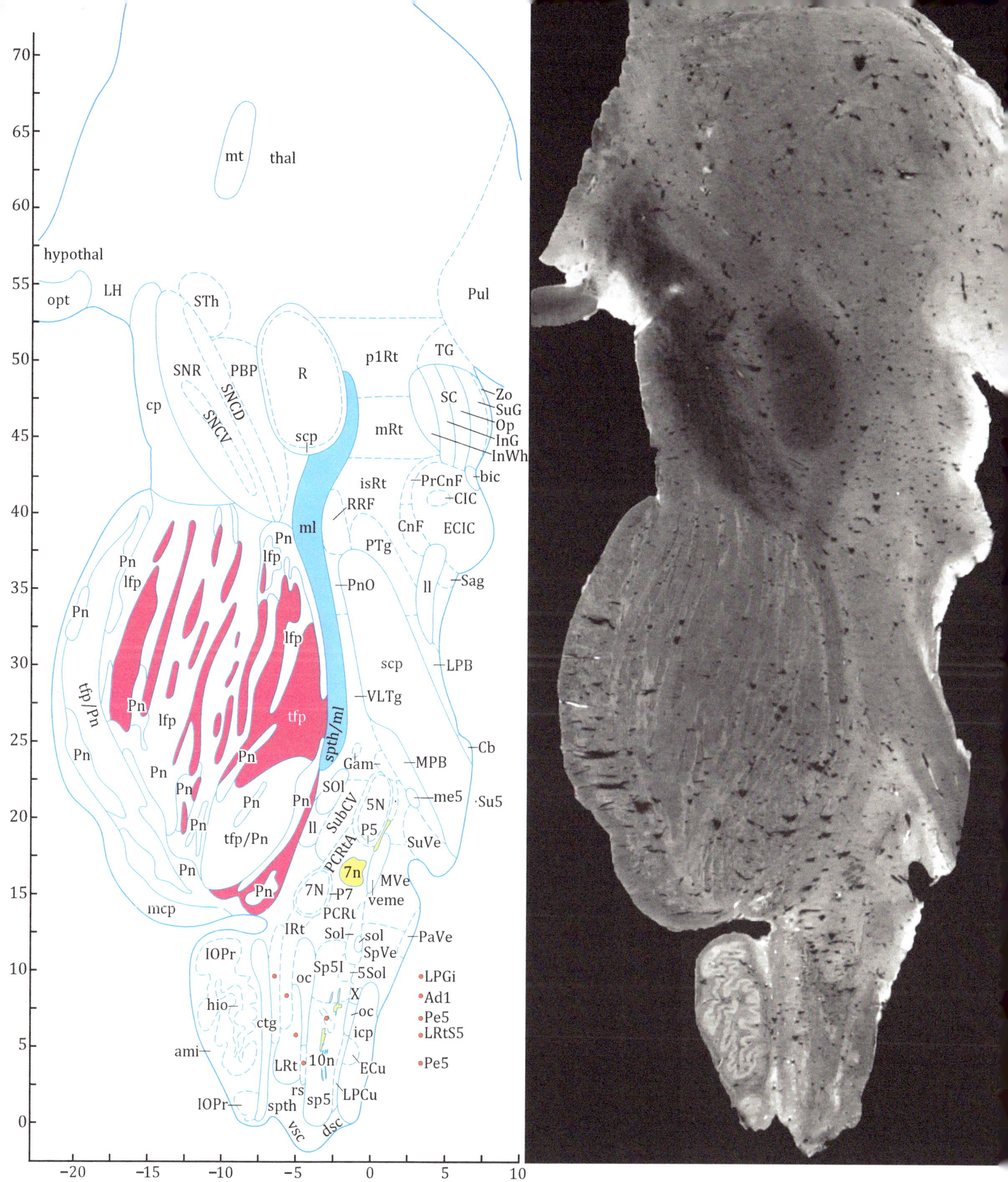

PaVe paravestibular nucleus
PBP parabrachial pigmented nu
PCRt parvicell reticular nu
PCRtA parvicell reticular nu, alpha
Pe5 peritrigeminal nucleus
Pn pontine nuclei
PnO pontine reticular nu, oral
PrCnF precuneiform area
PTg pedunculopontine tegmental
Pul pulvinar nuclei
R red nu
RRF retrorubral field
rs rubrospinal tract
Sag sagulum nu
SC superior colliculus
scp superior Cb peduncle
SNCD s nigra, compact, dors tier
SNCV s nigra, compact, vent tier
SNR s nigra, reticular
SOl superior olive, lateral
sol solitary tract
Sol nu of solitary tract
sp5 spinal trigem tract
Sp5I spinal trigem nu, interpolar
spth spinothalamic tract
spth/ml spinothal tract/med lemniscus
SpVe spinal vestibular nu
STh subthalamic nu
Su5 supratrigeminal nu
SubCV subcoeruleus nu, ventral
SuG superficial gray sup coll
SuVe superior vestibular nu
tfp transverse fibers pons
tfp/Pn transv fibers pons/pontine nu
TG tectal gray
thal thalamus
veme vestibulomesencephalic tr
VLTg ventrolateral tegmental area
vsc ventral spinocerebellar tr
X nu X
Zo zona layer sup coll

Fig 74b
Obex +6.5mm
Sagittal 14

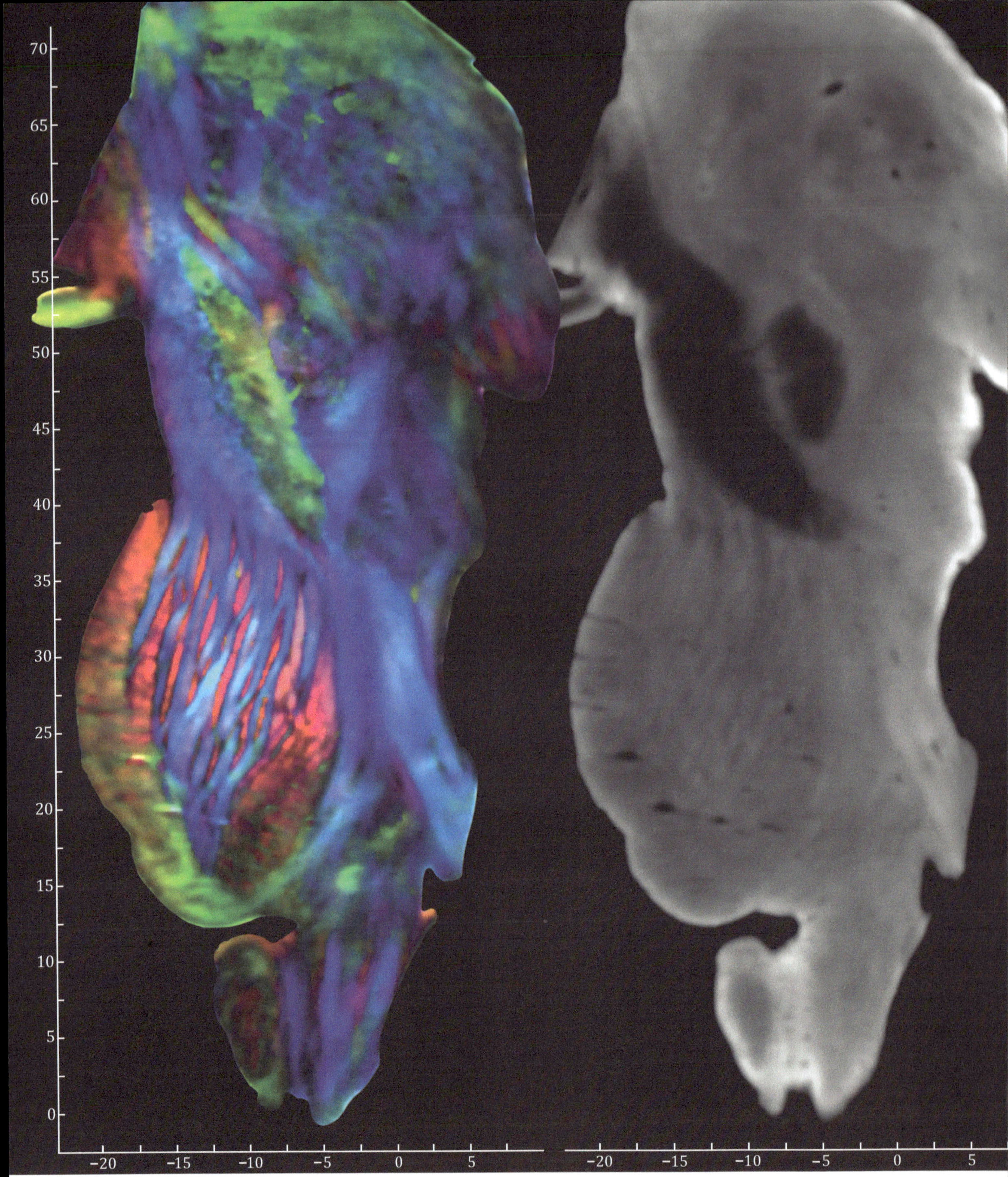

Fig 75a
Obex +7.0mm
Sagittal 15

10n vagus nerve
5N motor trigeminal nu
5Sol trigeminal-solitary trans
7n facial nerve
7N facial nu
Ad1 Ad1 adrenalin cells
ami amiculum of the olive
APT anterior pretectal nu
bic brachium inf colliculus
BIC nu brachium inf colliculus
bsc brachium superior colliculus
Cb cerebellum
CnF cuneiform nu
cp cerebral peduncle
ctg central tegmental tract
DLL dorsal nu lateral lemniscus
dsc dorsal spinocereb tract
ECIC external cx inferior coll
Gam gamma pontine nucleus
ic internal capsule
icp inf cerebellar peduncle
InG intermed gray layer SC
InWh intermediate white layer SC
IOPr inf olive, principal nu
isRt isthmic reticular formation
KF Kölliker-Fuse nu
lfp longitudinal fasciculus pons
LH lateral hypothalamic area
ll lateral lemniscus
LPB lateral parabrachial nu
LPCu lateral pericuneate nucleus
LRt lateral reticular nu
LRtS5 lat reticular nu, subtrigeminal
mcp middle cerebellar peduncle
me5 mesencephalic trigem tract
MiTg microcell tegmental nu
ml medial lemniscus
MPB medial parabrachial nu
mRt mesencephalic retic form
mt mammillothalamic tract
MVe medial vestibular nu
oc olivocerebellar tract
Op optic nerve layer sup coll
opt optic tract
p1Rt prosomere 1 reticular formation
P5 peritrigeminal zone

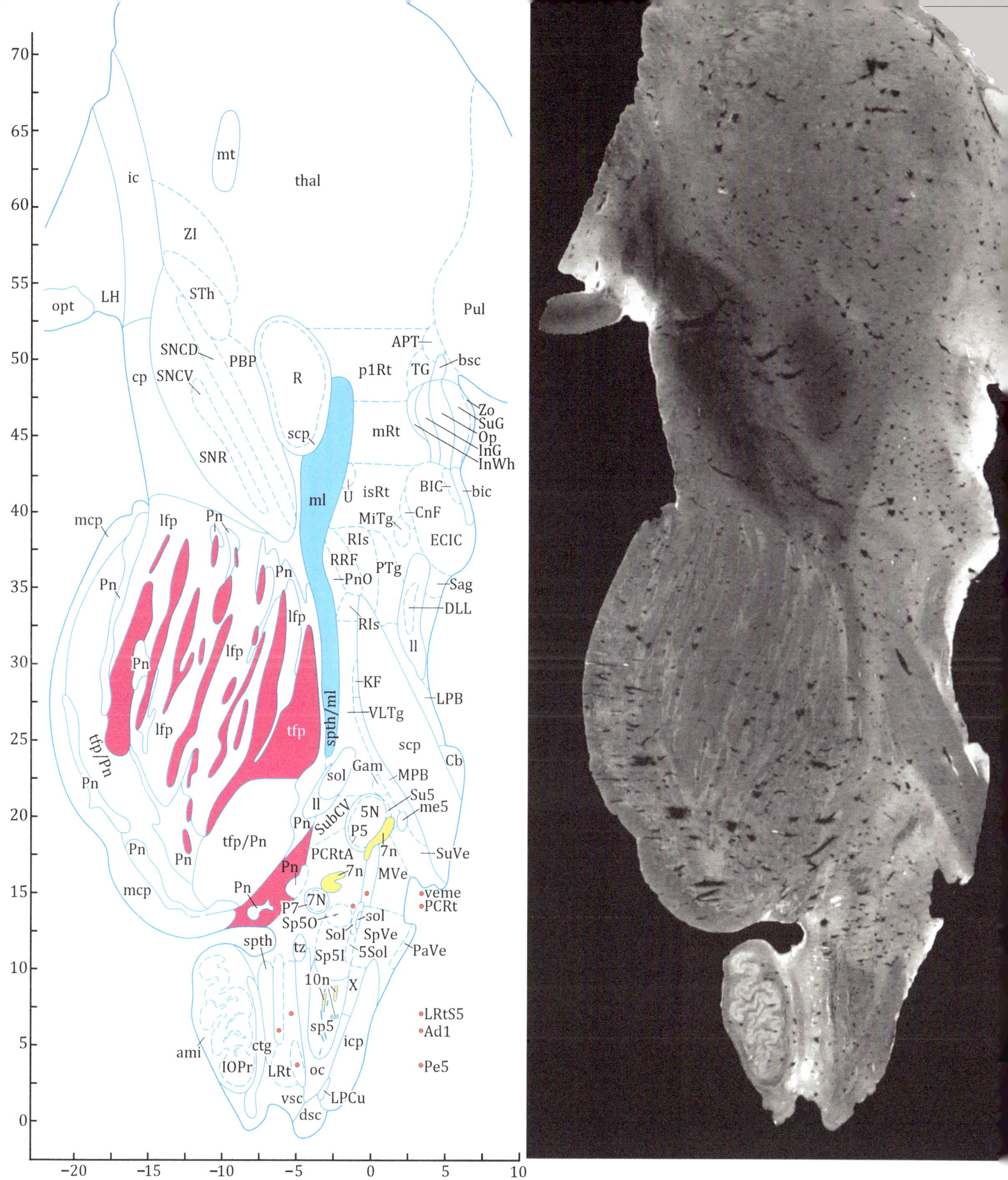

P7 perifacial zone
PaVe paravestibular nucleus
PBP parabrachial pigmented nu
PCRt parvicell reticular nu
PCRtA parvicell reticular nu, alpha
Pe5 peritrigeminal nucleus
Pn pontine nuclei
PnO pontine reticular nu, oral
PTg pedunculopontine tegmental
Pul pulvinar nuclei
R red nu

RIs retroisthmic nucleus
RRF **retrorubral field**
Sag sagulum nu
scp superior Cb peduncle
SNCD s nigra, compact, dors tier
SNCV s nigra, compact, vent tier
SNR s nigra, reticular
sol solitary tract
Sol nu of solitary tract
sp5 spinal trigem tract
Sp5I spinal trigem nu, interpolar

Sp5O spinal trigeml nu, oral
spth spinothalamic tract
spth/ml spinothal tract/med lemniscus
SpVe spinal vestibular nu
STh subthalamic nu
Su5 supratrigeminal nu
SubCV subcoeruleus nu, ventral
SuG **superficial gray sup coll**
SuVe superior vestibular nu
tfp **transverse fibers pons**
tfp/Pn **transv fibers pons/pontine nu**

TG tectal gray
thal thalamus
tz trapezoid body
U nucleus U
veme vestibulomesencephalic tr
VLTg ventrolateral tegmental area
vsc ventral spinocerebellar tr
X nu X
ZI zona incerta
Zo zona layer sup coll

Fig 75b
Obex +7.0mm
Sagittal 15

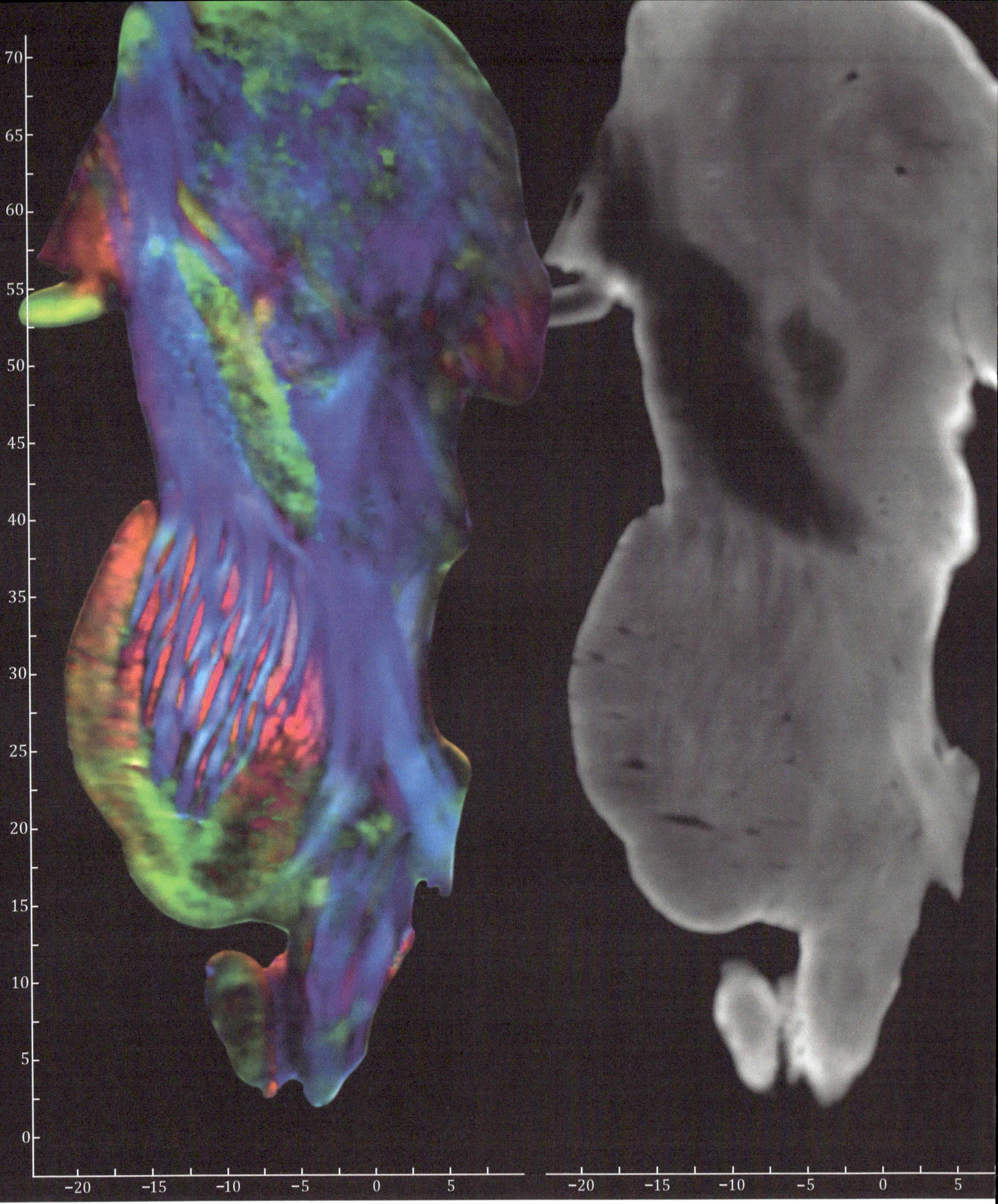

Fig 76a
Obex +7.5mm
Sagittal 16

10n vagus nerve
5N motor trigeminal nu
7n facial nerve
ami amiculum of the olive
APT anterior pretectal nu
bic brachium inf colliculus
BIC nu brachium inf colliculus
bsc brachium superior colliculus
Cb cerebellum
cp cerebral peduncle
ctg central tegmental tract
DLL dorsal nu lateral lemniscus
dsc dorsal spinocereb tract
ECIC external cx inferior coll
EnR endorestiform nucleus
ic internal capsule
icp inf cerebellar peduncle
InG intermed gray layer SC
IOPr inf olive, principal nu
isRt isthmic reticular formation
KF Kölliker-Fuse nu
lfp longitudinal fasciculus pons
ll lateral lemniscus
LPB lateral parabrachial nu
LRt lateral reticular nu
LVe lateral vestibular nu
m5 motor root trigeminal nerve
mcp middle cerebellar peduncle
MiTg microcell tegmental nu
ml medial lemniscus
MPB medial parabrachial nu
MPBE med parabrachial nu ext
mRt mesencephalic retic form
mt mammillothalamic tract
MVe medial vestibular nu
oc olivocerebellar tract
Op optic nerve layer sup coll
opt optic tract
p1Rt prosomere 1 reticular formation
P5 peritrigeminal zone
PaVe paravestibular nucleus
PBP parabrachial pigmented nu
PCRtA parvicell reticular nu, alpha
Pe5 peritrigeminal nucleus
PLi posterior limitans thal nu
Pn pontine nuclei

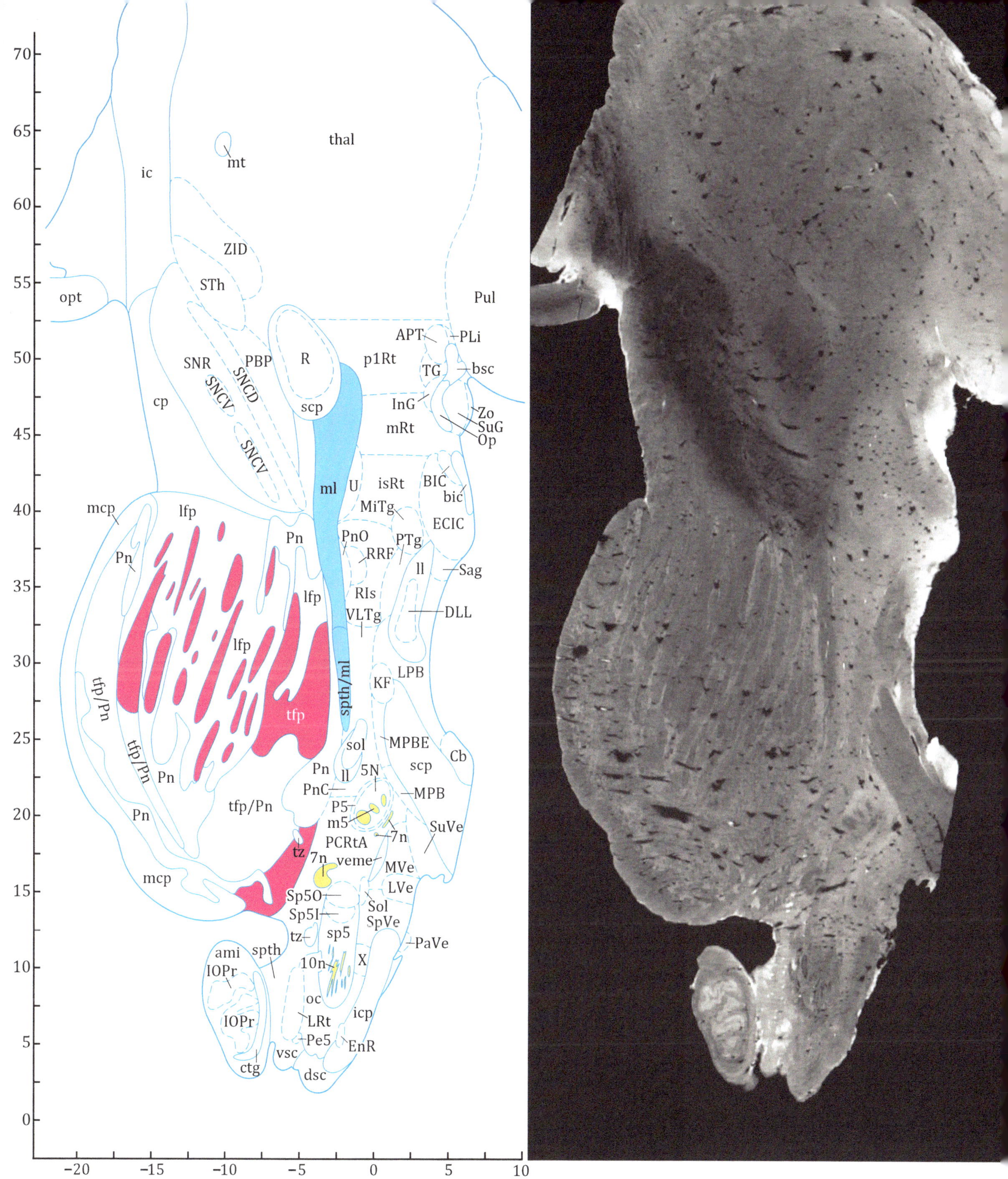

PnC pontine reticular nu, caudal
PnO pontine reticular nu, oral
PTg pedunculopontine tegmental
Pul pulvinar nuclei
R red nu
RIs retroisthmic nucleus
RRF retrorubral field
Sag sagulum nu
scp superior Cb peduncle
SNCD s nigra, compact, dors tier
SNCV s nigra, compact, vent tier

SNR s nigra, reticular
sol solitary tract
Sol nu of solitary tract
sp5 spinal trigem tract
Sp5I spinal trigem nu, interpolar
Sp5O spinal trigeml nu, oral
spth spinothalamic tract
spth/ml spinothal tract/med lemniscus
SpVe spinal vestibular nu
STh subthalamic nu
SuG superficial gray sup coll

SuVe superior vestibular nu
tfp transverse fibers pons
tfp/Pn transv fibers pons/pontine nu
TG tectal gray
thal thalamus
tz trapezoid body
U nucleus U
veme vestibulomesencephalic tr
VLTg ventrolateral tegmental area
vsc ventral spinocerebellar tr
X nu X

ZID zona incerta, dorsal
Zo zona layer sup coll

Fig 76b
Obex +7.5mm
Sagittal 16

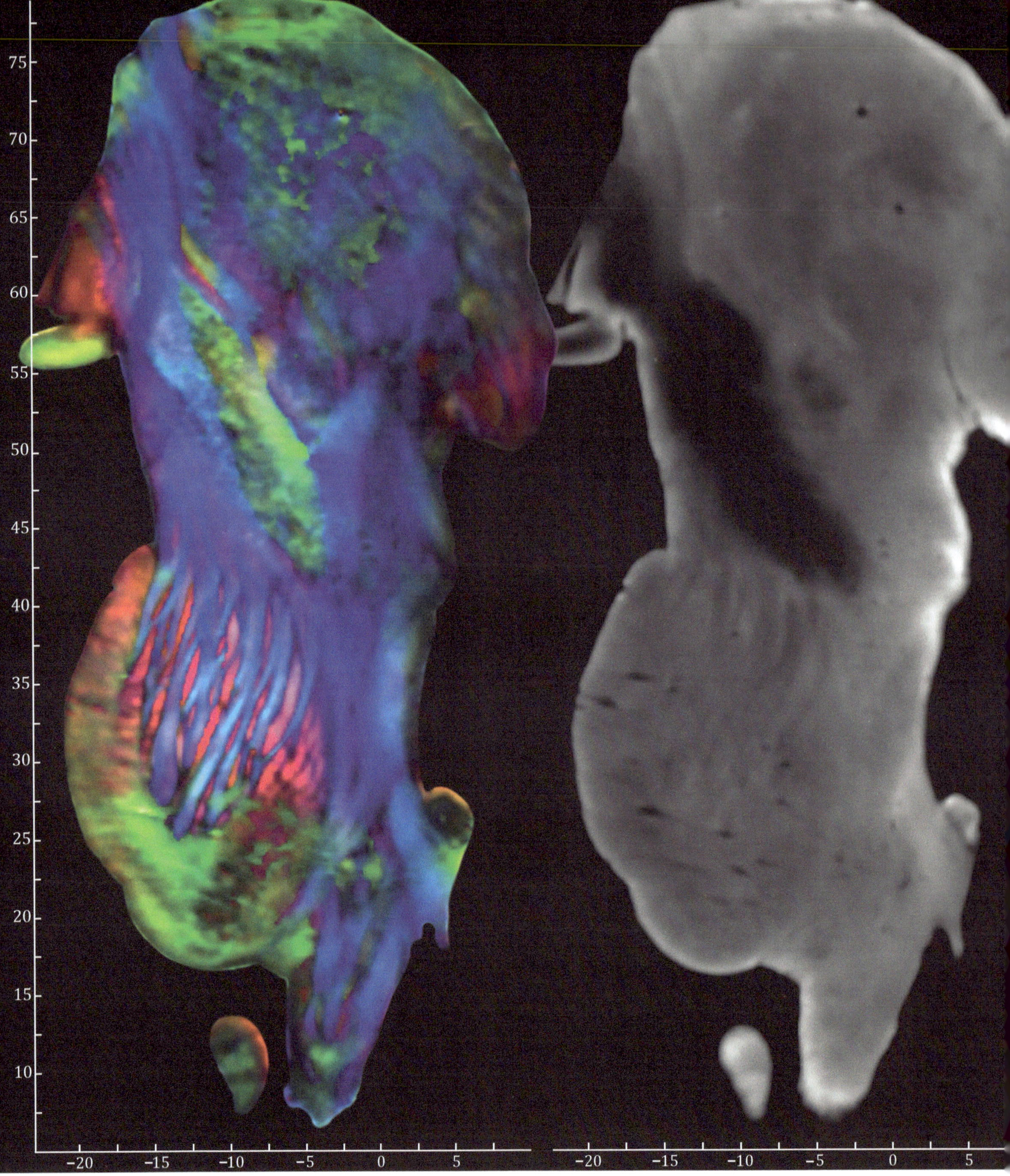

Fig 77a
Obex +8.0mm
Sagittal 17

5N motor trigeminal nu
5Tr trigem transition zone
7n facial nerve
ami amiculum of the olive
APT anterior pretectal nu
bic brachium inf colliculus
BIC nu brachium inf colliculus
bsc brachium superior colliculus
Cb cerebellum
cp cerebral peduncle
ctg central tegmental tract
DLL dorsal nu lateral lemniscus
ECIC external cx inferior coll
EnR endorestiform nucleus
Gam gamma pontine nucleus
ic internal capsule
icp inf cerebellar peduncle
IF5 interfascicular trigeminal nu
ILL intermed nu lat lemniscus
IOPr inf olive, principal nu
isRt isthmic reticular formation
KF Kölliker-Fuse nu
lfp longitudinal fasciculus pons
ll lateral lemniscus
LPB lateral parabrachial nu
LRt lateral reticular nu
LVe lateral vestibular nu
m5 motor root trigeminal nerve
mcp middle cerebellar peduncle
MiTg microcell tegmental nu
ml medial lemniscus
MPB medial parabrachial nu
MPBE med parabrachial nu ext
mRt mesencephalic retic form
MVe medial vestibular nu
oc olivocerebellar tract
opt optic tract
p1Rt prosomere 1 reticular formation
PBP parabrachial pigmented nu
PCRt parvicell reticular nu
PCRtA parvicell reticular nu, alpha
Pe5 peritrigeminal nucleus
PLi posterior limitans thal nu
Pn pontine nuclei
Pr5 principal sensory trigem nu
Pul pulvinar nuclei

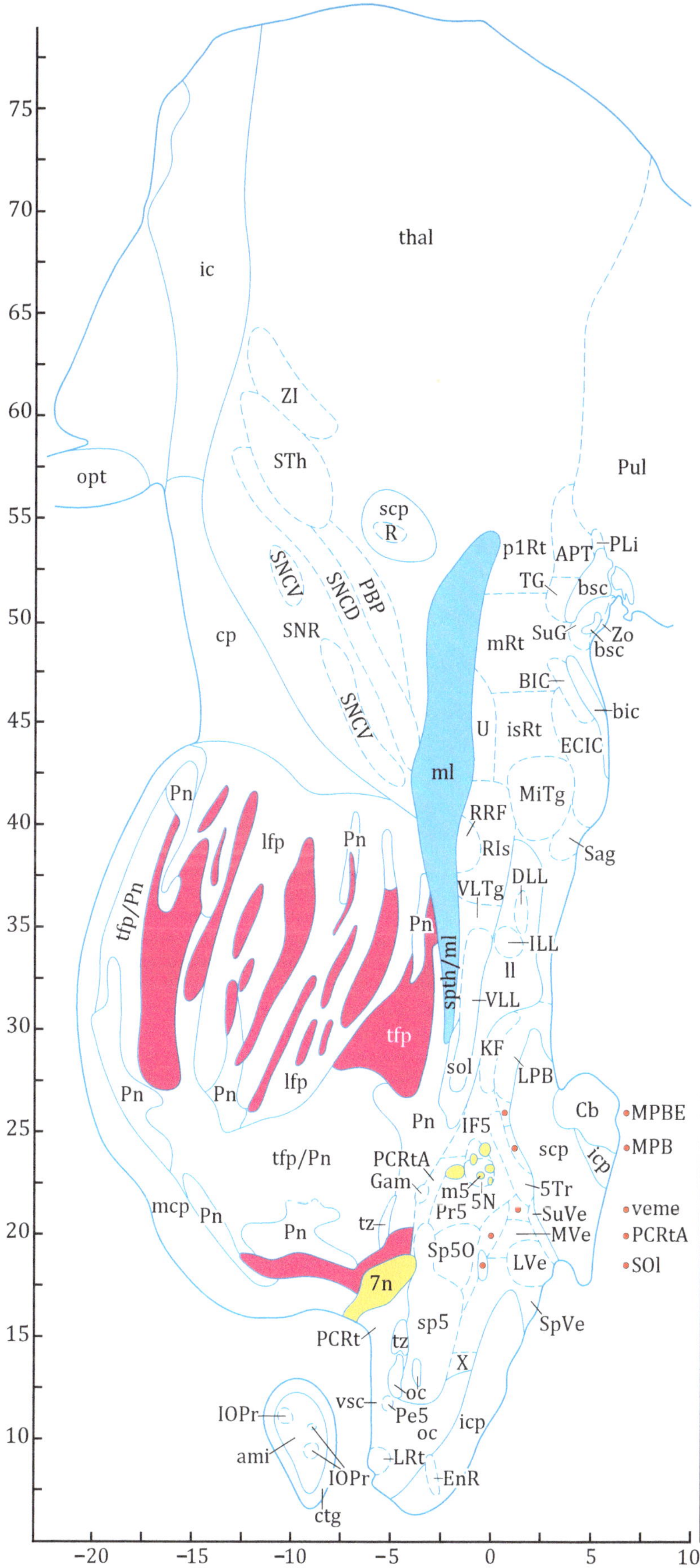

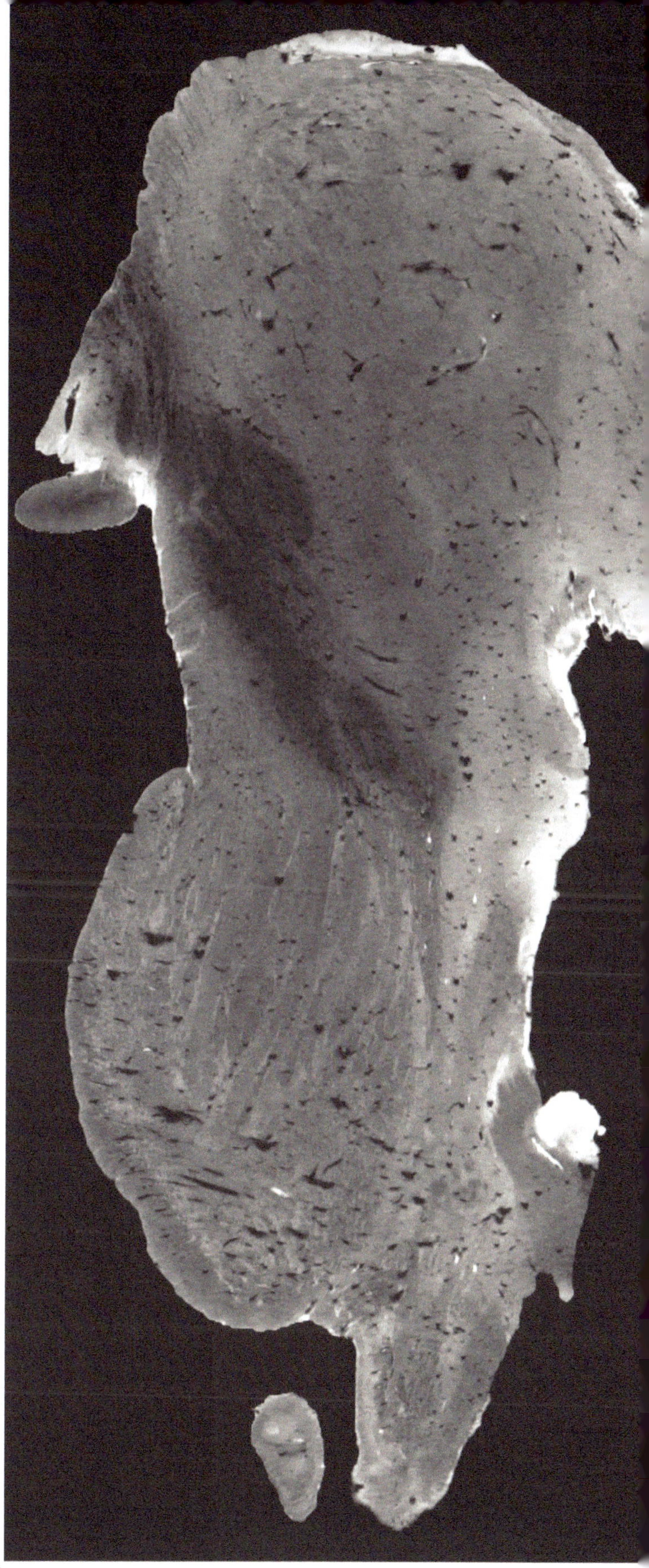

R red nu
RIs retroisthic nucleus
RRF retrorubral field
Sag sagulum nu
scp superior Cb peduncle
SNCD s nigra, compact, dors tier
SNCV s nigra, compact, vent tier
SNR s nigra, reticular
sol solitary tract
SOl superior olive, lateral
sp5 spinal trigem tract
Sp5O spinal trigeml nu, oral
spth/ml spinothal tract/med lemniscus
SpVe spinal vestibular nu
STh subthalamic nu
SuG superficial gray sup coll
SuVe superior vestibular nu
tfp transverse fibers pons
tfp/Pn transv fibers pons/pontine nu
TG tectal gray
thal thalamus
tz trapezoid body
U nucleus U
veme vestibulomesencephalic tr
VLL ventral nu lat lemniscus
VLTg ventrolateral tegmental area
vsc ventral spinocerebellar tr
X nu X
ZI zona incerta
Zo zona layer sup coll

Fig 77b
Obex +8.0mm
Sagittal 17

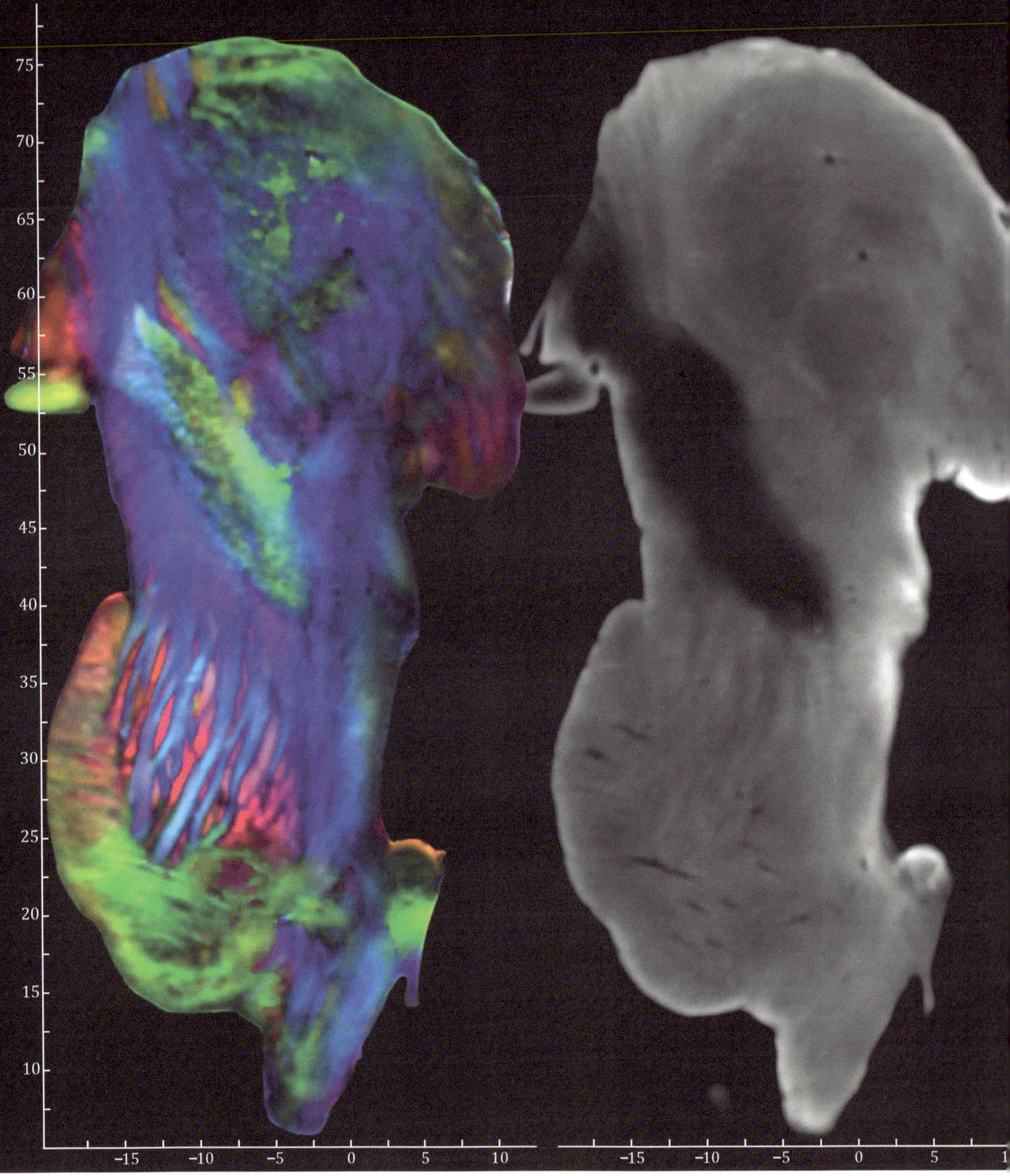

Fig 78a
Obex +8.5mm
Sagittal 18

5Tr trigem transition zone
7n facial nerve
APT anterior pretectal nu
bic brachium inf colliculus
BIC nu brachium inf colliculus
bsc brachium superior colliculus
Cb cerebellum
cp cerebral peduncle
ECIC external cx inferior coll
Gam gamma pontine nucleus
ic internal capsule
icp inf cerebellar peduncle
ILL intermed nu lat lemniscus
isRt isthmic reticular formation
KF Kölliker-Fuse nu
lfp longitudinal fasciculus pons
ll lateral lemniscus
LPB lateral parabrachial nu
LVe lateral vestibular nu
m5 motor root trigeminal nerve
mcp middle cerebellar peduncle
MiTg microcell tegmental nu
ml medial lemniscus
mRt mesencephalic retic form
MVe medial vestibular nu
oc olivocerebellar tract
opt optic tract
p1Rt prosomere 1 reticular formation
PBP parabrachial pigmented nu
PCRtA parvicell reticular nu, alpha
PLi posterior limitans thal nu
Pn pontine nuclei
Pn/tfp pontine nu/transv fibres pons
Pr5 principal sensory trigem nu
Pul pulvinar nuclei
Sag sagulum nu
scp superior Cb peduncle
SNCD s nigra, compact, dors tier
SNCV s nigra, compact, vent tier
SNR s nigra, reticular
sol solitary tract
sp5 spinal trigem tract
Sp5O spinal trigeml nu, oral
spth spinothalamic tract
SpVe spinal vestibular nu
STh subthalamic nu

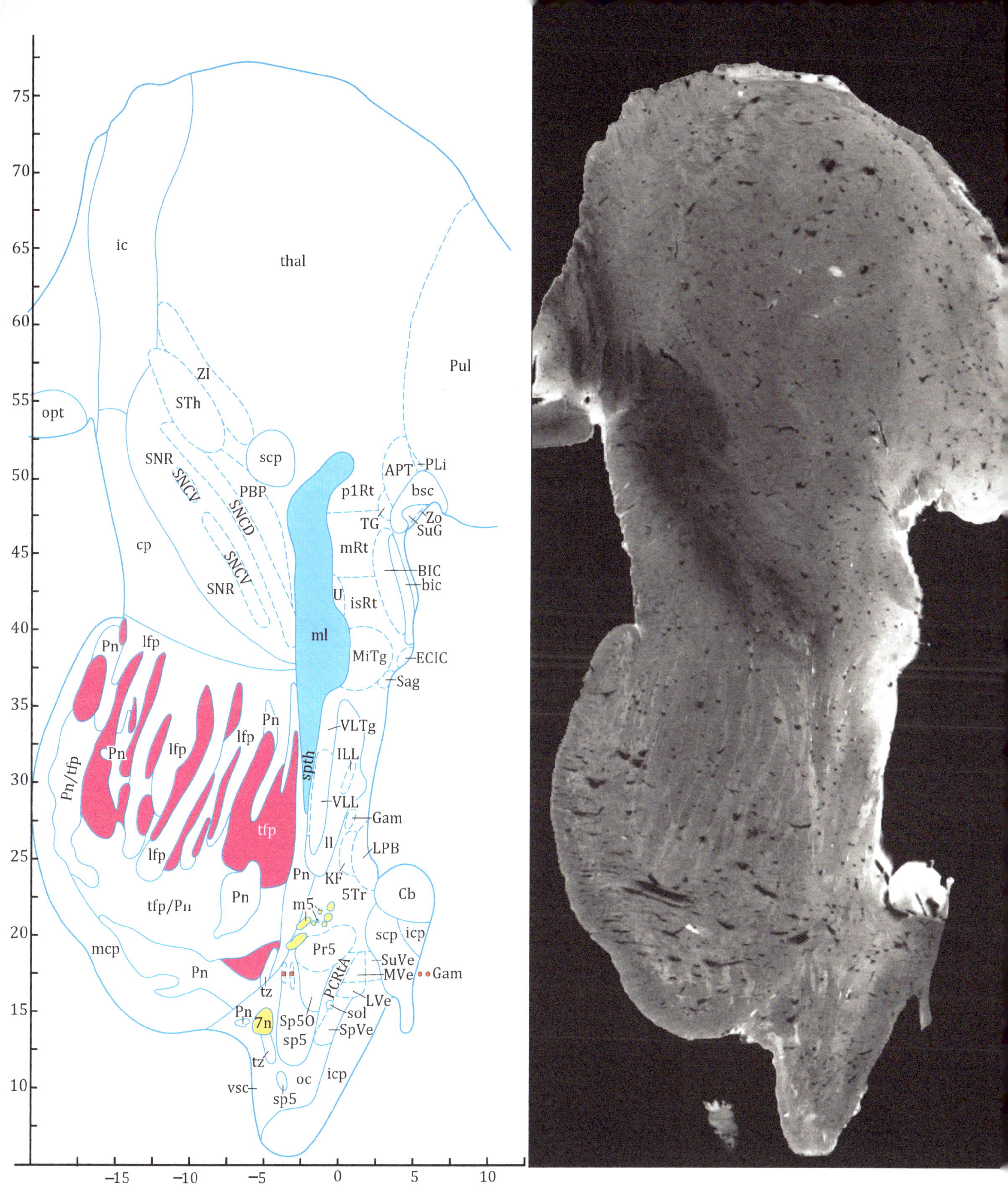

SuG superficial gray sup coll
SuVe superior vestibular nu
tfp transverse fibers pons
tfp/Pn transv fibers pons/pontine nu
TG tectal gray
thal thalamus
tz trapezoid body
U nucleus U
VLL ventral nu lat lemniscus
VLTg ventrolateral tegmental area
vsc ventral spinocerebellar tr

ZI zona incerta
Zo zona layer sup coll

Fig 78b
Obex +8.5mm
Sagittal 18

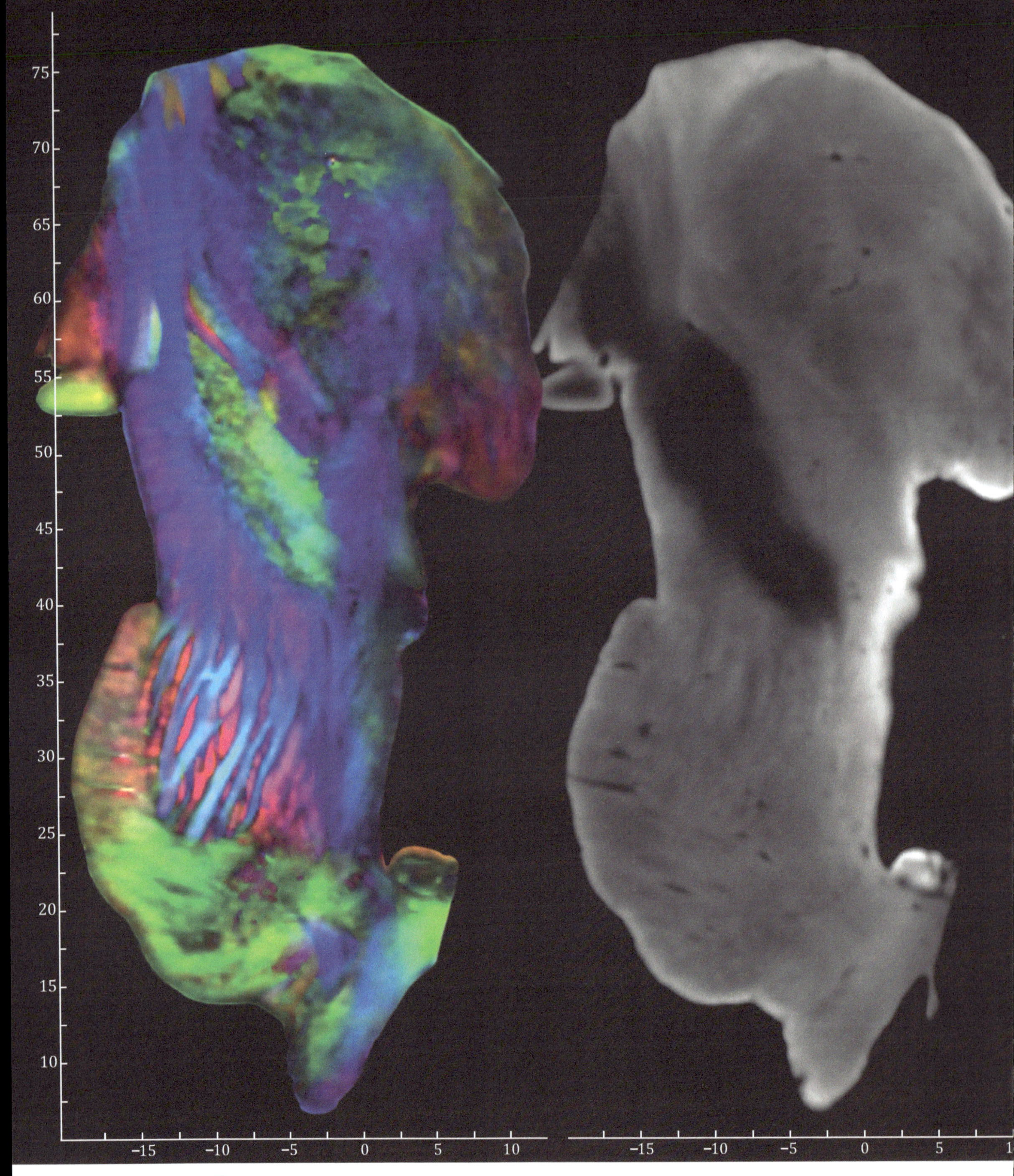

Fig 79a
Obex +9.0mm
Sagittal 19

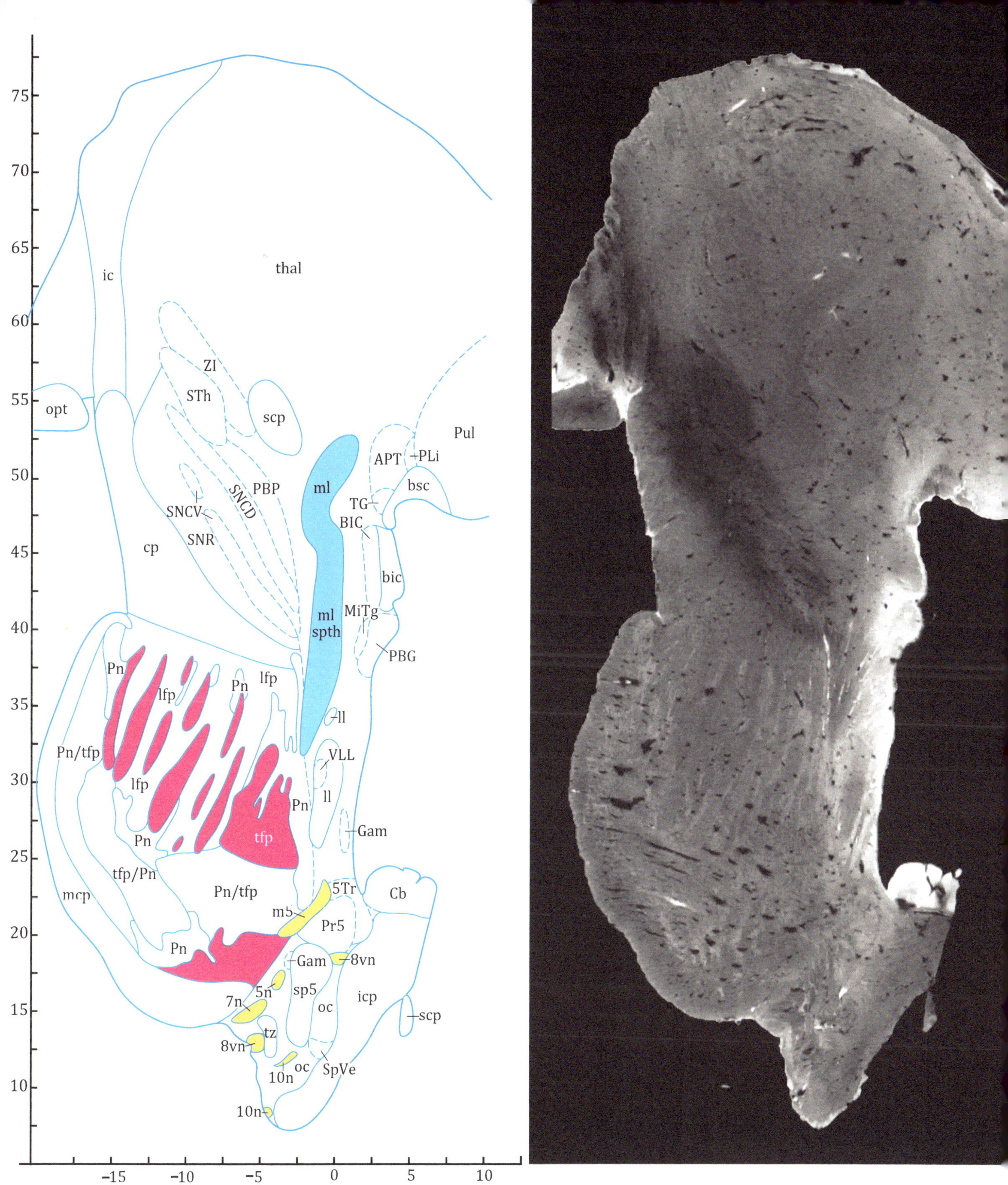

10n vagus nerve
5n trigeminal nerve
5Tr trigem transition zone
7n facial nerve
8vn vestibular root of 8th nerve
APT anterior pretectal nu
bic brachium inf colliculus
BIC nu brachium inf colliculus
bsc brachium superior colliculus
Cb cerebellum
cp cerebral peduncle
Gam gamma pontine nucleus
ic internal capsule
icp inf cerebellar peduncle
lfp longitudinal fasciculus pons
ll lateral lemniscus
m5 motor root trigeminal nerve
mcp middle cerebellar peduncle
MiTg microcell tegmental nu
ml medial lemniscus
oc olivocerebellar tract
opt optic tract
PBG parabigeminal nu
PBP parabrachial pigmented nu
PLi posterior limitans thal nu
Pn pontine nuclei
Pn/tfp pontine nu/transv fibres pons
Pr5 principal sensory trigem nu
Pul pulvinar nuclei
scp superior Cb peduncle
SNCD s nigra, compact, dors tier
SNCV s nigra, compact, vent tier
SNR s nigra, reticular
sp5 spinal trigem tract
spth spinothalamic tract
SpVe spinal vestibular nu
STh subthalamic nu
tfp transverse fibers pons
tfp/Pn transv fibers pons/pontine nu
TG tectal gray
thal thalamus
tz trapezoid body
VLL ventral nu lat lemniscus
ZI zona incerta

Fig 79b
Obex +9.0mm
Sagittal 19

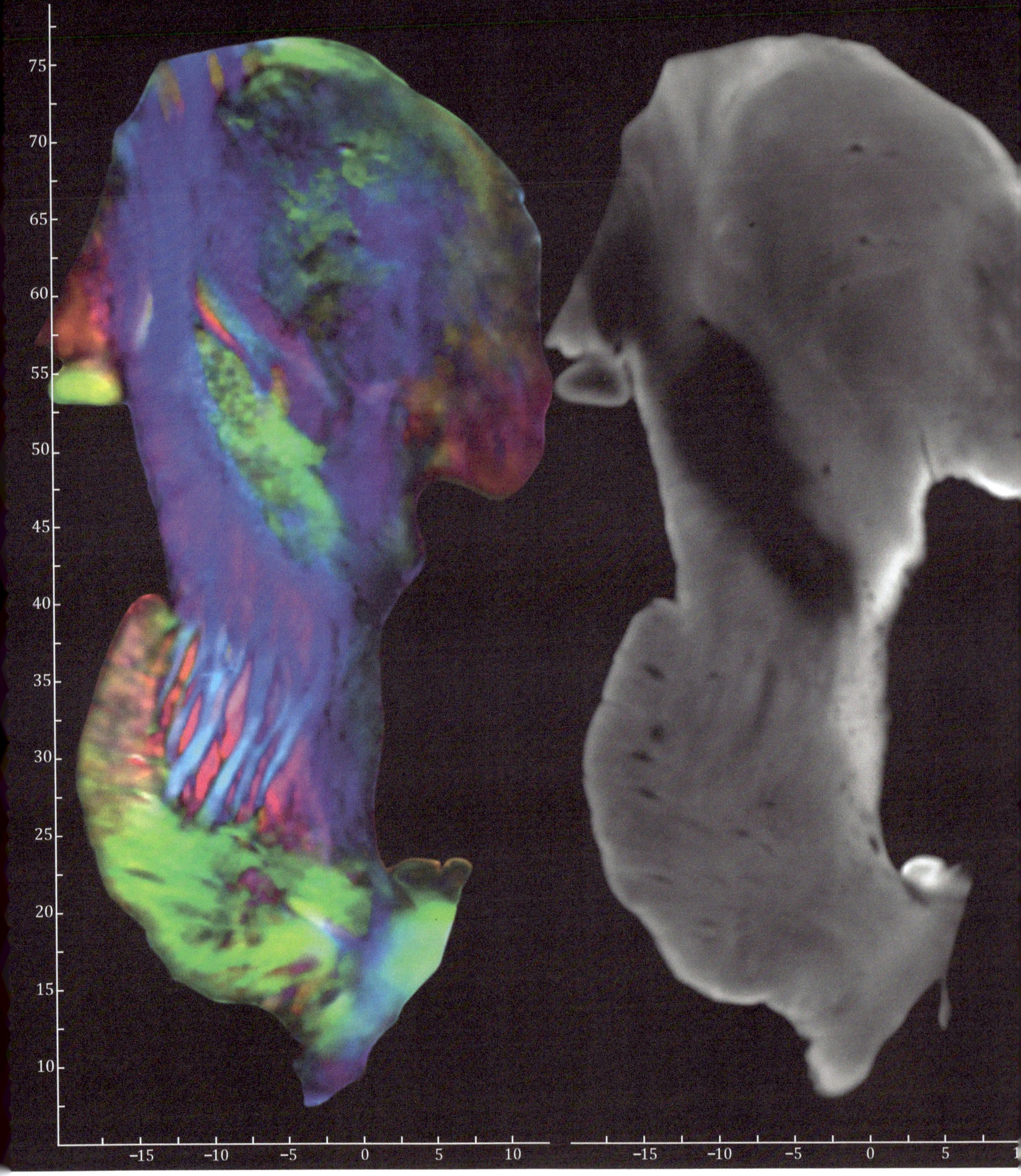

Fig 80a
Obex +9.5mm
Sagittal 20

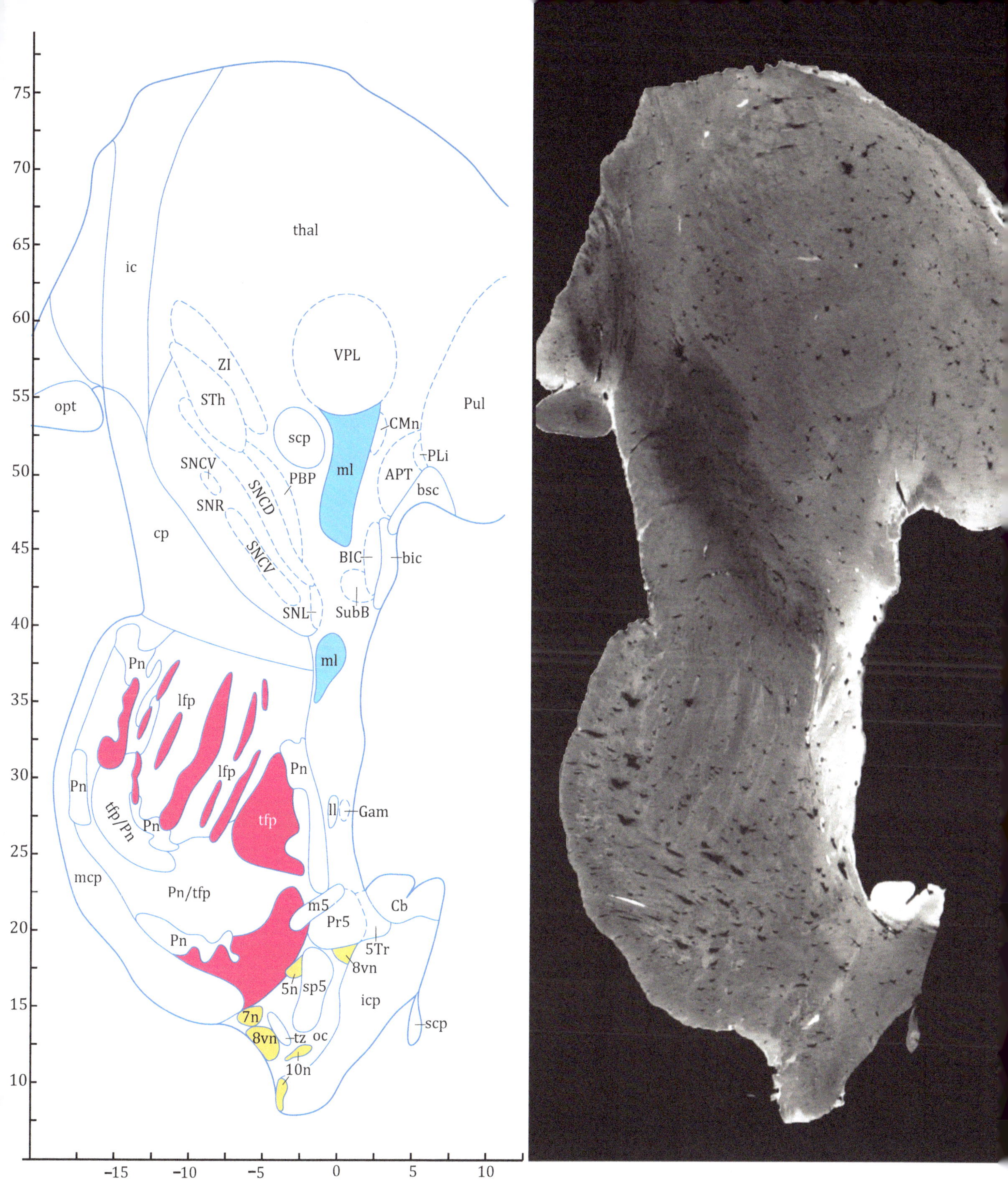

10n vagus nerve
5n trigeminal nerve
5Tr trigem transition zone
7n facial nerve
8vn vestibular root of 8th nerve
APT anterior pretectal nu
bic brachium inf colliculus
BIC nu brachium inf colliculus
bsc brachium superior colliculus
Cb cerebellum
CMn centromedian thalamic nucleus
cp cerebral peduncle
Gam gamma pontine nucleus
ic internal capsule
icp inf cerebellar peduncle
lfp longitudinal fasciculus pons
ll lateral lemniscus
m5 motor root trigeminal nerve
mcp middle cerebellar peduncle
ml medial lemniscus
oc olivocerebellar tract
opt optic tract
PBP parabrachial pigmented nu
PLi posterior limitans thal nu
Pn pontine nuclei
Pn/tfp pontine nu/transv fibres pons
Pr5 principal sensory trigem nu
Pul pulvinar nuclei
scp superior Cb peduncle
SNCD s nigra, compact, dors tier
SNCV s nigra, compact, vent tier
SNL s nigra, lateral
SNR s nigra, reticular
sp5 spinal trigem tract
STh subthalamic nu
SubB subbrachial nu
tfp transverse fibers pons
tfp/Pn transv fibers pons/pontine nu
thal thalamus
tz trapezoid body
VPL ventral posterolat thal nu
ZI zona incerta

Fig 80b
Obex +9.5mm
Sagittal 20

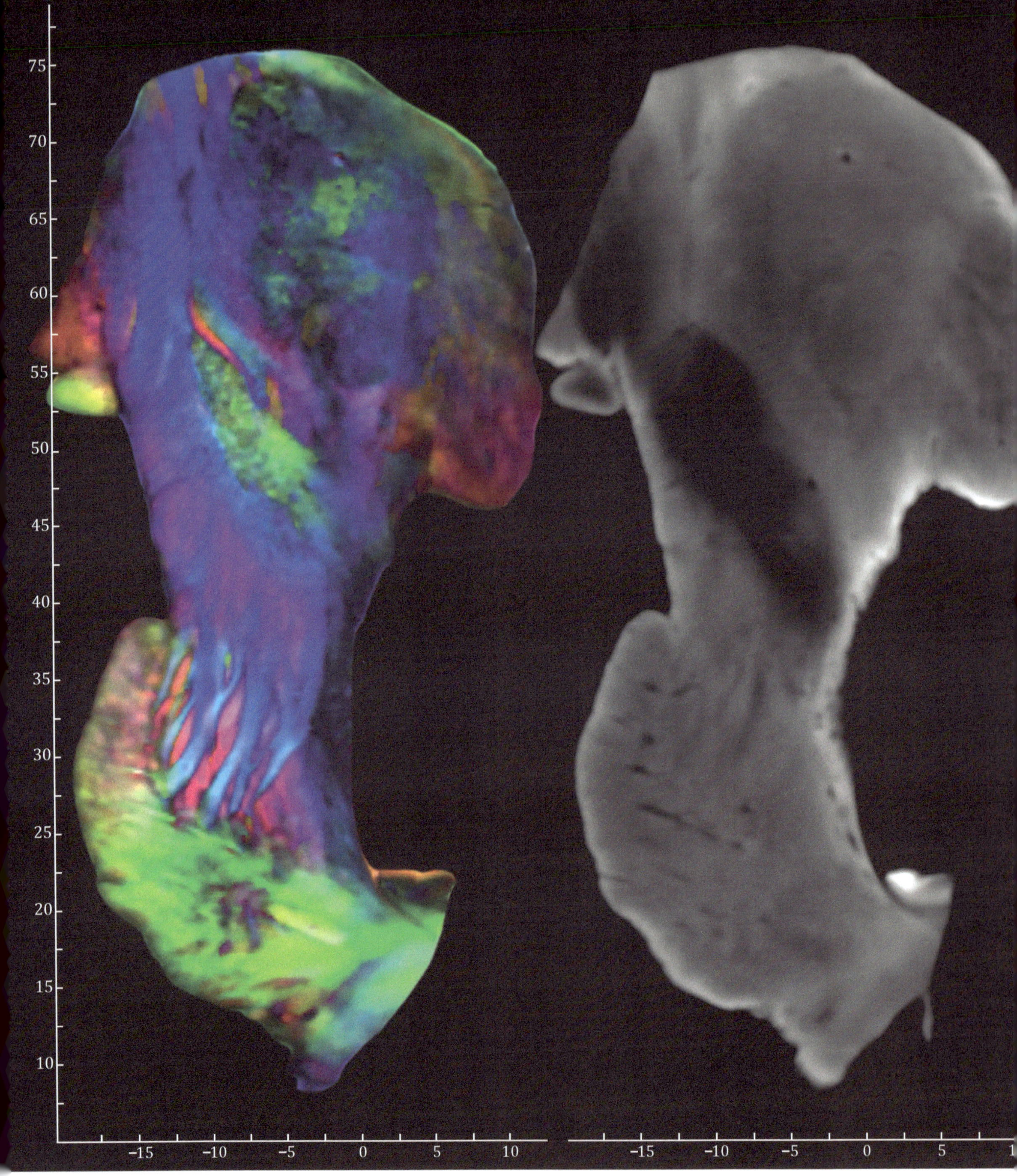

Fig 81a
Obex +10.0mm
Sagittal 21

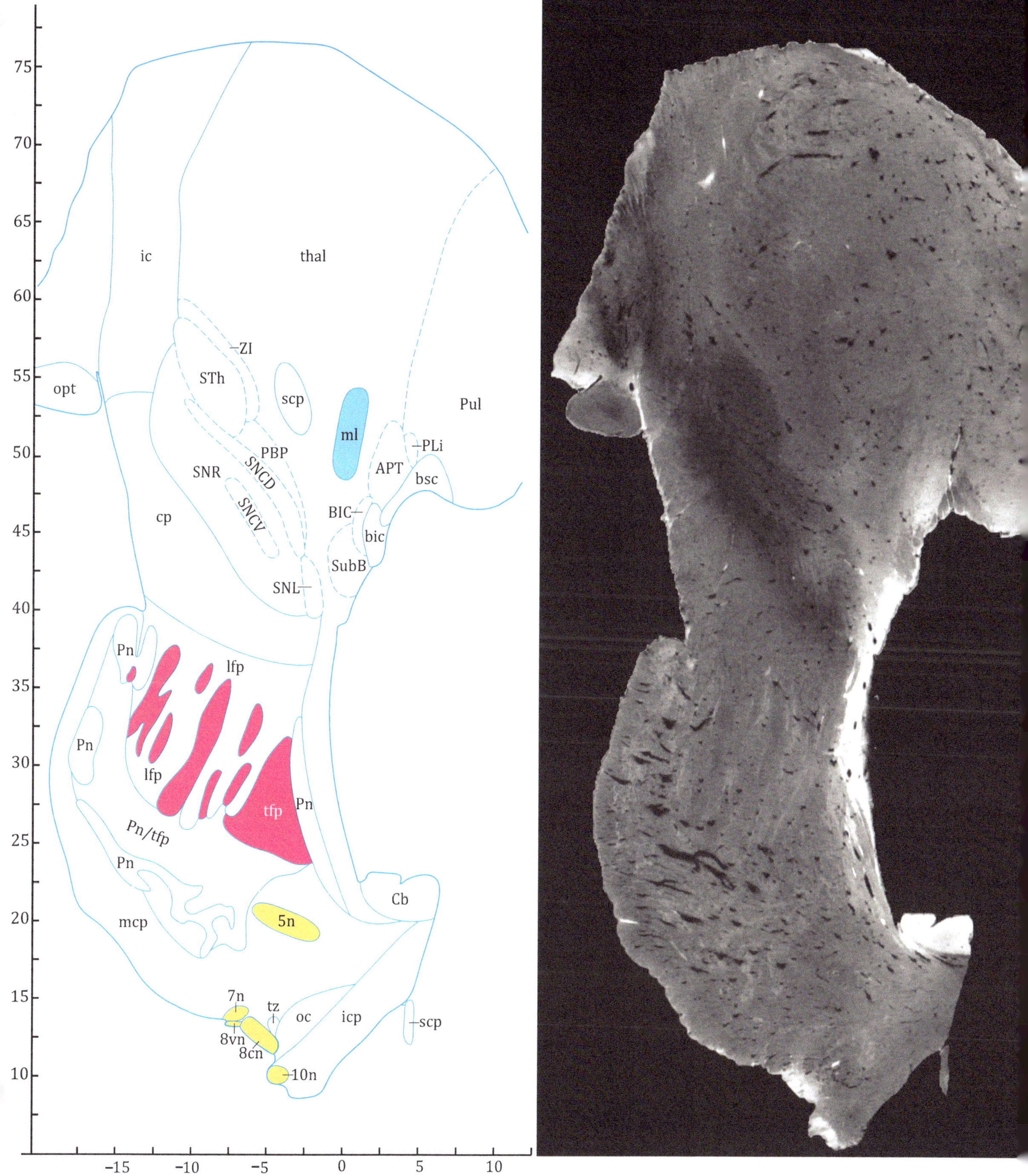

10n vagus nerve
5n trigeminal nerve
7n facial nerve
8cn cochlear root 8th nerve
8vn vestibular root of 8th nerve
APT anterior pretectal nu
bic brachium inf colliculus
BIC nu brachium inf colliculus
bsc brachium superior colliculus
Cb cerebellum
cp cerebral peduncle
ic internal capsule
icp inf cerebellar peduncle
lfp longitudinal fasciculus pons
mcp middle cerebellar peduncle
ml medial lemniscus
oc olivocerebellar tract
opt optic tract
PBP parabrachial pigmented nu
PLi posterior limitans thal nu
Pn pontine nuclei
Pn/tfp pontine nu/transv fibres pons
Pul pulvinar nuclei
scp superior Cb peduncle
SNCD s nigra, compact, dors tier
SNCV s nigra, compact, vent tier
SNL s nigra, lateral
SNR s nigra, reticular
STh subthalamic nu
SubB subbrachial nu
tfp transverse fibers pons
thal thalamus
tz trapezoid body
ZI zona incerta

Fig 81b
Obex +10.0mm
Sagittal 21

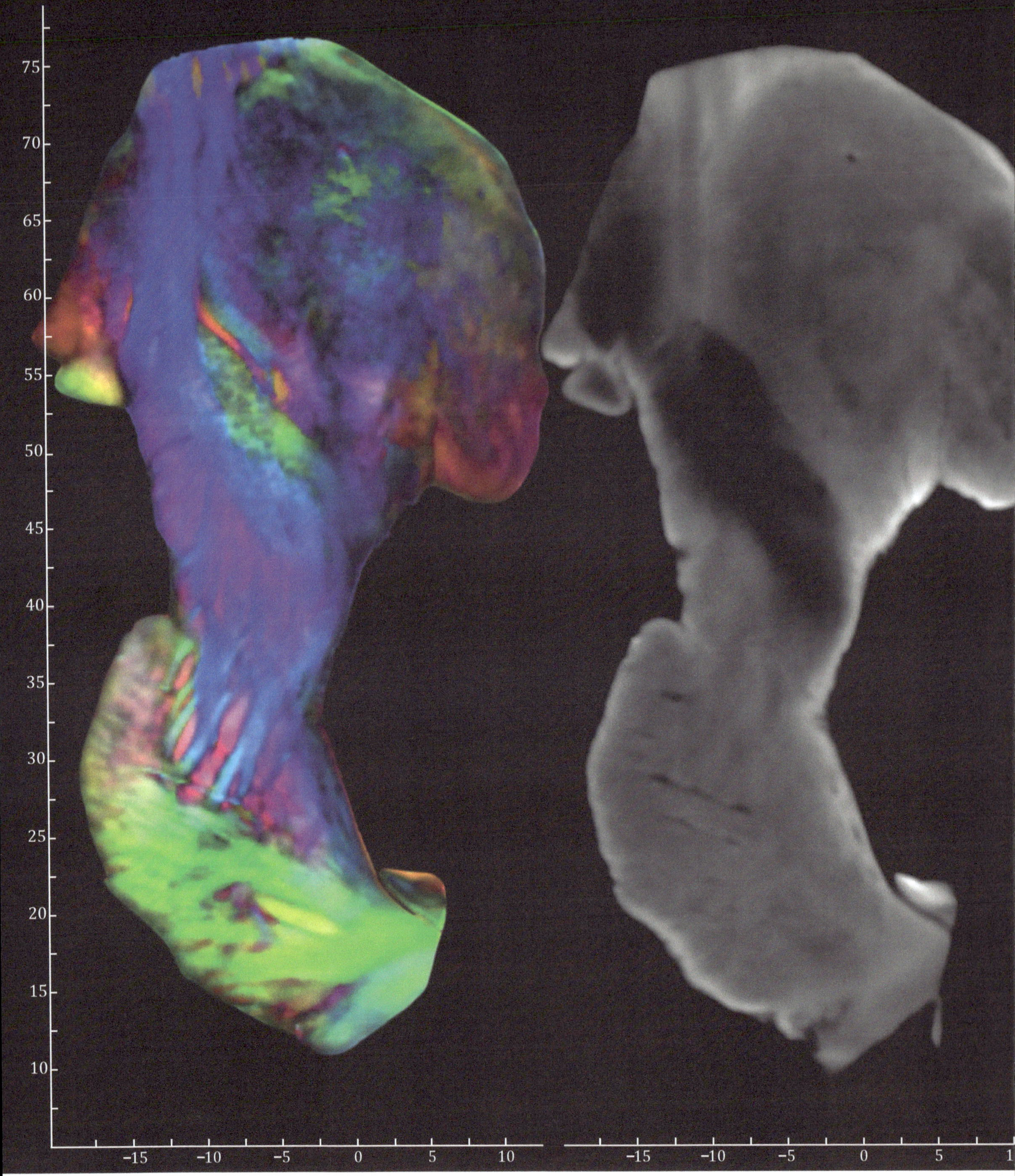

Fig 82a
Obex +10.5mm
Sagittal 22

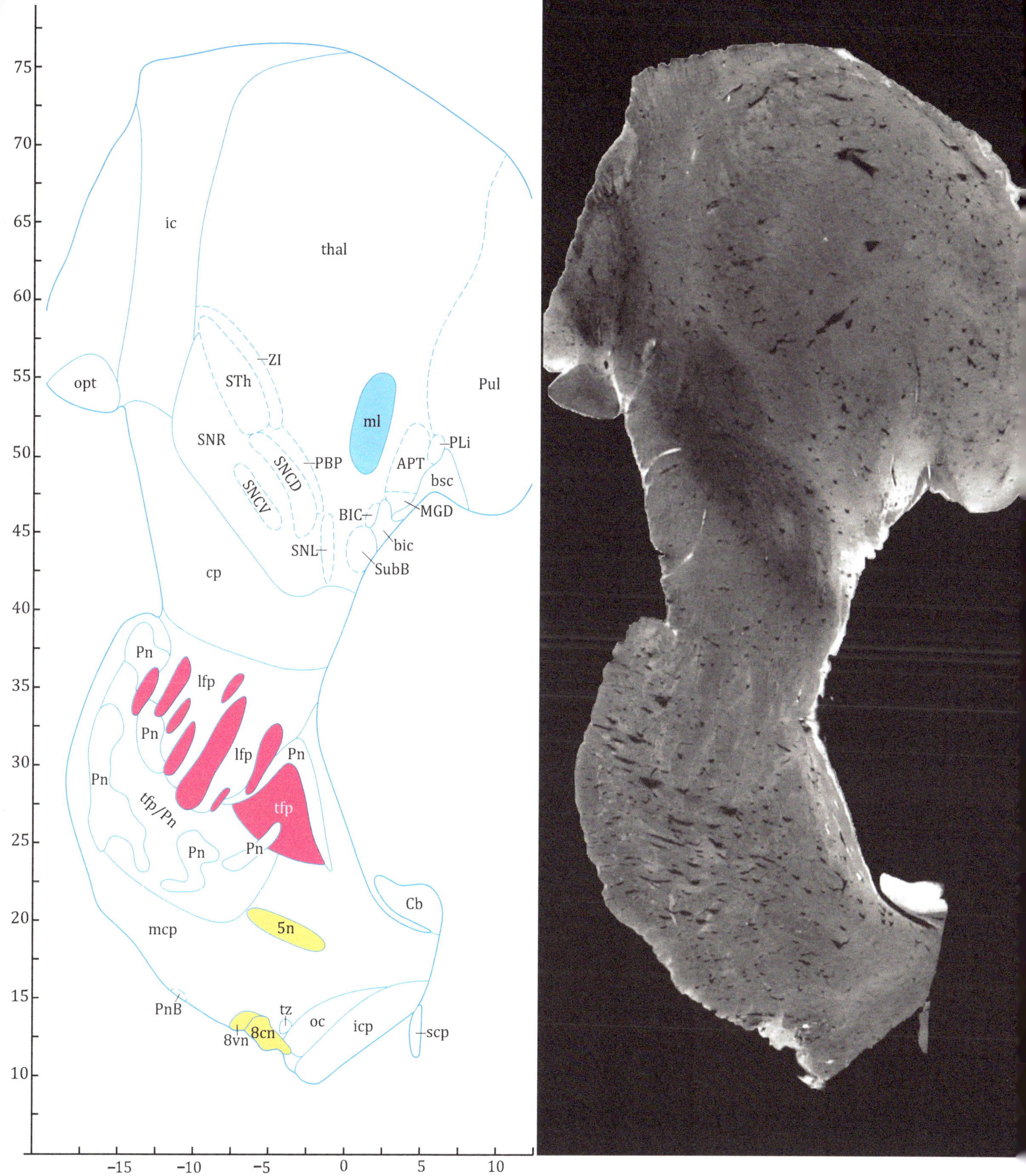

5n trigeminal nerve
8cn cochlear root 8th nerve
8vn vestibular root of 8th nerve
APT anterior pretectal nu
bic brachium inf colliculus
BIC nu brachium inf colliculus
bsc brachium superior colliculus
Cb cerebellum
cp cerebral peduncle
ic internal capsule
icp inf cerebellar peduncle
lfp longitudinal fasciculus pons
mcp middle cerebellar peduncle
MGD med geniculate nu, dorsal
ml medial lemniscus
oc olivocerebellar tract
opt optic tract
PBP parabrachial pigmented nu
PLi posterior limitans thal nu
Pn pontine nuclei
PnB pontobulbar nucleus
Pul pulvinar nuclei
scp superior Cb peduncle
SNCD s nigra, compact, dors tier
SNCV s nigra, compact, vent tier
SNL s nigra, lateral
SNR s nigra, reticular
STh subthalamic nu
SubB subbrachial nu
tfp transverse fibers pons
tfp/Pn transv fibers pons/pontine nu
thal thalamus
tz trapezoid body
ZI zona incerta

Fig 82b
Obex +10.5mm
Sagittal 22

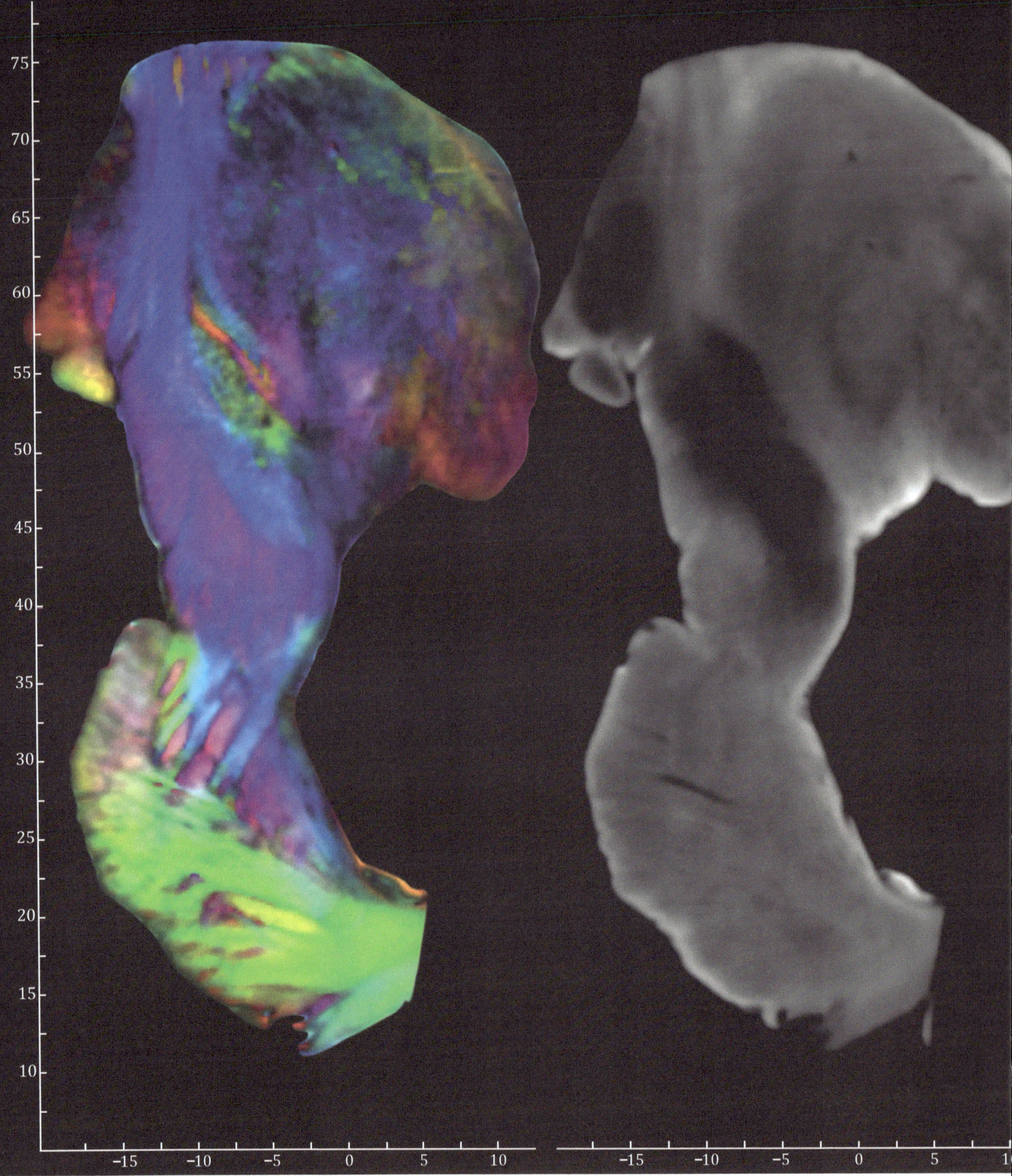

Fig 83a
Obex +11.0mm
Sagittal 23

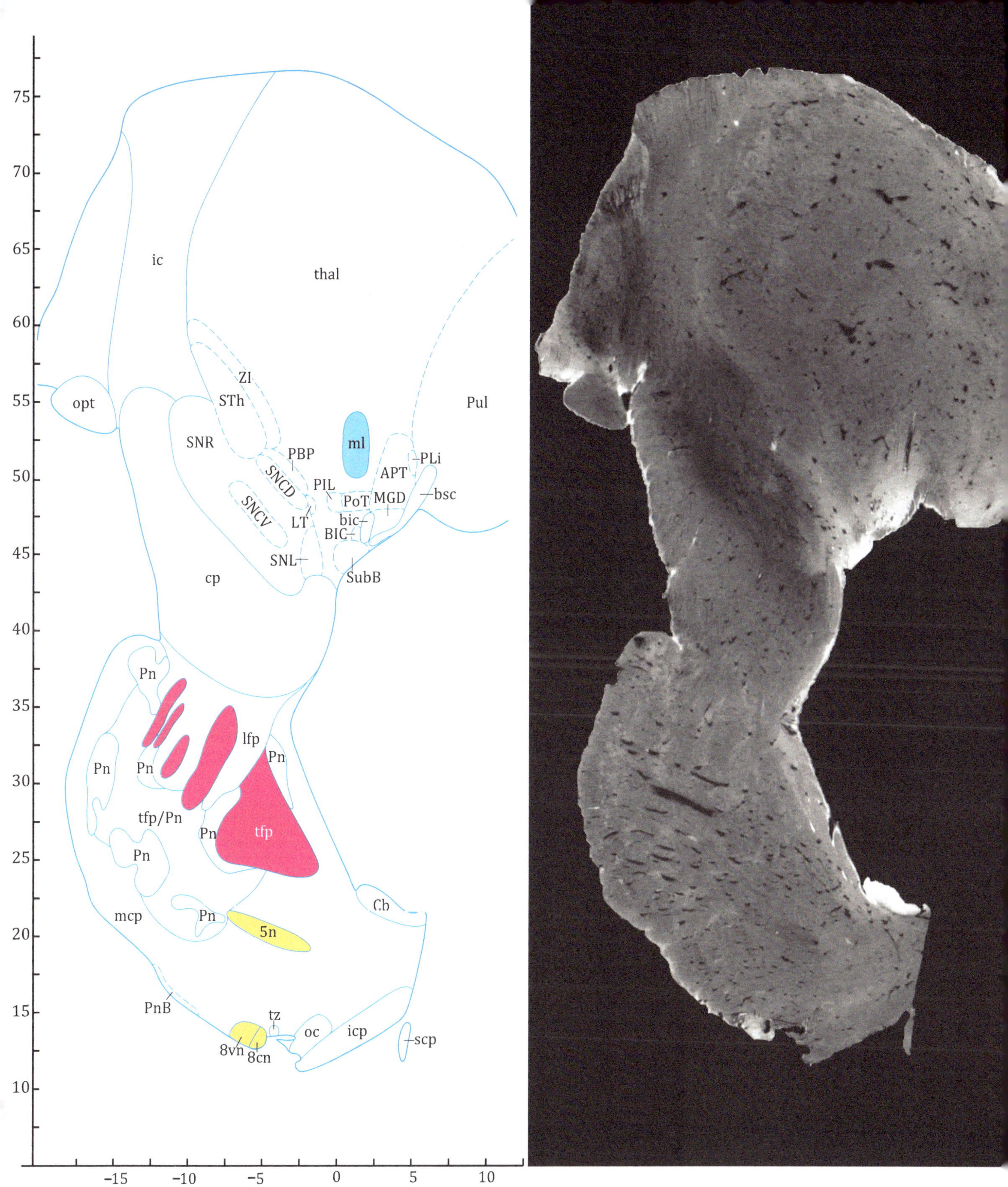

5n trigeminal nerve
8cn cochlear root 8th nerve
8vn vestibular root of 8th nerve
APT anterior pretectal nu
bic brachium inf colliculus
BIC nu brachium inf colliculus
bsc brachium superior colliculus
Cb cerebellum
cp cerebral peduncle
ic internal capsule
icp inf cerebellar peduncle
lfp longitudinal fasciculus pons
LT lat terminal nu acc optic tr
mcp middle cerebellar peduncle
MGD med geniculate nu, dorsal
ml medial lemniscus
oc olivocerebellar tract
opt optic tract
PBP parabrachial pigmented nu
PIL post intralaminar thal nu
PLi posterior limitans thal nu
Pn pontine nuclei
PnB pontobulbar nucleus
PoT posterior thal nu, triangular
Pul pulvinar nuclei
scp superior Cb peduncle
SNCD s nigra, compact, dors tier
SNCV s nigra, compact, vent tier
SNL s nigra, lateral
SNR s nigra, reticular
STh subthalamic nu
SubB subbrachial nu
tfp transverse fibers pons
tfp/Pn transv fibers pons/pontine nu
thal thalamus
tz trapezoid body
ZI zona incerta

Fig 83b
Obex +11.0mm
Sagittal 23

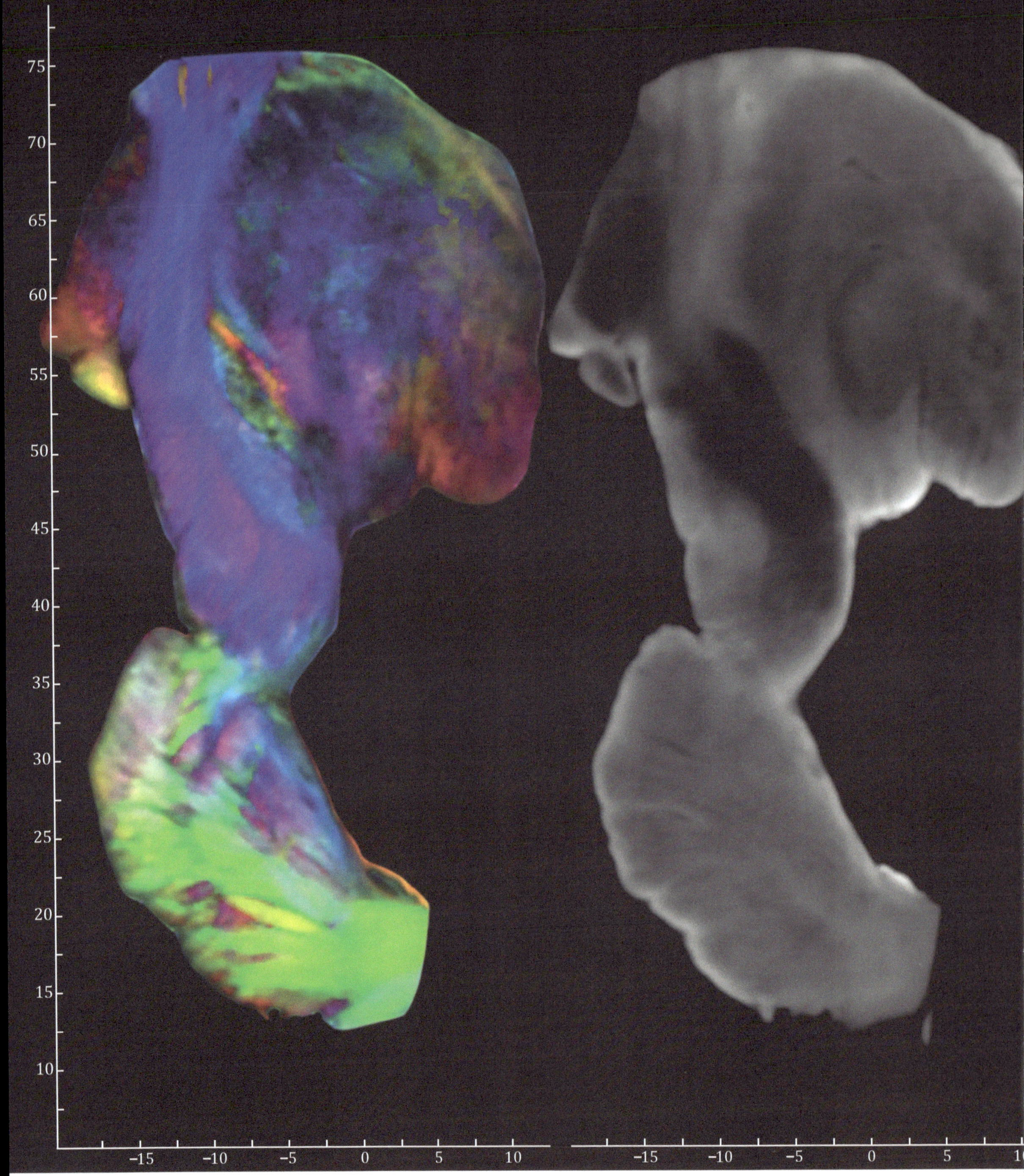

Fig 84a
Obex +11.5mm
Sagittal 24

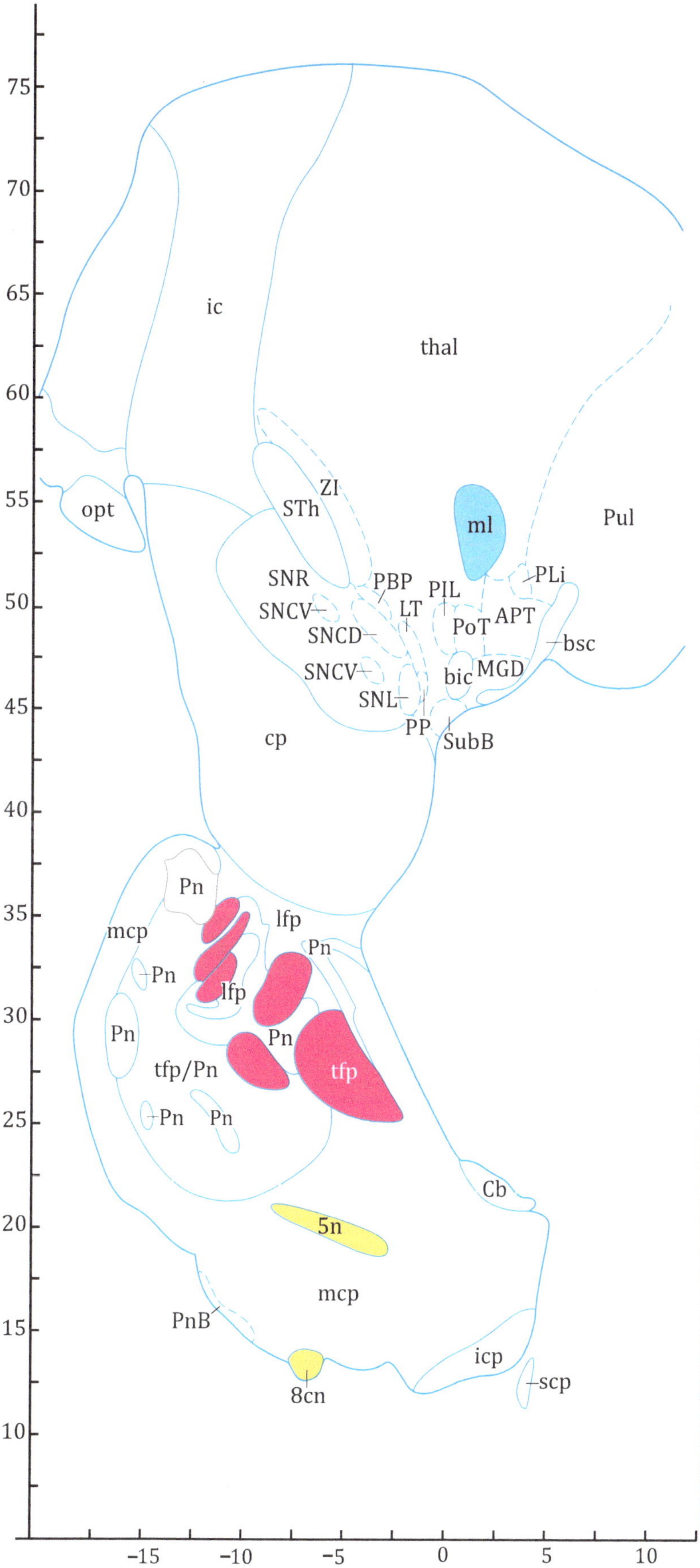

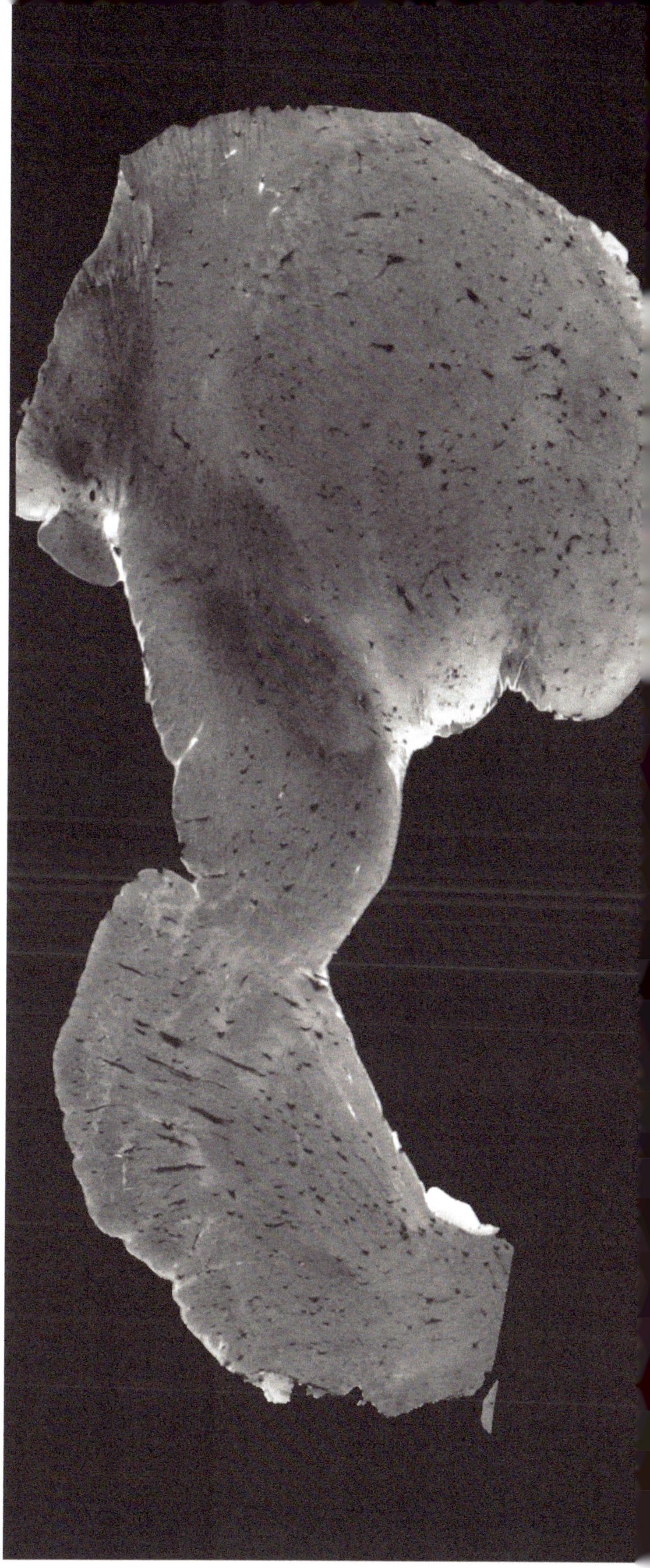

5n trigeminal nerve
8cn cochlear root 8th nerve
APT anterior pretectal nu
bic brachium inf colliculus
bsc brachium superior colliculus
Cb cerebellum
cp cerebral peduncle
ic internal capsule
icp inf cerebellar peduncle
lfp longitudinal fasciculus pons
LT lat terminal nu acc optic tr
mcp middle cerebellar peduncle
MGD med geniculate nu, dorsal
ml medial lemniscus
opt optic tract
PBP parabrachial pigmented nu
PIL post intralaminar thal nu
PLi posterior limitans thal nu
Pn pontine nuclei
PnB pontobulbar nucleus
PoT posterior thal nu, triangular
PP peripeduncular nu
Pul pulvinar nuclei
scp superior Cb peduncle
SNCD s nigra, compact, dors tier
SNCV s nigra, compact, vent tier
SNL s nigra, lateral
SNR s nigra, reticular
STh subthalamic nu
SubB subbrachial nu
tfp transverse fibers pons
tfp/Pn transv fibers pons/pontine nu
thal thalamus
ZI zona incerta

Fig 84b
Obex +11.5mm
Sagittal 24

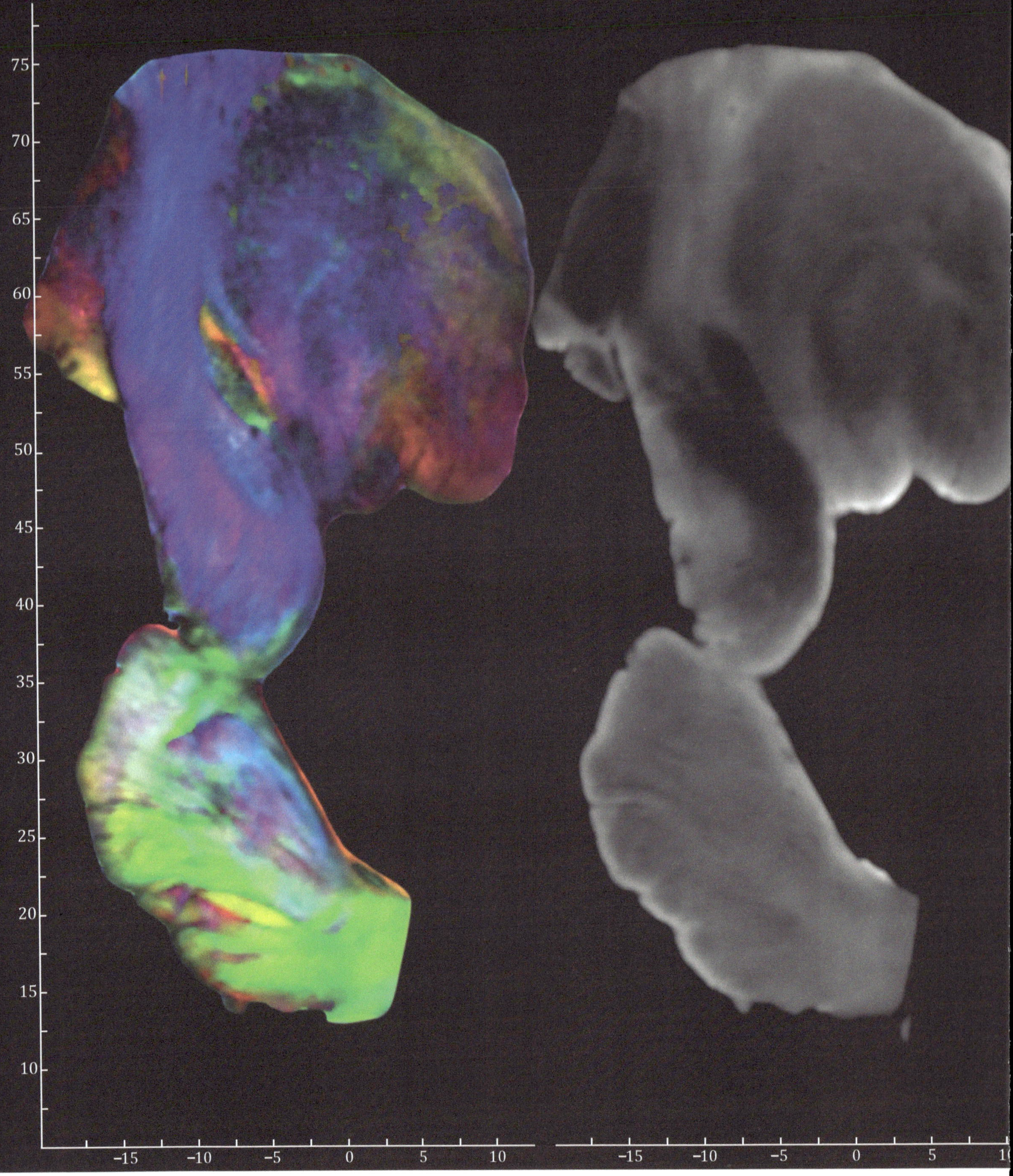

Fig 85a
Obex +12.0mm
Sagittal 25

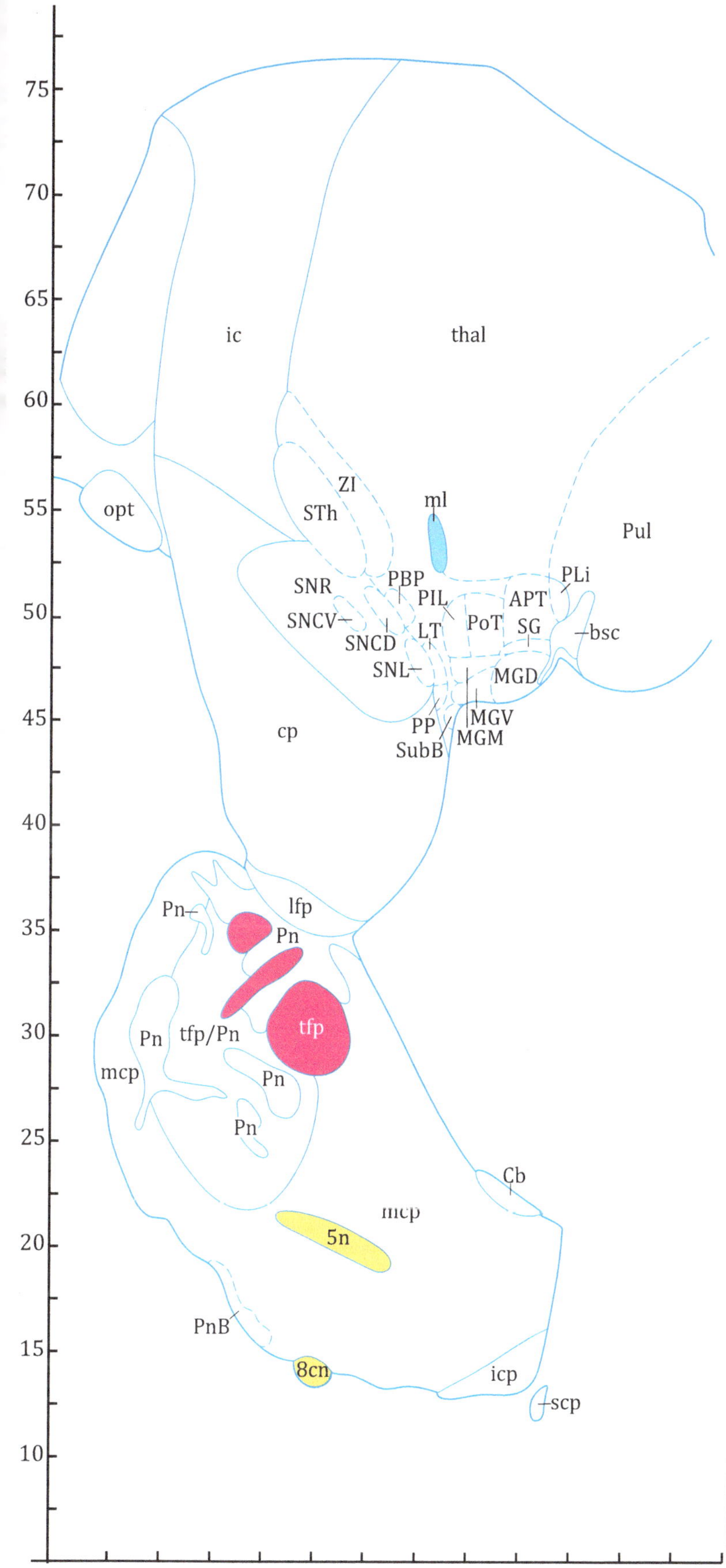

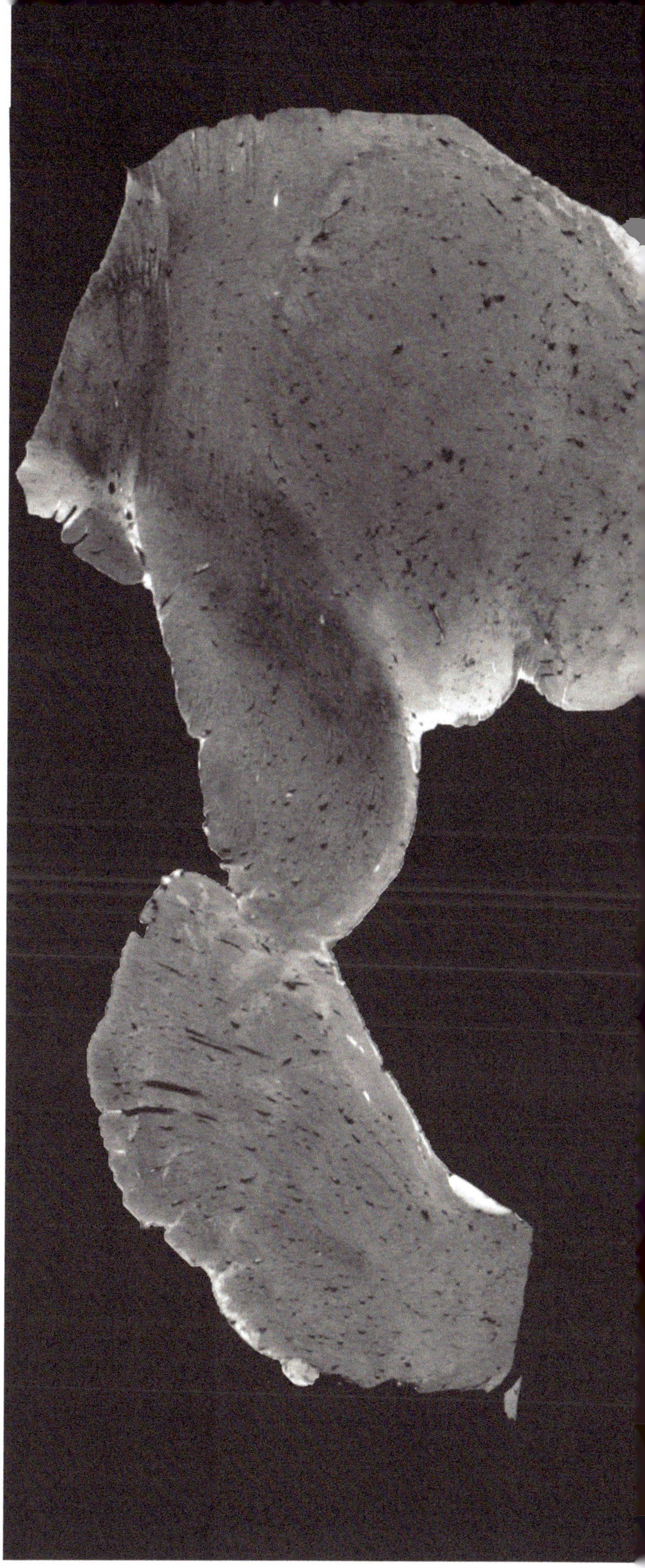

5n trigeminal nerve
8cn cochlear root 8th nerve
APT anterior pretectal nu
bsc brachium superior colliculus
Cb cerebellum
cp cerebral peduncle
ic internal capsule
icp inf cerebellar peduncle
lfp longitudinal fasciculus pons
LT lat terminal nu acc optic tr
mcp middle cerebellar peduncle
MGD med geniculate nu, dorsal
MGM med geniculate nu, medial
MGV med geniculate nu, ventral
ml medial lemniscus
opt optic tract
PBP parabrachial pigmented nu
PIL post intralaminar thal nu
PLi posterior limitans thal nu
Pn pontine nuclei
PnB pontobulbar nucleus
PoT posterior thal nu, triangular
PP peripeduncular nu
Pul pulvinar nuclei
scp superior Cb peduncle
SG suprageniculate thal nu
SNCD s nigra, compact, dors tier
SNCV s nigra, compact, vent tier
SNL s nigra, lateral
SNR s nigra, reticular
STh subthalamic nu
SubB subbrachial nu
tfp transverse fibers pons
tfp/Pn transv fibers pons/pontine nu
thal thalamus
ZI zona incerta

Fig 85b
Obex +12.0mm
Sagittal 25

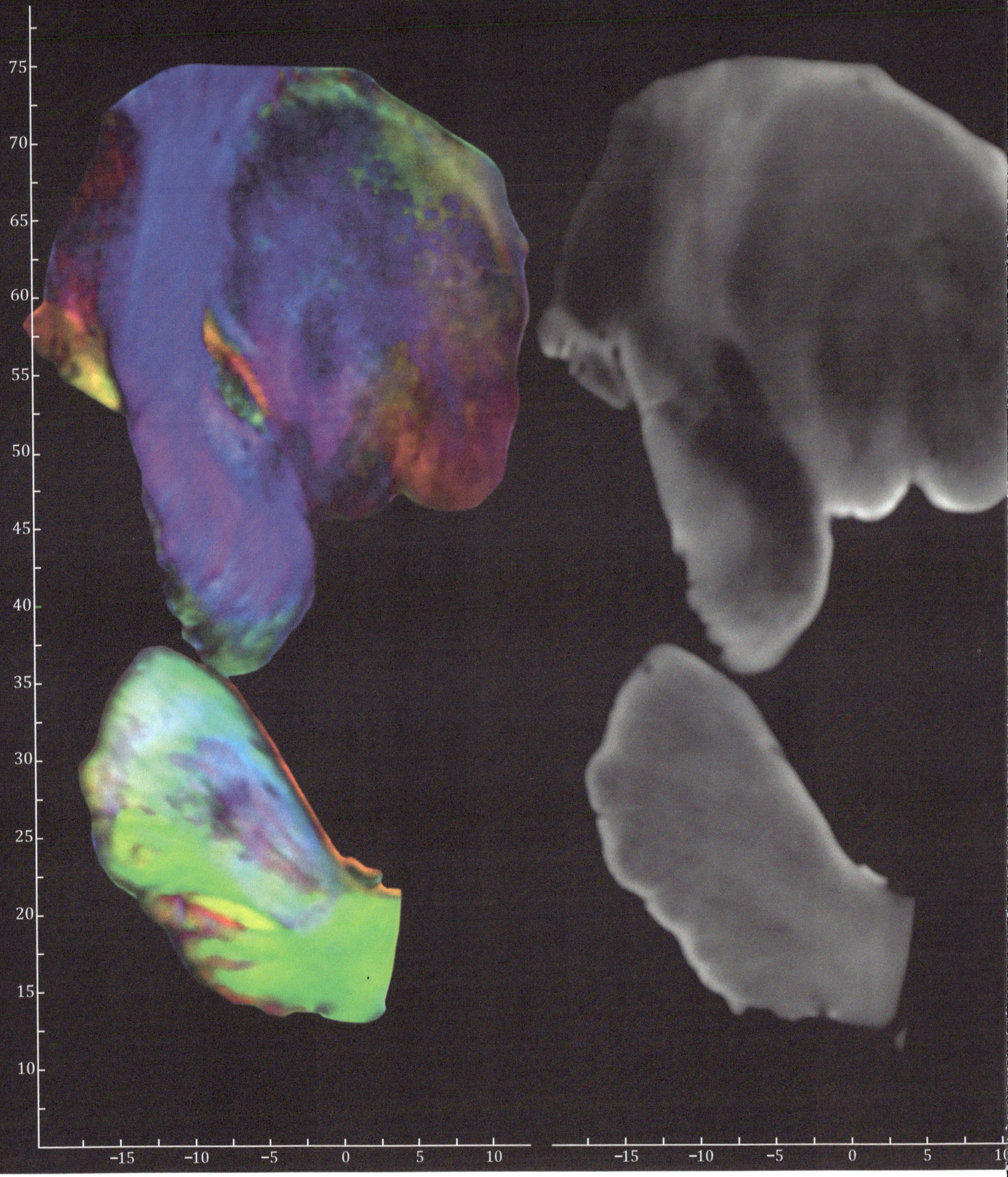

Fig 86a
Obex +12.5mm
Sagittal 26

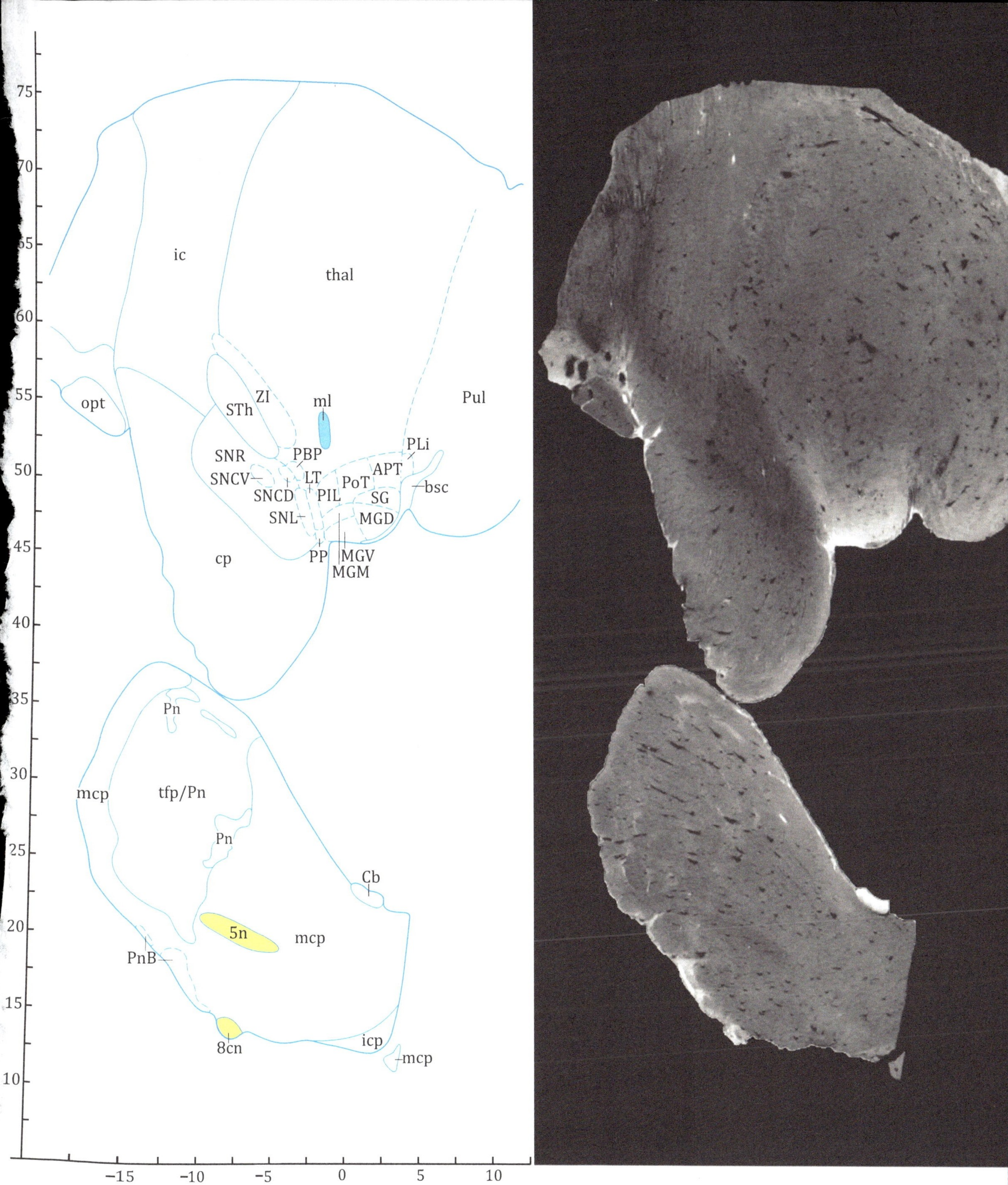

5n trigeminal nerve
8cn cochlear root 8th nerve
APT anterior pretectal nu
bsc brachium superior colliculus
Cb cerebellum
cp cerebral peduncle
ic internal capsule
icp inf cerebellar peduncle
LT lat terminal nu acc optic tr
mcp middle cerebellar peduncle
MGD med geniculate nu, dorsal
MGM med geniculate nu, medial
MGV med geniculate nu, ventral
ml medial lemniscus
opt optic tract
PBP parabrachial pigmented nu
PIL post intralaminar thal nu
PLi posterior limitans thal nu
Pn pontine nuclei
PnB pontobulbar nucleus
PoT posterior thal nu, triangular
PP peripeduncular nu
Pul pulvinar nuclei
SG suprageniculate thal nu
SNCD s nigra, compact, dors tier
SNCV s nigra, compact, vent tier
SNL s nigra, lateral
SNR s nigra, reticular
STh subthalamic nu
tfp/Pn transv fibers pons/pontine nu
thal thalamus
ZI zona incerta

Fig 86b
Obex +12.5mm
Sagittal 26